T0141997

Lecture Notes in Networks and Systems

Volume 189

Series Editor

Janusz Kacprzyk, Systems Research Institute, Polish Academy of Sciences,
Warsaw, Poland

Advisory Editors

Fernando Gomide, Department of Computer Engineering and Automation—DCA,
School of Electrical and Computer Engineering—FEEC, University of Campinas—
UNICAMP, São Paulo, Brazil

Okyay Kaynak, Department of Electrical and Electronic Engineering,
Bogazici University, Istanbul, Turkey

Derong Liu, Department of Electrical and Computer Engineering, University
of Illinois at Chicago, Chicago, USA; Institute of Automation, Chinese Academy
of Sciences, Beijing, China

Witold Pedrycz, Department of Electrical and Computer Engineering,
University of Alberta, Alberta, Canada; Systems Research Institute,
Polish Academy of Sciences, Warsaw, Poland

Marios M. Polycarpou, Department of Electrical and Computer Engineering,
KIOS Research Center for Intelligent Systems and Networks, University of Cyprus,
Nicosia, Cyprus

Imre J. Rudas, Óbuda University, Budapest, Hungary

Jun Wang, Department of Computer Science, City University of Hong Kong,
Kowloon, Hong Kong

The series "Lecture Notes in Networks and Systems" publishes the latest developments in Networks and Systems—quickly, informally and with high quality. Original research reported in proceedings and post-proceedings represents the core of LNNS.

Volumes published in LNNS embrace all aspects and subfields of, as well as new challenges in, Networks and Systems.

The series contains proceedings and edited volumes in systems and networks, spanning the areas of Cyber-Physical Systems, Autonomous Systems, Sensor Networks, Control Systems, Energy Systems, Automotive Systems, Biological Systems, Vehicular Networking and Connected Vehicles, Aerospace Systems, Automation, Manufacturing, Smart Grids, Nonlinear Systems, Power Systems, Robotics, Social Systems, Economic Systems and other. Of particular value to both the contributors and the readership are the short publication timeframe and the world-wide distribution and exposure which enable both a wide and rapid dissemination of research output.

The series covers the theory, applications, and perspectives on the state of the art and future developments relevant to systems and networks, decision making, control, complex processes and related areas, as embedded in the fields of interdisciplinary and applied sciences, engineering, computer science, physics, economics, social, and life sciences, as well as the paradigms and methodologies behind them.

Indexed by SCOPUS, INSPEC, WTI Frankfurt eG, zbMATH, SCImago.

All books published in the series are submitted for consideration in Web of Science.

More information about this series at http://www.springer.com/series/15179

Daniela Soitu · Šárka Hošková-Mayerová ·
Fabrizio Maturo

Editors

Decisions and Trends in Social Systems

Innovative and Integrated Approaches of Care
Services

 Springer

Editors
Daniela Soitu 🆔
Department of Sociology and Social Work
Faculty of Philosophy and Social-Political
Sciences
Alexandru Ioan Cuza University of Iași
Iași, Romania

Šárka Hošková-Mayerová 🆔
Department of Mathematics and Physics
Faculty of Military Technology
University of Defence
Brno, Czech Republic

Fabrizio Maturo 🆔
Department of Mathematics and Physics
University of Campania "Luigi Vanvitelli"
Caserta, Italy

ISSN 2367-3370   ISSN 2367-3389   (electronic)
Lecture Notes in Networks and Systems
ISBN 978-3-030-69093-9   ISBN 978-3-030-69094-6   (eBook)
https://doi.org/10.1007/978-3-030-69094-6

© The Editor(s) (if applicable) and The Author(s), under exclusive license to Springer Nature Switzerland AG 2021
This work is subject to copyright. All rights are solely and exclusively licensed by the Publisher, whether the whole or part of the material is concerned, specifically the rights of translation, reprinting, reuse of illustrations, recitation, broadcasting, reproduction on microfilms or in any other physical way, and transmission or information storage and retrieval, electronic adaptation, computer software, or by similar or dissimilar methodology now known or hereafter developed.
The use of general descriptive names, registered names, trademarks, service marks, etc. in this publication does not imply, even in the absence of a specific statement, that such names are exempt from the relevant protective laws and regulations and therefore free for general use.
The publisher, the authors and the editors are safe to assume that the advice and information in this book are believed to be true and accurate at the date of publication. Neither the publisher nor the authors or the editors give a warranty, expressed or implied, with respect to the material contained herein or for any errors or omissions that may have been made. The publisher remains neutral with regard to jurisdictional claims in published maps and institutional affiliations.

This Springer imprint is published by the registered company Springer Nature Switzerland AG
The registered company address is: Gewerbestrasse 11, 6330 Cham, Switzerland

Preface

From a systemic perspective, things can be understood and related in a network by engaging balanced resources and actors to assure the functionality. The systemic perspectives bring us the whole, in an integrative and unitary approach. It offers us tools to analyse and interpret inter-unities and interpersonal relations, functional schemes to observe and explore the real world, without rejecting it. The systemic perspective stresses the individual, institutional and organisational structures, micro, mezzo and macro levels, in their interrelations. They are open to other approaches, to proposals and suggestions, and adapted to the real world. The systemic theory sustains contexts for understandings, interpretations and actions, accepts and suggests adapted way of changes, involving even participant actors as changing agents. The systemic approach evolves behaviours and human relations, policies and interinstitutional frameworks, sometimes from an ecological lens and sometimes from a constructivist perspective. The individual in his/her environment, the potential of any actor to change, to adapt, to improve the things and the world around, and the impact of relations and environment are all underlined systemic paths used for analyses and proposals in this volume.

The first part, named "*Innovative and Integrated Approaches in Long-Term Care*", starts with an innovative example of the collaborative economy, i.e. "*time banks*", consolidated in recent years and presented by Alberto Sarasola-Fernández. As the author underlines, "*with time banks, citizens can achieve many benefits while covering part of their basic needs.*" Thus, there are economic benefits, both for the consumer who receives service at a low price or free, and producer who provides the service. The chapter underlines not only the economic but also the emotional benefits for users.

If sometimes citizens are finding solutions by themselves, there are also systems involving administration to support them. It is the example of Italy, where, starting from 2004, the state, represented by local authorities, intervenes in preventive and also curative support actions. Used usually for elderly and disabled, the tutor, as a person who is responsible for taking care of someone being unable to take care of her/himself, can be institutional support for many other categories being in incapacity or in risk situation of inability to be able to decide and administrate his or her goods or to decide. As Gabriele Di Francesco mentions, Italian law empowers the guardian judge to appoint a person as a curator or personal assistant for elderly, disabled persons, those needing long-term care, but also for alcoholics and other drug-addicted people. The chapter analyses the effects and social consequences for people in need of long-term care.

In the same country, i.e, Italy, from a legal perspective, something can still be done. Mariateresa Gammone draws our attention to dangerous discourses of common sense related to the relaxed approach of the results explained by: *"to err is human"*. She invites the reader to learn from experience, *"to fight mistakes and abuses"* by *"not ignoring or denying their existence but by building an institutional system that allows for learning from adverse events, making abuses and mistakes visible, traceable, and burdensome"*.

The invitation to learn, to prevent and manage better the individual health events in later life comes from the next chapter wrote by Daniela Soitu. The author underlines a shared responsibility for healthy ageing: it *"can be done individually, in the family, but also at the level of the educational, economic, environmental, cultural system."* The chapter makes significant contributions through the proposals it makes for three Romanian national programs as prerequisites for an excellent long-term care system: education and awareness of health determinants throughout life; regular assessment of the health status of adult citizens; services that support an active and healthy life.

Going back to the Italian context, we have the opportunity to read and analyse the *"recent transformations of social and welfare needs"* and a structure for the planning of social services according to emergent needs. As Vicenzo Corsi reveals, these are both framed by the population ageing phenomena and capacity of the governance to give answers for now and the future in the long-term care areas.

One explanation for living longer and healthy is found in the Mediterranean diet. It explains the numerous searches on Google about the miracle of this nutritional approach, even in the Iberian peninsula. Caruso and Fortuna provide us *"deep insights into population behaviour"* analysing the interest for Mediterranean diet through functional data analysis.

Active and healthy ageing suppose un open access to the labour market of older persons, even of retirees. Mihaela Rădoi, Gabriela Irimescu and Adrian Lupu stress the need for changing current perceptions of people's ageing and the need to invest in active policies for healthy ageing. Using the social determinants of health lens, authors stress the need for individual and systemic preparation. Otherwise, as European Commission emphasised (2012), all these processes will result in an increased demand for long-term health and care services, higher poverty rates, more significant social exclusion, and addiction among the elderly soon.

Preparing, preventing, but also intervention constitutes ways to improve older people's quality of life. Marius Neculaeş, Paul Lucaci and Mihaela Raluca Onose study how physical therapy and medical rehabilitation are increasing the quality of life of people diagnosed with chronic lumbar pain and arthroses. Specific interventions, adapted to individual's needs, increased the level of autonomy and gave back to the *"active world"* those 43, respectively 24 patients who received treatments and medical procedures.

People are living longer and healthier, but not always and not in any contexts. Life expectancy and healthy life expectancy index shows us geographical, cultural and individual differences. For the older adults with comorbidities living in Moscow, for instance, the geriatric syndromes are analysed by Sharashina, Runikina, Eruslanova and Tkacheva. Being all doctors, the authors measured, diagnosed and intervened according to people's needs and for their better life.

An integrated and innovative approach requests investments in an adequate training of human resources and especially of managers of care centres. It may concern daycare centres, facilities, home care centres, socio-medical institutions, public or private, with more or fewer beneficiaries or patients. Conţiu Şoitu and Daniela Soitu make "*An analysis of social care services for the elderly in Romania, focused on the particularities of top management from two perspectives: self-declared training needs and legal, administrative regulations*". Considering the "*significant change of the traditional Romanian model of intergenerational family care*", the chapter makes "*radiographs the current realities, but it is and can be a reference document for persons or institutions involved in (re) thinking the training offer.*"

The chapter is a natural link to the second part of the book: "*Actors in Social System. Decisions and Trends*". There is a need to know what is happening in the system: at a large scale: how are decisions being made; on a medium scale: how categories of beneficiaries, of professionals, of institutions, of administrations are acting and reacting; at a small scale: how individuals, families, small centres and communities are acting, deciding, promoting and integrating new visions and trends. Trends consist of analysis and solutions. In a postmodern world, many researchers and analysts are active actors, being involved in the transformative process, using transformative research. Each chapter, from this second part of the book, ends with solutions and proposals related to their finding data.

Nathalie Burnay performs a critical analysis of the transformations of the Belgian welfare model through a *marketisation process*. It supposes a "*dual process of privatisation of the home care sector and transformation of the benefits provided to recipients*" which will provoke a "*true paradigm shift through the introduction of a neoliberal logic that undermines the principles of the Belgian welfare state while shifting the burden of assistance to the family, and particularly to women.*"

The actorness of changes in the social care sector comprises institutional participation and institutional responsibilities. Daniela Soitu reveals data from an analysis performed over structural, financial, institutional, and legislative challenges from variate sources: documents, researches and analysis, opinions of specialists in the field of long-term health care. She also makes proposals for future policies, for a legislative act dedicated to long-term care, for other working funding mechanisms between institutions and care fields.

In the welfare sector, there are both seen and unseen actors. Antonella Sciarra compares the state' role, as a visible actor, with the market, having lighter spots but also hidden. A dilemma faced here starts from the principle of solidarity versus profit needed in the market. The author is advancing a solution to this conflictual relationship through a "*rational model of the ecological decision*".

A visible and stressful trend is that of assuring the quality of social services and care, in general. In Romania, this is an ongoing process from some years, and results start to appear: the beneficiaries are happier, dignify, felts respected by the providers, sensing the increasing quality of life. The accreditation and licensing of social and healthcare providers are significant steps in this process. Simona Bodogai analyses concrete decisions made to the certification, and licensing process of residential centres for the elderly in Romania, focusing on minimum qualitative standards care providers have to reach and drawing possible improvements. As Adina Daniela Rebeleanu and Paula Cristina Nicoara underlined, the accreditation process means "*compliance of the institution with a set of explicit criteria and, especially, known by the two parties—service provider and beneficiary.*" The authors performed research looking for an answer to an important question:

Is accreditation a sufficient and necessary approach for quality of elderly care? They revealed the social providers' perceptions concerning the quality of long-term care for the elderly, from an accreditation perspective and ended their contribution with critical conclusions.

Working in a residential centre/nursing homes/facility for the elderly is not the most straightforward job. Employees from these facilities can encounter, beyond work satisfaction and joy, different risk factors: depression, anxiety, burnout, emotional distress, and grief. In Spain, residential centres and nursing homes for the elderly are new trends and not so old administrative decisions who switch the paradigm from the Mediterranean model of care in the families towards public and private care in institutions. Ana Andrada Vallejo, Jose Luis Sarasola Sanches-Serano, Evaristo Barrera Algarin, and Francisco Caravaca Sanches have tested a battery of teste in ninety centres to make a single questionnaire adequate for studies concerning the quality of working life for the employees in nursing homes. In the following chapter, the four authors develop the study of the one identified risk factor: the mourning and the influence on burnout syndrome among nursing home workers in Spain.

Romania was, until some years ago, a specific model of informal care of all vulnerable persons, but mainly elderly and disabled, in the family. Demographic, economic and health challenges of last decades and years or months, have destroyed this model and raise new questions about state and administrative involvement in social and healthcare area. Luise Mladen Macovei and Andra Bertha Sănduleasa develop, in their chapter, innovative suggestions, to support services and funding mechanisms for informal care in Romania. These request involvement of new actors and insurances paths now and in a very near future.

Changes are related to perceptions and attitudes concerning ageing and older people. Mihaela Ghența and Aniela Matei present the results of qualitative research about this topic and find that there is a significant interest for the elderly and their image in Romania.

With the chapter signed by Ana Andrada Vallejo, Jose Luis Sarasola Sanches-Serano and Evaristo Barrera Algarin, the second part of the volume ends, having a double perspective on the mourning impact: institutional and among workers and their necessities. Having a massive emotional impact, the mourning in a residential context influence interpersonal and professional relations. Knowing the factors resulting from the survey performed by the authors, it may help in developing the support actions for all who experience such situations.

To have functional social systems and decisions, some prerequisites are needed on vertical, horizontal and multi-network levels. A few of the angles for the analyse contexts and to transform policies—at a local, regional or glocal levels—are investments in living standards, in training and education, in human resources and labour market, in the economy. At the European Union's level, such challenges came from new social and economic phenomena: (i)migration, financial crises, new pandemic situation, etc. In all these "*endogenous poli-crises*", European Union, as Bogdan Ştefanachi mention, have to manage the "*recalibration of the international balance of power*" and to re-legitimise in front of her citizens.

In the same regional and cultural contexts, Pedro Antonio Martín Cervantes, Nuria Rueda López, and Salvador Cruz Rambaud perform in 20 European countries "An Analysis of the Relative Importance of Social, Educational and Environmental Expenditures on Life Expectancy at Birth. Evidence from Europe". They find that social expenditure has the most significant relative importance on explaining life expectancy, whereas public educational expenditure has the least relative significance. Finally, there is no evidence demonstrating the extent to which environmental cost contributes to improving population health.

Education and actors involved in this process are significant. Franco Blezza, in his chapter "Who is the professional Pedagogist and How He Practices" took a synthetic examination of some "*toolbox*" the Pedagogist can use in its professional practice, in the context of a composite and plural culture.

A global challenge comes from new technologies and their omnipresence. Ayten Özsavaş Akçay, Fatma Baysen and Nermin Çakmak, using a qualitative methodology, investigate "Architecture Students' Smartphone Use in Design Studio". The chapter shows that students use smartphones frequently and would like to use more often to access on-point information, quickly, and rapidly without putting in much effort. The authors conclude on the needed guidance and motivation for students while using smartphones for Design Studio a close with recommendations for smartphone use.

Gender and its associated perspectives is a current debate in Europe and around the world. Simona Vrânceanu, in her chapter "Gender Equality Between Romanian Difficult Path in Women Political Representation and EU Perspective", proposes a theoretical approach of the pro-active engagement of women and some declarative reforms of women status. The chapter brings consistent terminology and request for standard European policies and actions for fulfilling the gender equality objectives.

Social biases and boundaries may be encountered in nursing homes. Javier Mesas Fernandez, Barrera Algarin and Ana Vallejo Andrada, in their chapter "Senior Citizen Centres and Sexual Affective Diversity: Homophobia and Residents", detailed the results of a study in 23 nursing homes for the elderly. Using a quantitative methodology, i.e. a questionnaire starting from Homophobia Scale, the authors find that residents reveal homophobic attitudes and believe when their opinion about homosexuality is investigated.

"Promoting Social Inclusion in Romanian Schools" is the next chapter written by Cristina Ispas and Alina Vişan. The authors present and analyse the dimensions of social inclusion in schools of disabled children. The paper underlines the opinion of the teaching staff from Romanian educational system regarding the characteristics of an inclusive school, i.e. "*the degree of promoting the inclusive school in the community, as well as the involvement degree of the family in the school situation improvement of the students with difficulties in learning*".

Old practices are kept in the societies for cultural, religious or traditional reasons. This is the case of "Societal Perception and Factors Promoting Female Genital Mutilation in Oyo State, Nigeria"—chapter written by Bolanle Misitura Oyundoyin, Fatai Adebayo and Tunbosun Soetan. Using a sample of 150 respondents, authors observe that there was a significant association between age, sex, educational level and perception of the parents on FGM. Considering the adverse effect of FGM on girls and women health, the authors recommend sensitive campaign to the grass root and inculcating consequences of female genital mutilation into the curriculum of primary and secondary schools, new policies and collective actions of all governmental and non-governmental organisations to eliminate this life-threatening act.

Traditions and historical paths remain in language, names and places. An interesting case study wrote by Theodora Flaut reveals it for a Romanian region in the chapter "The Etymology of Romanian Place Names. Case Study: The Vrancea Region". Discussing several family names that are common in Vrancea area, the author emphasises how these have turned into toponyms (anthroponyms) and hydronyms, revealing the relationship between people and places: "*the level of concern, affection, attachment and gratitude felt towards one's environment, as well as the friendly coexistence of the inhabitants, irrespective of their ethnic origin; they thus provide a unique insight into the lives and history of dwellers of this region*".

The next chapter focuses on research, and case study, but to an extent European level. A group of six authors, i.e. Maria Incoronata Fredella, Roberto Jannelli, Paola Materia, Maria Grazia Olivieri, Juan Jose Dañobeitia and Massimo Squillante, present the research "Performance Management Risk-Based Approach in the European Research Infrastructures (RIs): An Introduction to the Integrated Perspective. A Case Study at EMSO ERIC". They recommend a combined methodology for the management of research infrastructures, using Risk Management System, Internal Control Process and Management Accounting, together, into the governance structures and processes.

To a different scale, Irena Tušer and Jiří Jánský underline the issues of security management in the emergency medical services. Risks also include accidents at work of employees. As part of the prevention of accidents, in the chapter "Security Management in the Emergency Medical Services of the Czech Republic—Pre-Case Study", the authors suggest measures that would prevent the occurrence or repetition of the occurrence of an injury with the same cause or source.

In summary, the book *Decisions and Trends in Social Systems. Innovative and Integrated Approaches of Care Services* comprises new and innovative researches, policies and practices, models and theories for understand and influence the social systems. Due to the variate number of interests of the papers collected in this volume, the latter is addressed, in equal measure, to educators, sociologists, philosophers, social workers, health practitioners, policy-makers, stakeholders and more generally to scholars and specialists of different social sciences, humanities and health and life sciences.

Iaşi, Romania   Daniela Soitu
Brno, Czech Republic   Šárka Hošková-Mayerová
Caserta, Italy   Fabrizio Maturo

Contents

Contents

About the Editors

Daniela Soitu is a specialist in social sciences, Ph.D. in sociology, habilitation in social work, Professor in the Department of Sociology and Social Work, Faculty of Philosophy and Social-Political Sciences, Alexandru Ioan Cuza University of Iași, Romania. She is an active member of the Scientific and Organising Committee of the International Conferences: *Integrated Systems of Long-Term Care (ISOLTC), Recent Trend sin Social Sciences (RTSS)*, 10th edition of *The European Conference for Social Work Research*. Professor Daniela Soitu teach, among others, *Life course approach and Lifelong well-being, Social Work with and for the Elderly, Older Families, Social work System, Counselling in Social Work, Counselling for Personal Development*, etc., and coordinate students for Applied social work and preparing final dissertation. She is the author and co-author of over 80 articles, studies and book chapters. From 2008, she is leading, as Editor in Chief, the peer-reviewed journals: *Scientific Annals of Alexandru Ioan Cuza University of Iași, Sociology and Social Work Section*, and for five years, the quarterly bilingual *Journal of Social Economy / Revista de Economie Sociala*. As a manager or team member in over 50 educational, research and developmental projects, Prof. Daniela Soitu has international experiences on social sciences, social and healthcare policies, practices and researches. She supports the intersectoral policies and integrative practices for vulnerable persons and groups. Currently, she is leading a working group focused on Good Health and Well-being in the *European Universities Alliance: EC2U – European Campus of City-Universities* and is coordinating a team from her university in a partnership with European Ageing Network and other training providers in a European project aiming to *modify the educational module finalised with a European certificate for long-term care providers for seniors.*

Šárka Hošková-Mayerová is full Professor in the Department of Mathematics and Physics at the University of Defence, Brno, Czech Republic. She is co-author of the monograph *Quality of Spatial Data in Command and Control System* and co-editor of four books published in Springer publishing house. She is author or co-author of several chapters of books and papers in valuable journals (e.g. *Soft Computing, Computers & Mathematics with Applications; Analele Stiintifice ale Universitatii*

Ovidius Constanţa-Seria Matematica; and Quality & Quantity). She is Chief Editor
of the journal *Ratio Mathematica* and Associate Editor of the *Journal of Intelligent
and Fuzzy Systems* published by IOS Press. Moreover, she is a member of the editorial
board of various journals, e.g. *Italian Journal of Pure and Applied Mathematics and
Advances in Military Technology.* Her areas of interest are mathematical modelling,
decision-making process, algebraic hyperstructures and fuzzy structures.

Fabrizio Maturo has a PhD in economics and statistics and is a researcher in
statistics at the Department of Mathematics and Physics of the University of
Campania Luigi Vanvitelli in Caserta, Italy. He is a lecturer in statistical learning
and experimental research designs at the international bachelor's degree in data
analytics. Starting from April 2019, Fabrizio is qualified to function as an associate
professor in statistics. Previously, he has been a researcher in statistics at the
University G. D'Annunzio of Chieti-Pescara (Italy) and then a researcher in
biostatistics at the National University of Ireland Galway. Fabrizio is the author of
about 70 publications in international journals and books. His main research
interests are FDA, biostatistics, IRT, classification, data streaming, fuzzy set theory,
econometrics, R programming and statistical models for business and finance. He
is the creator and maintainer of the R packages BioFTF and AnnuityRIR. Fabrizio
participates in international research groups in Ireland, Italy, France, Spain,
Romania and the Czech Republic. He is the chief of Ratio Mathematica – Journal
of Mathematics, Statistics and Applications, an associate editor of the Journal of
Intelligent & Fuzzy Systems and the Journal of Statistics and Management
Systems, review editor of International Journal of Public Health and a member of
the editorial board of Italian Journal of Pure and Applied Mathematics. Fabrizio
has served and works as a referee for many journals in statistics, biostatistics and
mathematics.

Innovative and Integrated Approaches
in Long-Term Care

Time Banks as an Example of Collaborative Economy

Alberto Sarasola-Fernández ⓘ

Abstract In recent decades we have witnessed the growth of a series of activities that are part of a new vision of the economy. We refer to the collaborative economy, extending its areas of action projects arising in different aspects. Within these areas we will focus on the banks of time, an initiative that has been consolidated in recent years, partly because of the situation of sustained global economic crisis. With time bank, citizens can achieve a number of benefits while covering part of their basic needs. Thus, we can say that shows economic benefits, both for the consumer who receives a service at a low price or free as part of the producer who provides the service. We also found emotional benefits users. The quantitative technique and the online questionnaire was used.

Keywords Time banks · Elderly people · Social work · Active aging · Old people · Collaborative economy

1 Introduction

In recent years we have witnessed the consolidation of various initiatives that although developed in the economic sphere, have characteristics that differentiate them from activities that fit into the conventional economy. We refer to those that are part of the collaborative economy, which is part of a new economic model booming worldwide, including various initiatives and practices.

According Cañigueral [1], it is an economic model that "empowers citizens (who are no longer mere consumers to become producers of value in many areas) and has economic, social and environmental benefits."

García [5], p. 4 defines it as a movement with a clear social orientation and whose foundations are "cooperation, innovation, proximity, sustainability, community, solidarity, responsible production and consumption, and even, ethical finance".

A. Sarasola-Fernández (✉)
Department of Social Work and Social Services, Pablo de Olavide University, Utrera Road, 1, 41013 Seville, Spain
e-mail: Alberto-sgs@hotmail.com

© The Author(s), under exclusive license to Springer Nature Switzerland AG 2021
D. Soitu et al. (eds.), *Decisions and Trends in Social Systems*,
Lecture Notes in Networks and Systems 189,
https://doi.org/10.1007/978-3-030-69094-6_1

3

In both definitions we can see that clearly shows characteristics that differentiate it from conventional economics: the role of citizens, social, economic and environmental sustainability …. In short, the economy is considered as a means to achieve the integral development of individuals and communities, considering that resources are limited and assessing cooperation over competition.

For various reasons, such as economic crisis, the greater possibilities of communication that gives technology, cultural changes …, citizens have found other ways to interact with the market, different to buy, such as selling, exchange, loan, rent … that is, the consumer happens to be a taxable person to be an asset subject:

- On the one hand, he discovers he can make a profit by sharing their goods and skills with others (e.g., car, house or cooking skills)
- On the other hand, you can access the temporary use of products or services that can not or will not acquire.

It's what the American writer Alvin Toffler called prosumer in his book "The Third Wave" [8], word that sums up in one word the words producer and consumer, and that is part of the essence of collaborative economy.

To get an idea of the importance of the collaborative economy, we can say that the National Commission of markets and competition and in 2014 became aware of its growing importance and conducted a study on the new models of service delivery posed by technologies information and in particular on the so-called collaborative economy. This model of consumption promotes exchange between individuals of goods and services which hitherto were idle or underused. The growth of this new model of consumption has been boosted by the economic crisis and global awareness that is not as important property as the use made of goods and services.

2   Time Banks

As discussed, Time Banks are an innovation in the field of collaborative economy that responds to both social and economic needs. Within this framework, people have found an ideal place to socialize and improve their quality of life, to feel active and useful to society space, and where you can feel fulfilled and valued.

These experiences can lead to new forms of collaboration between citizens of a town, promoting the relations of solidarity, common good, cultural development, etc.

Therefore, we believe that Time Banks are presented as a new proposal whose many benefits deserve special attention, not only from professional practice, but also through the development of studies and research that address in depth the positive impact that participation in these initiatives can have on the quality of life of people.

These are initiatives that can not only be a tool with great potential for professionals working in the social sphere, but may provide new intervention scenarios that in turn generate new job opportunities.

Del Moral [3] provides the following definition of Time Bank:

A Time Bank is a network exchange support, skills and knowledge in which the currency is time. People put their time available to/os others and hope to have the time of/os other/os members/os to meet the daily needs.

According to Gisbert [6], we can define Time Banks as: "An economic experience of local character that provides the community where it implants of information services that its members can be interchanged using a coin or own currency for transactions whose value is agreed by themselves and whose name is characteristic, recording all transactions between them."

Main feature should be noted that in the Time Bank (BDT) exchange unit is the hour, regardless of the service offered or received. Thus, as valued the work of a person performing household chores, that of an economist who makes a tax settlement.

Usually only services are exchanged, not products, although some BdT are also integrating the exchange of products, such as BdT of Burgos.

They are also characterized by being essentially local nature experiences since exchanges are made between the inhabitants of the area in which it is located. So far they have not been successful experiences of regional or national BdT. This is because it is very important confidence among participants and for anyone to feel cheated or have intent to defraud is important that exchanges are made on a small scale for so you can keep personal contact. Although we can not forget that the more participants there are, the greater the possibilities of exchange and variety of services.

They run through a promoter and management team, which can be neighborhood associations, cultural associations, municipalities, schools, universities or other institutions (homes for the elderly, prison, etc.).

It is a tool that fosters communally services cooperation and solidarity among people. Thus improving community health and quality of life for them.

The BdT have a number of basic principles [2]:

- Equality: among people as in the BdT all services are valued according to the time unit.
- Confidence: among users of BdT belonging to a community bonds of trust and solidarity narrow.
- Multiple reciprocity: which arises because all participants providing and receiving services. Multilateral exchanges are such that the service does not have to be returned to the person who has borrowed, but any other participant BdT.
- Flexibility and voluntariness: the exchanges do not have to be regular but more often they are specific in nature and arise spontaneously, because we must not forget that providers are particular because it is not professional services.
- Accountability: Each user must respond BdT quality of the service provided. In addition, each enrollee has a liability insurance. The BoT is responsible for monitoring compliance with the regulations and standards adherence partners.

Time Banks are already realities we have been finding in many towns and villages of our country, integrating them to a lot of people using, and generate a number of benefits such as:

- Promote social integration, since the BdT appreciates the service provided by every person equally, regardless of age, education level, gender, social class or disability.
- economic benefits, both from the point of view of the consumer, which gets the provision of a service to a free or low price, as the producer or user can provide a service.
- Avoid isolation and loneliness. According to Fernández and Ponce de León [4], p. 94, Time Banks are a resource that "offers the possibility of greater social contact, avoid isolation, prevent loneliness, activate self-esteem, allow intergenerational encounter, maintain activity, improve social cohesion and combat exclusion".

Thanks to the activities carried avoid people immerse themselves in isolation, so it offers the opportunity to share space and time with others through exchanges and group activities organized.

- Encourage support network of the person. Through entities BdT users can weave a web of intergenerational support, since in these projects involving people of all ages and social status, which makes improve coexistence, social peace and cultural development, both people themselves as municipalities or territories in which the work of BdT develops.

This benefit occurs regardless of the age of users of BdT is particularly noteworthy in the case of the elderly. "Social networks have a positive and protective effect. Those older people who have frequent contact with family, close friends and neighbors, tend to have better mental and physical health than those who are less involved. Moreover, greater involvement with the neighborhood and the community is associated with more social support, more physical activity and lower stress levels" [7].

- Emotional well-being. Users participating in BdT get a series of emotional benefits such as feeling useful, considered an integrated within the community in which their daily lives, feelings of mutual aid … all strengthens the bonds of trust and solidarity unfolds part between BdT members.

For older people involved in the BoT, they have a function or occupation within them, so that every person is valuable in offering assistance. This fact makes them feel special and useful to people, enabling them to share knowledge, facts and experiences, thus helping to improve the quality of life of the participants.

- Increased activity and exercise. Regular physical activity or exercise, especially in the elderly, contributes greatly to delay the onset of motor deficiencies and improve functional capacity, prevents morbidity and mortality cause many diseases, favoring autonomy, and therefore improving the quality of life (Spanish Society of Geriatrics and Gerontology, nd).

Finally, we can also see that a social paradigm shift is achieved and certain achievements alzanzan as can be:

- Ethnic barriers disappear when each group is seen as part of a whole.

- Age barriers fade with intergeneracionismo.
- Strengthen the bonds of community bonding, providing a social support network and support for individuals and families.
- Attention to vulnerable groups.

3 Methodology

This research is part of an extended investigation about Time Banks and ederly people. On this occasion, we decided to focus on the satisfaction of the people participating in the entity, with the following assumptions:

- People who participate in the time banks have a positive view on the Banks of Time.
- Older people have a high degree of satisfaction with the entity.
- There are optimal conditions for intergenerational relations between the participants.

With the aim to answer these concerns we have used a quantitative technique, the questionnaire was created using a Google Survey and social media was used to distribute and compile information.

To study in more detail the case of Seville and the importance of the Cross Project, we have carried out an investigation, in which we have worked with the data provided by a sample of 100 users of the Banco del Tiempo of Nervión-San Pablo.

The participants in this study are between the ages of 16 and 80. 3% of respondents are under 18, 45% are between 18 and 65, 49% are over 65, and 3% have not answered the question.

Regarding gender, 70% of the people surveyed were women, compared to 30% who were men.

Our research factors are (Table 1).

Table 1 Research factors

Code	Factor	Question
BDT001	Sex	1
BDT002	Age	2
BDT003	Studies	3
BDT004	Profession	4
BDT005	Volunteer in another organization	5
BDT006	Frequency of participation	6
BDT007	Satisfaction level	7

Source Author compilation

4 Results

In this section, we are going to focus on the results of our research and the possible relationships between the factors which could be interesting for our conclusions (Graph 1).

In this graph, we could see the level of satisfaction whit the services in the Time Bank. In this case, most of the participants show a very high level of satisfaction with their participation and interaction with other people.

Table 2 show as the level of satisfaction with respect to gender. Where we can appreciate that women have a greater degree of satisfaction than men.

By age groups, we observed that it is the elder and young groups that are most satisfied (Table 3). The group of adults includes a user not very satisfied, and presents a more similar distribution in the categories of extremely satisfied and satisfied (Graph 2).

It also highlights the degree of satisfaction shown by the users of the time bank in terms of the relationship between the participants, whether from the same age group or a different one.

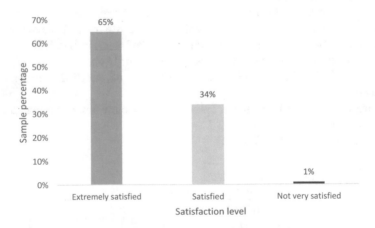

Graph 1 Satisfaction with the services rendered or received. *Source* Own elaboration

Table 2 Service given or received by gender

		Service given or received		Total
		Extremely satisfied	Satisfied	
Gender	Female	47	22	69
	Male	18	12	30
Total		65	34	99

Source Own elaboration

Table 3 Service given or received by age group

Age group	Service given or received			Total (%)
	Extremely satisfied (%)	Not very satisfied (%)	Satisfied (%)	
Adult	56.5	2.2	41.3	100
Elder	72.0		28.0	100
Young	75.0		25.0	100
Total	65.0	1.0	34.0	100

Source Own elaboration

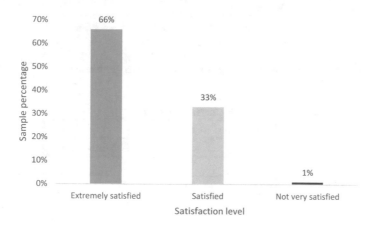

Graph 2 Relationship between participants. *Source* Own elaboration

By gender (Table 4), we appreciate that there are no significant differences because, although there is an association between the variables (Chi-square = 0.020), it is so small that we can talk about independence (Phi = 0.014) (Graph 3).

Regarding the service of the Time Bank in general also highlights the high degree of satisfaction, both by professionals who develop their work, as external staff invited on occasion, such as professionals from some sector, volunteers, etc. (Table 5).

If we perform the study by gender (Table 6), we again observe that both variables are practically independent (Chi-square = 0.429 and Phi = 0.065).

Table 4 Relationship between participants by gender

		Relationship between participants			Total (%)
		Extremely satisfied (%)	Not very satisfied (%)	Satisfied (%)	
Gender	Female	67.1	1.4	31.4	100
	Male	66.7		33.3	100
Total		67	1	32	100

Source Own elaboration

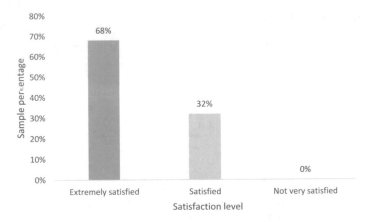

Graph 3 Level of satisfaction with the general service. *Source* Own elaboration

Table 5 Satisfaction with the service in general by gender

		Overall, the service		Total (%)
		Extremely satisfied (%)	Satisfied (%)	
Gender	Female	70	30	100
	Male	63.3	36.7	100
Total		68	32	100

Source Own elaboration

Table 6 Satisfaction with the service in general by age group

Age group	Overall, the service		Total (%)
	Extremely satisfied (%)	Satisfied (%)	
Adult	63	37	100
Elder	70	30	100
Young	100		100
Total	68	32	100

Source Own elaboration

By age groups, it is the young and elder who are extremely satisfied to a greater extent with the service provided and received within the organization (Graph 4).

Another question that we find interesting to know is the degree of satisfaction related to the level of attention received and the quality of the care provided (Fig. 10). In both cases 69 of the 100 users surveyed showed to be extremely satisfied, between 28% and 30% satisfied and less than 3% not very satisfied.

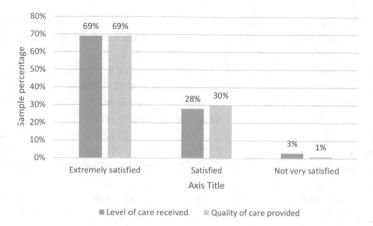

Graph 4 Satisfaction with the level of attention received and with the quality of the attention provided. *Source* Own elaboration

5 Discussion

After an analysis of the theory and the results provided by the investigation, we can observe that the people who participate in the Banks of the time are generally active people, as Cuenca has stated in its publication.

The elderly are a population group that for their large amount of free time, availability, willingness to feel useful, etc., are one of the basic pillars of time banks. It is not surprising that their level of satisfaction with the service is so high, due to the great time they provide and the work they perform.

This level of satisfaction, we can relate, as established Cañigueral, to the feeling of continuing to contribute things to society, making them valid and useful, as well as the possibility of interacting with people who, if it were not for this area, would not have the opportunity to do it.

Finally, it is very important to note that the high level of satisfaction of the participants is with the workers of the time bank, with the services they provide and receive, and with the rest of the users, which underscores the good atmosphere occur in these types of organizations.

6 Conclusion

In this research we wanted to reflect on one of the initiatives that are framed within the collaborative economy: we refer to the Banks of Time, in which objects and money are not exchanged but knowledge, skills and services.

With this type of initiatives we are moving from the donation of time to the exchange, so that this makes it possible for new people to work. Remember that in the Time Banks time is the unit of payment, the unit of measurement and also the deposit of value that allows moving current productions to future consumption (that is, allowing accumulation of savings).

Although we believe that these types of initiatives are not going to replace conventional economics, we believe that they can be complementary, allowing certain people to satisfy needs and provide services, so that the initial sense of economic functioning is recovered. Thus there should be no advantage for anyone. This type of initiatives remind us that the essence of economic exchanges lies in the fact that the members of a community need us, being the exchange a form of solidarity between those of us who dedicate ourselves to activities that do not generate food and the rest of goods and services that we need to live.

The exchange of activities with other people to solve situations of need of daily life, facilitate the expansion of contacts and resources, predispose us to trust others … are some of the benefits inherent in the collaborative economy and the Bank of time.

Adolescents, the unemployed, housewives, elderly people … who have no place in the formal economy, can become service providers in the Bank of time, which in this way recognize their skills and provide them with a second opportunity to feel useful and able to access to present and future services that could not otherwise afford.

For us to be aware of the importance of the Bank of time, we can say that there are currently more than 300 operated in Spain. That is to say, we already have a high number of experiences that, using the unit of time as payment of benefits and services, are revolutionizing the way of understanding solidarity and work in the community. And to show its potential, we have studied in more detail a Bank of time that has taken part in the last years of a European Project, in which we wanted to promote this activity of collaborative economy through the use of a digital platform that allows the relationship between providers and receivers of Bank of time services. After carefully studying the characteristics of the selected sample, we can say that in general terms the participants in the European project CROSS have felt satisfied or very satisfied with the experience, so it is to encourage the participation of new users in the future.

Likewise, we have detected that the online knowledge of the Bank of Time initiative is insufficient, which is why efforts should be made in the future, given the importance that new technologies play in today's society and it is foreseeable that they do so in the medium long term.

Finally, we want to comment that we firmly believe that this type of initiative, which has received a greater boost in times of economic crisis, will not disappear when a period of improvement of the economy arrives. Citizens have become aware of the need to stay out of the vagaries of the globalized economy, which at certain times expels a large group of citizens from the market. At the same time, these initiatives of collaborative economy have favored the strengthening of networks of mutual help, solidarity and empowerment, necessary feelings to face the challenges of daily life.

References

1. Cañigueral A (2014) Vivir mejor con menos: descubre las ventajas de la nueva economía colaborativa. Penguin Random House Grupo Editorial, S. A. U, Barcelona
2. Cuenca C (2007) La Gestión del tiempo y su intercambio: los Bancos del Tiempo. In: Esteban C, De la Torre I, Martín JI, Rodríguez JM, De Puelles I, Vega P, Fernández P-, Alia JC, Vidal R (eds) Gestión del Tiempo y Evolución de los usos del tiempo. Visión Net, Madrid, pp 121–134
3. Del Moral L (2014) Trueques e intercambios de tiempo: ¿respuestas inmediatas o propuestas de fondo frente a una crisis multidimensional y sistémica? Economista Sin Fronteras 12:15–18
4. Fernández T, Ponce de León L (2013) Envejecimiento activo: recomendaciones para la intervención social desde el modelo de gestión de casos. Portularia 13(1):87–97
5. García J (2014) Guía de economía social y solidaria para la administración local. Diputación de Barcelona, Barcelona
6. Gisbert J (2010) Bancos de Tiempo y monedas locales en la red. Confederación Española de Cajas de Ahorro, Imprenta
7. Lang G, Resch K, Hofer K, Braddick F, Gabilondo A (2010) Background document for the thematic conference on mental health and well-being among older. European Communities, Luxembourg
8. Toffler A (1980) The third wave. William Morrow (US), ISBN 0-517-32719-8

Long Term Care and Support Administration

Gabriele Di Francesco

Abstract Since 2004, Italian legislation has established and regulated the role of the Support Administration. The law responded to the growing needs of long-term care among the population, protecting more effectively those who find themselves unable, even partially or temporarily, to provide for their own interests, due to an illness or physical or mental impairment. The elderly and the disabled, but also alcoholics, drug addicts, terminally ill patients can obtain, even in anticipation of their own future incapacity, that the guardian judge appoints a person who takes care of their person and their heritage. The Support Administration aims to protect people who, due to their conditions, are totally or partially unable to perform the functions of daily life. The chapter aims to analyze the effects and social consequences of Support Administration for those who need long-term care.

Keywords Support administration · Italian legislation · Long-term care · Social consequences

1 Long Term Care and Conditions of Non-self-sufficiency

Long Term Care (LTC) means that set of interventions necessary to ensure adequate assistance to individuals in conditions of non-self-sufficiency, caused by injuries, diseases or disabling pathologies, but also only by aging. These interventions, provided by public or private institutions, generally can be included in the complex of healthcare services and/or in the context of the complex of social-assistance services,

---

It is more important to know what sort of person a disease has.
than to know what sort of disease a person has.
Hippocrates c400BC.

G. Di Francesco (✉)
Department of Business Administration, Faculty of Social Work, "G. d'Annunzio" University of Chieti-Pescara, via dei Vestini, 31, 66100 Chieti, Italy
e-mail: gabriele.difrancesco@unich.it

© The Author(s), under exclusive license to Springer Nature Switzerland AG 2021
D. Soitu et al. (eds.), *Decisions and Trends in Social Systems*,
Lecture Notes in Networks and Systems 189,
https://doi.org/10.1007/978-3-030-69094-6_2

15

both at home and in in care homes. "Attention to these forms of social security insurance is mainly due to the aging process of the population that affects all industrialized countries. This phenomenon creates serious financial and coverage problems in various sectors of the welfare state such as the pension system, the demand for health services and, more specifically, the demand for those long-term social and health benefits of which a person needs in the event of total or partial loss of their autonomy" [23, p. 43].

LTC is generally understood as a guarantee linked to problems of non-self-sufficiency. It is a protection and care solution that is activated in the event that you completely, permanently and irreversibly lose your ability to carry out the elementary activities of daily life independently, related to mobility, nutrition and personal hygiene [31, 36]. Although they concern a multiplicity of fragile subjects such as cancer patients, severely disabled people, mentally ill, suffering from degenerative diseases, etc., these situations more specifically concern the condition of the elderly people [30], pp. 196–197.

According to United Nations data, in 2010 the world population reached 6 billion and 895 million in 2010. At the same time, the percentage of people aged 65 and over increases in all parts of the planet [8].

The population of the European Union (27 members) as of January 1, 2010 was estimated at 501.1 million people. Of these 87.1 million were 65 years old and older. The two countries that lead the European ranking are Germany and Italy. As of January 1, 2010: 16.9 million people aged 65 and over lived in Germany while in Italy there were 12.2 million [8].

The over 80s in Europe were 27.3 million in 2016, 7 million more than ten years ago. The increase in elderly people in the EU (from 4.1% in 2006 to 5.4% in 2016) means that in 2016, one in every 20 people living in the EU was 80 years old and over. The aging of the population is, at least in part, the result of a growing life expectancy: at 80 years old it rose from 8.4 years in 2005 to 9.2 years in 2015. Although the percentage of women in the population of 80 years and more declined between 2006 and 2016, still accounted for about two thirds (64%) of older people in the EU [16].

Italy is among the countries with the highest percentage of elderly people. Currently, the life expectancy at birth of males is 80.1 years, while that of women is 84.7 years and consequently the elderly have become increasingly numerous. It is a challenge from both a health, economic and social point of view. Indeed, the progressive lengthening of life requires society to take on the responsibility of ensuring the elderly live as long as possible in good health [2].

Eurostat has drawn up a ranking of the oldest European regions, which sees at the top places eight Italian regions that have a population over 80 years old: Umbria, Marche, Abruzzo, Friuli-Venezia Giulia, Emilia Romagna, Molise, Piedmont and Liguria [16]. Europeans live longer and healthier than in the past, but at the same time the rate of not full self-sufficiency also increases.

The latest ISTAT Multiscopo survey on the "Health condition and use of health services" found that 18.5% of over 65s (about 2.5 million people) report a condition of total lack of self-sufficiency and it is estimated that in the 2050 will be over 5 million [34].

Further demographic and social changes taking place reinforce the need to take care of yourself. These include the drop in birth rates and the consequent reduction in households, the increase in the involvement of women in the world of work and the substantial decline in the working-age population, with a contraction in the number of employees [35].

The Italian LTC system has traditionally compensated for the shortcomings in the offer of direct services with monetary transfer mechanisms. The cost of these transfers, in particular for the accompanying allowance, has increased significantly in recent years reflecting the actual increase in need, also in relation to various social, cultural and organizational factors [34].

In reality, the whole European population lives longer and in better health than in the past, but it is good to underline how both the rate of not full self-sufficiency and the need to face assistance problems increase. The traditional model which largely entrusted the functions of care, assistance and moral support to families has definitely entered into crisis in the Mediterranean countries, which have also maintained this organizational structure longer than in other countries.

The family structure has gradually decreased in terms of components and organizational capacity. The new cultural models that impose themselves see the expansion of the mononuclear families, with only one component.

From this it follows that the organization of roles and functions has been totally changed. In fact, families are no longer able to provide valid support for people in difficulty. Indeed, there is an increase in social isolation which also affects the community level, where participation in social life becomes occasional if not entirely punctual, that is, limited to rare circumstances. In this scenario it is necessary to face the problems related to assistance, care and sociability.

2 Family Transformations and Care Functions

The families present in Italy today are about 25 million and 300 thousand. In twenty years the average number of components has fallen from 2.7 to 2.4. Profound demographic and social changes have affected our Country, triggering a gradual reduction in the family dimension, which has also affected our South, which historically has the primacy of large families. As a consequence single-member households, with only one component, have increased enormously, which from 20.5% of the two-year period 1995–96 now reach 31.6%, practically one family on three [37]. The large families, with five or more components, have declined, falling from 8.1% in 1995–96 to 5.4% today. Families with two components represent 27.3%, those with three 19.8% and those with four 16%. Otherwise it can be said that in comparison to the 31.6% of single-member households, 20.5% is represented by couples without children, 34.7% by couples with children, 9.7% by single-parent families (whether of the mother—more often—or of the father), 1.4% from families with two or more families and the remaining 2% from other types of families [7].

Already these first data indicate that the Country has been in a phase of full demographic decline since at least 2015. The natural movement of the population in 2017 shows a negative balance (between births and deaths) of about 200 thousand units, mainly due to Italian residents [37].

The natural movement of the population shows from 2017 to 2019 a negative balance (between births and deaths) of about 200 thousand units per year ($-191,000$ in 2017, $-187,000$ in 2018 and $-212,000$ in 2019), mainly due to Italian residents. However, Italy has a primacy over the average life expectancy at birth which is now close to 83 years, higher for women and a little less for men [5].

The elderly represent 22.6% of the total population; the old age index shows that today there are 168.9 elderly people per hundred young people, while the statistical dependency index reveals that compared to one hundred people of working age there are 56 of them in non-working age. These data place Italy in first place in Europe for the old age index and fourth for the dependency index. However, things change if we consider the indicator relating to years lived in good health and without limitations in activities from the age of 65: in this case, the Italian elderly are below the European average [17, 37].

According to the data, it would seem that marriage is affected by the loss of long-term planning capacity among the new generations. The reasons are manifold but overall it would seem that the choice of marriage is no longer central to social life: as recently noted by the director of Censis, Massimiliano Valeri, marriage is no longer the center of gravity of people's lives, it is not the primary reason leaving the family for young people, it precedes the experience of parenting less and less, it is no longer a mechanism of social ascension for women. In this sense, marriage becomes an option subordinated to the individualistic trait of the time. Certainly the cultural factor counts, which no longer recognizes the authority behind the institution, but also the instability of individuals who, more and more alone and less protected, no longer risk, "aware that every gamble would leave impressed deep scars on one's lonely personal biographies" [1, 37, p. 10].

In 2006, 7.2% of Europeans lived in a condition of social isolation: the data represents the share of those who declare that they have no contact with friends or relatives not even once a year. It is a fact that Eurostat researchers rightly consider an important indicator of social exclusion the most obvious and burdensome expression of relational deprivation is just not being able to get any help from family or friends [27].

Italy stands out among other European countries and, while taking into account the possible exceptions and distortions of the data, certainly the indication that comes from such high rates is that of not neglecting this dimension of poverty [14].

In the face of these transformations, it is noted that there is a marked increase in longevity. We live a greater number of years, but these are years in which health is precarious and conditions of fragility and disability have increased.

Over the years, a change has also developed in the epidemiology of emerging diseases: we have gone from a prevalence of infectious and deficient diseases to a preponderance of chronic degenerative ones.

There is also a trend towards a contraction in services, probably a sign of the repeated economic crises over the period of time from 2008 to today and paradoxically increases the care-load of families. Consequently, problems related to assistance, care and, not least, sociality and interpersonal relationships arise [8, 11].

3   Problems Related to Disability, Health, Care and Sociability

Due to aging and chronic degenerative diseases, disability is worryingly increasing with a rate of 6.7% (about 4.1 million people) in 2010, of 7.9% in 2020 (equal to 4.8 million of people). "According to the Censis estimate [4], disability will grow to 10.7% in 2040 (equal to 6.7 million people) an increase destined to create a very strong increase in the demand for services" [29].

From the chapter "The welfare system" of the 53rd Censis Report on the italian social situation of the Country/2019, it is noted that in Italy non self-sufficient people have grown by +25% since 2008. The great majority of them are elderly: 80.8% are over 65 years old. 20.8% of the elderly are not self-sufficient. The public responses to a phenomenon destined to grow are insufficient and inadequate, given the progressive aging of the population.

The burden of non-self-sufficiency falls directly on families, called to count on their own economic and care forces. For 33.6% of people with a non-self-sufficient component in the family, welfare costs weigh heavily on the family budget, against 22.4% of the total population [6].

A homogeneous and coherent model of support for non self-sufficiency is not present always and everywhere in the Italian regions. On the contrary, in some areas there are different types of support and not a few disadvantages for resident citizens.

In particular, four support models can be identified:

- the social-welfare residential type advanced in some northern municipal areas;
- the high-intensity care solution for elderly beneficiaries of continuous assistance services in some central and northern regions;
- the low-growth model of the residential network, a high rate of beneficiaries of the accompanying allowance and a tendency to monetize services (central southern cash for care);

- the medium/low care intensity model where the spread of *Integrated Home Care* (Italian ADI) and *Home Care Service* (Italian SAD)[1] is slightly lower than the national average [23].

For socio-welfare purposes, moreover, the identification of a person experiencing situations of disability does not follow clear and unambiguous definitions. We note the use of the term "handicapped" or the term "disabled" or "incapacitated" as a synonym for a person who is not self-sufficient with extremely varied purposes and linked to different motivations and interventions over time. An example of what has been said can be found in the use of the term disability and invalidity: sometimes reference is made to the individual's ability to carry out the fundamental activities of human life independently; other times we refer to people who have reported biological damage regardless of self-sufficiency in daily life.

A further problem is related to the practice of using the concept of disability, intended as a reduction in the ability to independently carry out certain daily activities, in a regulatory framework in which public intervention finds partial support in the profit and non-profit private sectors.

For the purposes of protecting the disabled person, the rules also raise many uncertainties. For example, we distinguish rules in which the conditions of non-autonomy are reduced, evaluating the economic, social and welfare aspects from time to time, from others in which the need to integrate the health and social dimensions into the definition and assessment of disability is highlighted.

With reference to the international classification of disability, the World Health Organization (WHO) has paid attention to those environmental and personal factors that limit the ability to function at an individual level [23].

A critical point is the focus of fair and appropriate interventions within a welfare system, characterized by a significant asymmetry between the incomplete social and health care model.

Consequently, uncertainty ensues as to the data relating to the "recipients" of LTC services.

A 1999 ISTAT survey on "Health conditions and use of health services", developed with the Multi-Purpose Survey of 1999–2000, allows us to identify the three fundamental functions to define disability:

- the functions of daily life, which includes personal care activities (washing, dressing, eating etc.);
- the functions of mobility and locomotion (walking, going up and down stairs, etc.) which can be defined as a sort of confinement of the individual in a bed or chair or home;
- the functions of communication, including speech, sight and hearing [22].

---

[1]Integrated Home Care means the intervention at home of the patient by health professionals, doctors, nurses, rehabilitation therapists and/or staff assigned to patient hygiene according to a personalized care plan. The Home Care Service aims to help the person in the handling of daily activities such as cleaning rooms, preparing meals, errands, etc.

In this context, ISTAT operated on the sample subject to estimate, defining the person who reported a total lack of autonomy for at least one essential function of daily life as "disabled" [23].

A model of intervention is therefore identified with regard to the disabled person, which includes a whole and heterogeneous set of highly specialized services: medical and nursing support, personal care, help domestic, social support and social secretariat [21]. This care model is not only aimed at chronic or disabled patients, but also at people with serious addictions, terminally ill and elderly who are not self-sufficient or suffering from age-related diseases, with the aim of improving their quality of life.

LTC services should also be included among the various interventions, which are also attributable to home and residential care. In fact, it is believed that domicile is the privileged form of assistance to guarantee a better quality of life and contain costs for assistance. With home care it is possible to maintain a certain degree of independence and self-sufficiency of people in a way that is more functional to their needs, since "the home is a place of physical and emotional associations, memories and comfort" [21]. In Italy, the LTC includes three types of assistance: home or residential interventions (with a strong health and social component) and financial services (the accompanying allowance, care allowance and vouchers for the purchase of social and healthcare benefits within accredited facilities) [29].

The complex of interventions listed does not fully cover long term care policies. In particular, most of the elderly resort to informal assistance provided by relatives, friends, acquaintances, volunteers for the care of the person and the home environment.

Family members who perform the task of caring for their loved ones are subjected to strong physical, psychological, emotional, social and financial stress. They are most often subjected to the management of the emotional burden and the patrimonial and economic management of the relatives. Very often they are also legally designated as their Support Administrators.

4   The Role of the Support Administration in Italian Legislation

The Support Administration within the Italian legal system was provided for and regulated by law no. 6/2004 [18]. This law provides that the person who, due to an illness or a physical or mental impairment, is unable, even partially or temporarily, to provide for their interests, can be assisted by a Support Administrator, appointed by the tutelary judge of the place where it has its residence or domicile.

The Support Administration law has introduced a deeply innovative logic in the area of personal and private law as regards the protection tools that apply to weak subjects, according to a protection model inspired by the need to satisfy the daily life needs of weak subjects, as they are not self-sufficient [3, p. 621].

The law has the purpose of protecting, with the least possible limitation of the ability to act, the person deprives in whole or in part of autonomy in the performance of daily life functions, through interventions of temporary or permanent support.

In the area of possible action of the Support Administration, there is an extremely heterogeneous map of subjects to manage and pathways of rights to be guaranteed: the fragility of minors or the elderly, physical or mental handicaps, behavior prob lems that touch the memory or the psychiatric pathology, people alone or living in structured situations, in economically sustainable situations or in socio-economic or cultural poverty-marginality. The intention of the legislator was to first of all operate a cultural and operational reversal along the lines of the new conception of disease (and in particular of mental illness) which shifted attention from illness to the person and his suffering [28].

This regulatory institution has strengthened the personal protection of the weak subject, creating an *ad personam* defense, capable of limiting his capacity as little as possible.

The protection function of the institute of the Support Administration is not limited only to the plan of the patrimonial interests of the weak subject, but also to the personal interests [25], and no longer in marginal terms as required for the interdiction.

The law establishing the figure of the Support Administrator is more flexible than in previous institutions in protecting the person by paying attention to the individual, his identity and freedom of choice, to his existential interests.

It should be noted that with the expression *existential interests* we want to refer not to patrimonial acts, but to those that are an expression of the individual's personality, of *individual freedoms* [32].

It has been stated [9] that "the tutelary judge, unlike the interdiction and inca-pacitation judge, does not move *with a view to ascertaining the inability to act* of the interested party, but rather in order to provide support to those who you find it impossible to take care of your own interests. In light of this consideration, it has been possible to recognize how the attribution of functions to the administrator constitutes the essential content of this legal institution [9].

It was also pointed out that "The flexibility of the institution of the Support Administration and its applicability in favor of a greater range of beneficiaries is further reflected in the analysis of the terms *infirm* and *impairment*, used in the law. The first noun derives from the Latin infirmus, that is *not firm, not stable, weak*, referring to the noun *impaired*, that is *decrease*; the second derives from *menomo* or *minimus*, superlative of small" [28].

It should also be noted that "the daily adjective, from the Latin *cotidie*, refers to facts or events that are repeated daily in an individual's life. This adjective acquires full meaning alongside the word life, a concept that carries within itself the very essence of man in all its aspects and areas (...). Therefore, it is not possible to lead (and reduce) the protection of the weak subject to an area of merely patrimonial relevance by excluding, limiting or even denying their personal aspect" [28].

"Again, the term *administration* derives from the Latin *administrare* and means *to supervise, to cure*; it is composed of *ad* and *ministrare* and derives from the word *minister*, or *helper* and *servant*, which is connected to the term *minor*, that is *minor*, placing the emphasis on this latter meaning" [28].

In the final analysis, the functions and duties of the Support Administration are aimed at enhancing the skills of weak people, at supporting frailties, at helping to achieve the aspirations of those who, alone, elderly, physically or psychically disabled, alcoholic, drug addict, cannot do it.

All this must be included in a network of family, social and institutional relationships aimed at ensuring that the human person, with difficulty, is not abandoned to himself, and, even more, that the family that must take care of helping the family member in this situation should not be left alone.

The Support Administration has provided an instrument of concrete help to families and also to the reception facilities for the sick, nursing homes, RSAs[2] and all other structures responsible for the reception and daytime and night-time residence of sick, disabled people fragile.

Finally, it allowed to solve the problem of identifying the subject with which to relate and share the decisions relating to the therapeutic choices of the individual administered.

The following words, written by the signatory of the Law, Professor Paolo Cendon, give full meaning to the legal figure: "The Support Administrator is a presence to be conceived, on paper, as something not very distant from a sort of older brother (a little manager and a little home handyman), but also is a guardian angel, one who must perceive, who must have the ability to listen to the person and must be able to transform needs, even small ones, into a legal response" [38].

5   Conclusion

The institute of the Support Administration therefore provides for a complex of interventions characterized by a notion of *soft protection*, careful to enhance the subject's residual physical and intellectual abilities in an effort of continuous integration into the social fabric of the disabled person.

This legal figure seems to be particularly coherent with the new community welfare system, precisely because of its characteristic of solidarity and the assumption of responsibility by volunteers towards people who live situations that jeopardize their autonomy. Due to the high value of this figure, the Public Service can only support this legal institution, a true goal of civilization, of social and cultural.

In 2012, a research carried out within the Voluntary Service Center (CSV) of Bologna revealed some considerations which it seems right to report here with reference to the voluntary connotation of this regulatory institution. The attention stopped on the fact that as a form of volunteering the Support Administration does not pass through the organization and public service circuits, but from an autonomous and aware taking in charge in favor of those people who can't do it alone; a form

---

[2]Italian RSA is a health care residence for the elderly.

of joint commitment that recognizes in the role of the Support Administrator not only managerial and organizational tasks, but also the communicative and relational power to build proximity with the beneficiary; form of voluntary work motivated to improve their skills, abilities and knowledge in which everyone provides time, experience (also professional and technical), expresses strong motivational depth and civil responsibility [10].

For those who work in the world of volunteering, these results can be a reflection on changes in the forms of active citizenship, it represents a strong incentive to support and promote this figure and to make it interact with other resources in the area.

Moreover, the figure of the Support Administration constitutes a concrete response to family and community problems, in the light of current trends in demographic, social, economic and health terms.

In the context of Long Term Care policies it seems to be able to play a fundamental role in ensuring a response to a radical social need, more effectively than in the past due to the possibility of greater inclusiveness and empowerment of fragile people. The objective remains to improve or renew the social skills and relationships of non self-sufficient people by determining a better use of goods and resources with a view to innovation in social services.

This theory would usefully fit into those which the *Bureau of European Policy Advisers, BEPA* [19] calls social innovations: "*new ideas (products, services and models) that simultaneously meet social needs (more effectively than alternatives) and create new social relationships or collaborations. In other words they are innovations that are not only good for society but also enhance society's capacity to act*" configurate "*in both their ends and their means*" [31, p. 3].

In reality, these are social challenges in the face of human needs and problems of collective relevance and which need new solutions and solutions especially for the LTC in order to allow the universality of access to services (social and health), the achievement of a high quality of assistance and ensuring its economic and financial sustainability based on the rational use of resources [24].

In the so-called *European Pillar of Social Rights* proclaimed in 2017 by the European Commission, Parliament and Council in Chap. 3—dedicated to "Social protection and inclusion"—point 18 states that every person has the right to long-term care services quality and affordable, in particular to home care services and local services [15].

In this regard, the European Commission has also reiterated several times how guaranteeing citizens a high level of protection from the risks of disease and addiction is a crucial objective for the Member States and for the European Union [12]. It is expected the adoption of a policy based on new forms of multi-actor and multi-level collaboration, socio-health integration, home and community care, rather than institutional, support of informal caregivers and qualification of formal ones, prevention and rehabilitation.

The growth in the demand for LTC (long-term care) requires lifelong prevention policies and the promotion of the ability of non-self-sufficient people (elderly and non-elderly) to live as independently as possible.

According to the report Adequate social protection for long-term care needs in an aging society [33] these are immense challenges that generate expectations and uncertainties. "To bridge the growing gap between supply and demand for care and to mitigate the risks of overloading family responsibilities, degrading the quality of care and unsustainability of public budgets" it is suggested to move from a reactive approach, focused on problems that cannot be controlled, to a proactive one, aimed at preventing future situations, trends or problems in order to plan the appropriate actions in advance [20].

These are mostly "age-friendly" adaptations to domestic spaces, prevention, rehabilitation and re-enablement measures, work-life balance measures to support family caregivers, etc. [26].

In this context, the Support Administration is well placed. It can and, in fact, is called to carry out its mandate of protection and management of the life of people, that need to be protected not only materially (asset management), but also by attempting to restore social relations, significant at the family and community level of daily life.

References

1. Avvenire (2016) Sposarsi, la missione (im)possibile, intervista a Massimiliano Valeri, in "Avvenire" del 19 luglio 2016
2. Capparucci S et al (2012) Rapporto sulla Povertà a Roma e nel Lazio 2012. Comunità di Sant'Egidio, Francesco Mondadori, Milano
3. Cendon P (1987) Infermi di mente ed altri "disabili", in una proposta di riforma del Codice Civile, in Politica del diritto, 1987 p 621 ss
4. CENSIS—Centro studi investimenti sociali (2014) Il welfare familiare è in crisi: italiani costretti a rinunciare a prestazioni sanitarie e badanti. Rapporto 2014 «Welfare, Italia. Laboratorio per le nuove politiche sociali» di Censis e Unipol. Comunicato stampa. Roma: 9 luglio 2014 [online 03-01-2020]. Available online at: http://www.censis.it/7?shadow_comunicato_s tampa=120972
5. CENSIS—Centro studi investimenti sociali (2018) Convivenze, relazioni e stili di vita delle famiglie italiane, I Rapporto Auditel-Censis, sintesi dei principali risultati, Roma 25 settembre 2018
6. CENSIS—Centro studi investimenti sociali (2019) Il capitolo «Il sistema di welfare» del 53° Rapporto Censis sulla situazione sociale del Paese/2019 [online 03-01-2020] Available online at: https://www.censis.it/welfare-e-salute/il-capitolo-%C2%ABil-sistema-di-welfare% C2%BB-del-53%C2%B0-rapporto-censis-sulla-situazione
7. CENSIS e Confcommercio (2013) Outlook Italia. Clima di fiducia e aspettative delle famiglie italiane nei primi mesi del 2013, Roma, 18 aprile 2013
8. Comunità di Sant'Egidio (2012) Anziani nel Mondo, Europa e Italia. Le regioni europee in cui vive più a lungo [online 12-15-2019]. Available online at: https://www.santegidio.org/downlo ads/DATI%20ANZIANI%20-%20%20novembre%202012.pdf
9. Corte di Cassazione (2007) Cassazione, sez. I, sentenza 29 novembre 2006, n. 25366. [online 12-15-2019]. Available online at: https://www.minoriefamiglia.org/index.php/documenti/ giurisprudenza/347-corte-di-cassazione-sezione-prima-sentenza-n-25366-del-29-novembre-2006-sul-rito-per-la-richiesta-di-nomina-dell-amministratore-di-sostegno-3-11-06
10. Bologna CSV (2012) Dalla voce degli amministratori di sostegno. Indagine sulle opinioni e sui bisogni degli amministratori di sostegno, Vola bo—Bologna, 2012 [online 12-15-2019] Available online at: https://www.sostegno.bz.it/it/quali-doveri-dellamministratore-di-sostegno

11. Di Meglio E, Kaczmarek-Firth A, Litwinska A, Rusu C (eds) (2018) Living conditions in Europe—2018 edition ISBN 978-92-79-86498-8, doi: 10.2785/39876, Cat. No KS-DZ-18-001-EN-N. Eurostat, Unit F.4., Income and living conditions; Quality of life, Luxembourg: Publications Office of the European Union
12. EC (2008) Commission staff working document "Long-term Care in the European Union" Accompanying the document "Communication from the Commission to the European Parliament, the Council, the European Economic and Social Committee and the Committee of the Regions, Towards Social Investment for Growth and Cohesion—including implementing the European Social Fund 2014–2020". Luxembourg, Office for Official Publications of the European Communities, [online 02-10-2020] Available online at: http://ec.europa.eu/social/BlobSe rvlet?docId=2781&langId=en
13. EC (2013) Long-term care in ageing societies—Challenges and policy options, [online 02-10-2020] Available online at: http://ec.europa.eu/social/BlobServlet?docId=12633&langId=en
14. EC (2019) Social participation statistics#Social_isolatio, [online 02-10-2020] Available online at: http://epp.eurostat.ec.europa.eu/statistics_explained/index.php/Social_participation_statist ics#Social_isolatio
15. EC, European Parliament and European Council (2017) European Pillar of social rights, [online 02-10-2020] Available online at: https://ec.europa.eu/commission/sites/beta-political/files/soc ial-summit-european-pillar-social-rights-booklet_it.pdf
16. EUROSTAT (2012) Active ageing and solidarity between generations A statistical portrait of the European Union 2012 [online 02-10-2020]. Available online at: http://epp.eurostat.ec.eur opa.eu/cache/ITY_OFFPUB/KSEP11001/EN/KSEP11001EN.PDF
17. Fondazione IREF (2019) [online 03-01-2020]. Available online at: https://irefricerche.acli.it/ index.html
18. Gazzetta Ufficiale della Repubblica Italiana n. 14 del 9 gennaio 2004, Legge 9 gennaio 2004, n. 6, "Introduzione nel libro primo, titolo XII, del codice civile del capo I, relativo all'istituzione dell'amministrazione di sostegno e modifica degli articoli 388, 414, 417, 418, 424, 426, 427 e 429 del codice civile in materia di interdizione e di inabilitazione, nonché relative norme di attuazione, di coordinamento e finali" [onli-ne 02-04-2020] Available online at: https://www. camera.it/parlam/leggi/04006l.htm
19. Hubert A (ed) (2010) Empowering people, driving change. Social Innovation in the European Union, Luxembourg, Publication Office of the European Union, [online 04-05-2020] Available online at: http://ec.europa.eu/social/BlobServlet?docId=2781&langId=en
20. Kudlak A, Urban R, Hoskova-Mayerova S (2020) Determination of the financial minimum in a municipal budget to deal with crisis situations. Soft Comput 24(12):8607–8616. https://doi. org/10.1007/s00500-019-04527-w
21. ICN—International Council of Nurses (2010) Delivering quality, serving communities: nurses leading chronic care. International Nurses Day 2010, Geneva. [online 04-15-2020] Available online at: https://www.hrhresourcecenter.org/node/5094.html
22. ISTAT (2001) Condizioni di salute e ricorso ai servizi sanitari, Indagine Multiscopo 1999-2000, Roma
23. ISVAP (2007) La non autosufficienza degli anziani: il caso italiano alla luce delle esperienze estere, rapporto realizzato dall'ISVAP con la collaborazione del Censis, Roma [online 04-05-2020]. Available online at: https://www.ivass.it/pubblicazioni-e-statistiche/pubblicazioni/ altre-pubblicazioni/2007/anziani-non-autosufficienti/index.html
24. Maino F, Razetti F (2019) Long term care: riflessioni e spunti dall'Ue, fra innovazione e investimento sociale, in la Rivista delle Politiche Sociali/Italian Journal of Social Policy, 1/2019. [online 04-15-2020] Available online at: https://www.researchgate.net/publication/333 749148_Long_term_care_riflessioni_e_spunti_dall'Ue_fra_innovazione_e_investimento_soc iale_in_La_Rivista_delle_Politiche_Sociali_12019
25. Marcazzan G (2000) La tutela degli interessi patrimoniali dei minori e degli incapaci ed i provvedimenti autorizzativi di competenza del Giudice Tutelare, in Il processo civile minorile, Quaderni del Consiglio Superiore della Magistratura n. 109/2000

26. NNA—Network Non Autosufficienza (2017) l'assistenza agli anziani non autosuffi-cienti in Italia. 6° Rapporto 2017/2018. Il tempo delle risposte, Rapporto promosso dalla Fondazione Cenci Gallingani, Maggioli editore, Sant'Arcangelo di Romagna
27. OECD (2011) The future of families to 2030. Projections, policy challenges and policy options, A synthesis Report, Paris
28. Orlando N (2013) Gli interventi dell'amministratore di sostegno relativi agli atti di carattere personale del beneficiario, tesi di dottorato in Diritto Internazionale e Diritto Privato e del Lavoro. Indirizzo Diritto Privato nella Dimensione Europea—Ciclo XXV—Università di Padova. [online 04-15-2020] Available online at: http://paduaresearch.cab.unipd.it/5782/
29. Paolini I (2015) La percezione del carico assistenziale nel Caregiver familiare di pazienti seguiti dal servizio di assistenza domiciliare dell'ULSS 9 Treviso. Indagine quantitativa, Tesi di laurea in Infermieristica a.a. 2014/2015, Università di Padova [online 04-152020] Available online at: http://tesi.cab.unipd.it/52500/
30. Pasquinelli S, Rusmini G (2013) Badare non basta. Il lavoro di cura: attori, progetti, politiche, Roma, Ediesse, pp 196–197
31. Razetti F (2018) Long-term care e innovazione sociale: quali spunti dall'Europa?, [online 12-28-2020] Available online at: https://www.secondowelfare.it/edt/file/Razetti_WP_UR1_breve.pdf
32. Ruscello F (2004) Amministrazione di sostegno » e tutela dei disabili. Impressioni estemporanee su una recente legge, in Studium Iuris, 2004, 2, p 153
33. Social Protection Committee—Working Group on Ageing (SPC-WG-AGE) (2014) Adequate social protection for long-term care needs in an ageing society, Report jointly prepared by the Social Protection Committee and the European Commission, Luxembourg, Publications Office of the European Union; [online 04-15-2020] Available online at: http://ec.europa.eu/social/BlobServlet?docId=12808&langId=en
34. Solipaca A (a cura di) (2010) La disabilità in Italia. Il quadro della statistica ufficiale, ISTAT, Roma
35. Tagarelli F (2019) Long term care, la tutela per un futuro incerto, [online 04-15-2020]. Available online at: https://www.tagarelliassicurazioni.it/long-term-care-la-tutela-per-un-futuro-incerto/
36. Urban R, Hoskova-Mayerova S (2017) Threat life cycle and its dynamics. Deturope 9(2):93–109
37. Volpi F (2019) Il vero volto della famiglia italiana: un racconto attraverso i dati. Fondazione IREF, Roma
38. Vorano M (2014) L'amministrazione di sostegno, intervista a Paolo Cendon 7/10/2014 [online 04-16-2020] Available online at: www.personaedanno.it

Legality in Long-Term Care. Research in Progress

Mariateresa Gammone

Abstract The thesis of this chapter is that in the medical field, errors are very frequent, dangerous, and misunderstood. From the Hippocratic Oath, written between the fifth and third centuries BC, a dimension wider than negligence or incompetence has become clear: in the practice of medicine, illegal behavior is a possibility, in the strict criminal meaning of the term. The misunderstanding of the omnipresent possibility that "to err is human" is a perspective that has been implanted in the common way of seeing healthcare, with the well-known and devastating effects of defensive medicine. The best way to fight mistakes and abuses is by not ignoring or denying their existence, but by building an institutional system that allows for learning from adverse events, making abuses and mistakes visible, traceable, and burdensome. Error can become an important learning opportunity and is not just a synonym for negligence or incompetence.

Keywords Medicine · Sociology · Law · Errors · Abuses · Crimes · Epistemology

1 The Hippocratic Problem

The history of errors and abuses in medicine begins with the beginnings of medicine. The *Hippocratic Oath*, written between the fifth and third centuries BC, is full of references in this sense. In the oath we can read a specific reference to "intentional wrong-doing and harm, especially from abusing the bodies". The same notion concerning the necessity of an oath introduced a dimension which is wider than negligence or incompetence: in the practice of medicine there is the possibility of illegal behavior, in the strict criminal meaning. In the *Hippocratic Oath* there are many, various, explicit references to it, including the famous saying "First do no harm", which does not appear in the oath, but is existing in the Hippocratic school and then arrives to our days, in an infinity of versions and representations. "The Hippocratic problem" was so stringent and conspicuous that it required an oath.

M. Gammone (✉)
University of L'Aquila, Piazzale S. Tommasi n.1 Coppito, 67100 L'Aquila, Italy
e-mail: mariateresa.gammone@univaq.it

© The Author(s), under exclusive license to Springer Nature Switzerland AG 2021
D. Soitu et al. (eds.), *Decisions and Trends in Social Systems*,
Lecture Notes in Networks and Systems 189,
https://doi.org/10.1007/978-3-030-69094-6_3

29

With J.B. Bichat, according to Foucault, French medicine enters interiority of the body and surgically discovers "man", for the first time, in the scientific sense of the word: a breakthrough to "a more genuinely scientific empiricism" [12]. Precisely in French culture [6], from Marcel Proust to Georges Canguilhem[3–5], there are the moot sincere representations regarding medical abuses and mistakes. At the beginning of modern medicine, Western health care was a fairly free-form enterprise, full of do-it-yourself initiatives and innovative cures. The medical mistake could only be understood as an innocent and involuntary consequence of an attempt to save human lives: people were only saved by medicine. The Hippocratic problems remained in the background, but they were not unknown. An outstanding text on abuses in medicine is *Knock ou le Triomphe de la médecine*, a 1923 French satirical play written by Jules Romains, which had an extraordinary success through triumphal theatrical productions, performed by the famous Louis Jouvet (who then played the same role in a film with the same title, rewarded by another great success). In the story, the ambitious Dr. Knock arrives, as the local physician, in a rural French village and understands that most of the villagers are in good health. He therefore decides to make everybody believe they were actually far sicker than they actually were. He manages to establish a long-term therapy for everyone, doing business with the pharmacist and turning the town hall into a clinic. One of the most famous phrases in the text is: "Healthy people are sick people who ignore their real situation". The whole story revolves around the possibility that someone may practice medicine as a swindle, a trick, a racket: a manipulative technique that can be carried out using the same techniques as advertising and hidden persuasion. Some years after, in 1957, in the American postwar era, Vance Packard published his classic (albeit controversial) volume on consumerism motivational research and on the possibility to manipulate expectations and induce desire for useless or harmful products.

Fictional writers are not the only ones who have transmitted unexpected accounts on the medical mind, with descriptions ranging from the sense of omnipotence that can give the power on life and death, to the sense of annihilation that transmits the body in his corruption, decomposition, dissolution. Physicians know the stink of death and even know that life is always born *inter urinas et feces*. Many physicians have written fiction, among them Rabelais and Bulgakov with a taste for humor and fantasy. However some practicing physicians have been authors of sulphurous texts, which open a window on hell (as Louis-Ferdinand Céline did), or which portrayed human bodies morbidly, just disease-ridden organisms (as Gottfried Benn did), or which wrote on the gardens of human cabbages: miserable, grotesque, repugnant human beings watered with the fertilizer of injections (as António Lobo Antunes did). Céline, Benn, Lobo Antunes maintained a clinical practice, performing the profession: they knew that survival is not simple and that the attempt to heal life from its ills is not easy either. Errors were taken for granted.

Obviously, pessimistic writers were but a minority in the polyphonic and glorious literature of medical science (for instance, in *Der Zauberberg*, 1924, after Roentgen's discovery, Thomas Mann wrote a wonderful kind of epic on chest X-rays). There are errors in the field of medicine, but many mistakes have been beneficial; just thinking about some of these errors, in the interpretation of puerperal fever, Ignác

Fülöp Semmelweis has written pages that not only remain exceptional in the history of medicine, but are exceptional in the history of science and human thought. His case demonstrates in extreme magnitude the modest thesis of this brief text: in medicine the error is very frequent and dangerous. The misunderstanding of the omnipresent human possibility of making mistakes in good faith is a perspective that has been implanted in the common way of seeing medicine, with the known and devastating effects of defensive medicine. The best way to fight against mistakes and abuses is not to ignore or deny their existence, but to build an institutional system that allows learning from mistakes and the reduction of abuses number, making them visible, traceable, and burdensome. Medical students can see many and great models of integrity, compassion, altruism and wit.

2 Modern Medical Errors

In modernity, the first discussion on the abuse of medicine is traced back to Ivan Illich, in 1974. His book on *Medical Nemesis*, famously opened with the statement: "The medical establishment has become a major threat to health". Illich's intellectual influence peaked in the mid-1970s and, within the leftist area, was connected with the fascination for the Maoist Chinese revolution, which to many Western intellectuals seemed to indicate a new viewpoint. The volumes of Needham opened the first road in this sense, legitimizing the idea that there could be a different way, unknown and better than the Western one [20]. While in areas such as psychiatry this kind of approach was widespread, in the proper field of medicine, it remained only a suspicion, a possibility, a doubt, one of the most difficult conundrums for the political experts [1, 2].

Illich argument that modern medicine had little effect on the overall health of populations had been made by some epidemiologists. Like Thomas McKeown, Illich said that general factors (such as housing, education, nutrition and sanitation) were more important determinants of public health. In his perspective, not only doctors contribute little to the health of people, they probably do more harm than good. The cultural wars of the sixties shook the foundations of the most common beliefs, starting with intimacy and sexuality. In those years the visibility and fear of errors in health care began to spread. Ralph Nader's fight for rights took place in many sectors, including medicine [13]. The *1973 Patient's Bill of Rights*, mounted everywhere on US hospital walls, is a symbol of the long wave of the Sixties in this area of democratic concerns. Despite many measures of clinical transparency, health care safety remains a big problem. Complex technologies, multifarious drugs, audacious intervention, intensive treatment, and prolonged hospital stay contribute to adverse effect of healthcare [21]. Perfect patient safety has been the holy grail for many intrepid missionaries, long sought after yet never achieved.

There has been much concern about the errors of medicine. In estimates and evaluations, the number of probable, unknown, hidden, medical errors have grown in a hyperbolic manner. Atul Gawande, a surgeon and public-health researcher, wrote:

"In 1991, the *New England Journal of Medicine* published a series of landmark papers from a project known as the Harvard Medical Practice Study—a review of more than thirty thousand hospital admissions in New York State. The study found that nearly 4% of hospital patients suffered complications from treatment which either prolonged their hospital stay or resulted in disability or death, and that two thirds of such complications were due to errors in care. One in four, or 1% of admissions, involved actual negligence. It was estimated that, nationwide, upward of forty-four thousand patients die each year at least partly as a result of errors in care" [9]. In the US, clinical error is considered the nation's third leading cause of death, behind only cancer and heart disease. Researchers say that it is responsible for more than 250,000 deaths each year.

In our perspective, the analysis of errors is a tool to assess factors that help the improvement of the system. The vision of James Reason [15, 16] has meant a radical change in the conception of error, linked to the description of cognitive processes. While relying on previous speculations, he reinterprets the concept of medical error in a new perspective and insists on the need to study human behavior in relation to the contexts in which errors are found. James Reason has connected his approach with a theory on latent gaps: the occurrence of an accident is the result of a concatenation of events that occurred despite the barriers that had been put in place.

The sociological literature on the increasing complexity of modern systems helps us understand that the growth of errors is inevitable. An application of the concept of autopoiesis to social organizations can be found in many authors (Luhmann, for instance). Generally speaking, all systems must continually construct and reconstruct themselves through processing the distinction between system and environment, and self-reproduce themselves as final intent of their own basic elements [11].

A canonical example of an autopoietic system is the biological cell. A living system is a closed topological space that continuously generates its own organization and its own factors, and does this in an endless turnover of components. An autopoietic system is to be contrasted with an allopoietic system, which is Cartesian and which has a functioning goal different from relentless reproduction. A system as a hospital can use its components to maintain its structure, which is something other than its formal statute. A hospital, as a system, can include external components in its vital environment, as principal reason for its existence. Patients, doctors, nurses, supply chains, plants, equipments, dealerships, contracts, competitors, spare pieces, corpses, and so on, can become parts of a kind of living system, which could be considered autopoietic: interested above all in its reproduction, with all components embedded in a sensory-motor coupling of dynamic changes. In that kind of system, even the individual members feel entitled to behave in a parallel way: they become interested only in themselves and in their egoistic purposes. In a self-referential system, coding and corrections must be autopoietic: especially concerned in the reproduction of individualistic and internal purposes, rather than fulfilling statutory tasks: saving lives and battle suffering. Survival then is more relevant than errors performed on the shoulders of others.

3   The American Way of Making Mistakes

It is almost impossible to understand Western health care systems without reference to the United States, even if the profiteering, non-universal U.S. system is different from the non-profit universal health care systems of other Western countries. In medicine the American model wins all over the world, regardless of the difference (which in some respects is enormous) concerning the pre-eminence of public, or private, or half-public and half-private systems. Medicine is a hierarchical organization, filled with asymmetries, but democracy enters interiority of its institutional body, as medicine enters interiority of the physical body. American medical wars are democratic wars. The opioid epidemic and the crack cocaine epidemic, secondhand smoke and embryonic stem cell research, emergency contraception and abstinence-only sex education, birth control and abortion, transgender discrimination and legalizing drugs are heavily politicized issues. Medicine is an unlimited combat zone, concerning life and death—for life, as for love and war, is everything permitted?

Historically, it is very clear that health mixes with democracy and legality; this is evident in the United States from the beginning of twentieth century democracy. In 1906, with *The Jungle*, Upton Sinclair wrote a book that is a classic of democracy and which gave rise to the *Federal Food and Drugs Act*. Almost everything in America has been big: great mistakes, great democratic attempts to repair errors.

The American model is based on a specific way of understanding science and medicine, organizing hospitals, accepting the supremacy of the pharmaceutical industry and aiming at the same time on the Welfare State. The American model can be summarized through the reference to Fordism, which has been described as a model of economic expansion and technological progress, based on mass production and mass consumption: the manufacture of standardized products in huge volumes. Although Fordism was a method used to improve productivity in the automotive industry, its organizational principle has been applied to any kind of manufacturing process, including health industry. In the Fordist way, the intensive use of purpose machinery and unskilled labor does not result only in standardization of the product. Nothing is handmade, everything is included in protocols.

It was in the American 1930s that Fordist protocols were pushing doctors "to get through with a patient in five or 10 min". Subsequently patients grieved for the loss of "good old Doc" and experts denounced that the "*whole* patient" was lost. It was only the beginning: together with the old patients and the old doctors, the nineteenth-century medicine was lost. In strident contradiction with the enormous progress of biomedicine and biotechnology, the twenty first century patient feels "dehumanized" and hospitals are loaded up with technology in order to impress clients with "new and scientific" medical care [17].

Standardization would be impossible without classification originated during the bubonic plague in seventeenth century London. Rightly health historians focus on the successes upon terrible maladies and tell medical narratives of glorious progress. They rarely studied the immense sociological literature on the bureaucratic phenomenon, which emphasizes the tendency of every organization (as of any profession) to expand in a geometric way: the growth of the means becomes an end in itself, independently of the statutes and official goals (of the organization). The *Heterogonie der Zwecke*, heterogony of ends, is a theme as old as human culture [11].

Legality is an end and an instrument in democracy, in ways that are often confused. The *Health Maintenance Organization Act* of 1973, in the US, is a case in point. Although officially this Act was introduced with the noble intention to guarantee the best health care in the world, subsequently falls in the powerful hands of insurance and drug lobbies [8]. Their huge profits have been often used to finance politicians at the highest level. Insurance companies and pharmaceutical companies try to control the American health market.

During the first half of the twentieth century, the number of tabulated diseases naturally increased with better understanding of human health. For more than a century now, Western health industry has been a happy marketplace, hosting a shotgun wedding among consumers, medical providers and regulators (including insurance companies). The pharmaceutical lobbies, supported by the medical profession, endorsed a medicalization of Western societies. With the affirmation of Welfare State, after the Second World War, medicalization has invented (critics say) new diseases, from social anxiety disorder (shyness) to testosterone-deficiency syndrome (old age).

Long before health insurance, biomedicine, biotechnology, and the Internet were invented, medical science began to change. The burdens of standardization and pharmaceutical marketplace began to engulf health care duties, back in the 1940s, with expansion of Fordism in all organized sectors. Observers openly advised a professional mutation: the doctor should be "more a businessman and less a saint". The combination of science and consumer capitalism changed the fundamentals of many things [22].

In the second half of past century, in the Seventies, United States spending on healthcare was 8% of the GDP; it is now more than 17%. The United States spends vastly more than other high-income countries. Healthcare makes up 10% of the entire global economy. There are, of course, many democratic good intentions [7] that have produced the United States' record-breaking $3 trillion health care bill: excessive testing, runaway drug prices and sky-high charges for even the most basic medical interventions. Many Americans, including those who are insured, decide to decline treatment and prescriptions that they need. Consumers and layers, patients and bills, providers and insurance companies deploy against each other in an endless war over the virtuous procedures that must be undertaken and how much to pay for them. Caught in the crossfire, American healthcare has been intended as rotten: lack of transparency, profit maximization and greed have been feared as dominant criterions. In John Grisham's America, illegality and health go hand in hand. He is not a successful author only in the United States: he is one of the most widely read authors in the world.

In health care, legality problems are not just American problems. Every democratic country needs independent public health control, intended to dominate lobbyism, cronyism, and partisan politics. There is almost no country in the world whose government does not say to be the government of a democracy. And democratic governments are in trouble virtually everywhere [10].

An escape from insurance companies and a thriving medical tourism business soured in many Western nations, which have been produced layer upon layer of well-meaning efforts to guarantee the safety of health services. A dire perception of inadequacy has become dominant. Concern about medical errors has become a tremendous worry of the great majority of average Western citizens, whose lives are punctuated by a variety of ills. Much more visible and vociferous are wellness hunters, informed by the Internet and worried by their insurer's cash: they are uniquely well-educated, angry, and empowered consumers. Patients want a stake in the medical game, which has been the game of survival, therefore is the game of life, progressively becomes the game of happy lifestyles. Every clinical mistake seems an opportunity for a possible gain. The number of lawsuits against Western physicians has increased within the last decades, with a substantial impact on medical practice, insurance premiums, overdiagnosis and overtreatment [13].

4   The Italian Experience

As in the United States, in Italy the close link between public health, democracy and legality is equally evident, even if in a peculiar way. In every country the problems of legality are peculiar [18]. The Italian case is characterized by a considerable amount of investigations and legal cases in the health sector. There have been many scandals. Despite the difficult situation, in relevant international evaluations (including vulgarizations, as in *Sicko*, the M. Moore movie), the Italian health system is one of the best in the world, perhaps the best, certainly so if one considers only the north of the country. Concerning long-term care, the world statistics of longevity, presented by the World Health Organization, objectively report that, together with Japan, Italy has the highest number of centenarians and the best life expectancy.

Article 32 of the Italian Constitution identifies protecting health as a "fundamental right of the individual and an interest of the community", thus considering this right as one of the most important rights of the individual. This principle must always be taken into account when we consider Italian medical legality, because it is on this basis that the whole regulatory framework is centered.

Concerning long-term care, Italy has a tradition of assistance entrusted mainly to Catholic organizations. The history of long-term care and the glorious history of piety have walked together. With the advent of the welfare state, a complex mechanism (to fund public care) has been put in place, with enormous economic means, which have aroused insatiable appetites. As in many European countries, arrangements exist to at least partially fund private care as well. Population aging is a demographic and democratic force: powerful and transforming. The Italian welfare system has not yet established all-embracing programs, relying on a shared assortment of informal care-givers and formal services. Private initiatives are often pre-eminent to public ones, partly because the Italian pension system was built in the golden years of economic growth and is very generous. Female labor-market participation, population ageing, migrations from one side of the country to another, the affirmation of a narcissistic mentality, and troubles in family structures have created peculiar demand for new long-term care services. There are many welfare-state opportunities: in the public sector, individual revenue grows across different life stage. In general, the pension system and the high levels of personal savings (which were a feature of popular economic wisdom) cover individualized assistance. The necessities of frail, elder, dependent, people have been partially satisfied by immigrant women of different nationalities, who are hired to provide private care in people's homes. Italian attitudes towards the welfare state and the care-labor market are split in two tendencies: while the wealthiest has a predilection for the free-market approach, the poorest advocates a stronger role of public authorities in helping families cope with the burden of long-term assistance. Procedural legal expectations and substantive legitimate expectations are at stake.

In Italy, as in the whole Western context, there is a lot of attention to the quality of health services and to the issue of legality. Safety costs and patient safety have become crucial priorities that the national authorities aim to achieve. The clinical risk management process has become a model for understanding the genesis of errors, following an operational viewpoint. Changes in the Italian health system followed publication, on a world scale, of public reports on errors in health care that had a strong impact. Also in Italy, the approach in terms of clinical risk management become dominant: a set of proceedings that involve problem's recognition, the identification of critical areas, the activation of a monitoring system in order to reporting and analyzing adverse events, the guidelines for measures regarding broad containment and accurate anticipation. To this end, the Italian Ministry of Health established in March 2003 a Technical Commission on Clinical Risk, which one year later drafted the document "Risk management in healthcare", containing a series of reflections and recommendations. The document states that "in order to reduce errors it is necessary to identify a uniform organizational model for clinical risk management, to draw up guidelines for the uniform detection of errors and risks of errors in healthcare facilities, to promote events to spread the culture of error prevention, [...] experimenting, at company level, methods and tools for reporting errors, collecting and processing data". All good intentions and objectives are difficult to reach with decreasing economic means that question the (until now acceptable) performance of Italy's healthcare system. Italians are in need of integrated health networks, at regional, national and international level. In the West, the cost disease of the public health care sector are similar to or even somewhat more than in private health care sector. In Europe, from Italy to France, the rising cost of welfare and education, safety and security, cast a shadow on virtually every democratic government.

In a knowledge society, it is absolutely necessary to manage information but also to give financial support to the modernizing initiatives. Heightened awareness can only be assessed by careful institutional attention to the clinical context.

Beyond the formal aspects, investigations and scandals abound in the Italian sector. Concerning long-term care, the elderly are among the biggest victims of greed, inefficiency, and illegality. But the fight for the rule of law did not damage the overall functioning of the system. Indeed, individual health is seen in Italy as a primary value and it seems that the angry attention of investigators, magistrates, journalists, citizens has made the problems very visible, but has also made the transformation of public health more difficult as in many other countries, in an easy prairie for the most dangerous interests groups and treacherous individual citizens. As we have said, there is a dirty heap of illegality in the Italian health system, but there is also more investigation and more visibility than in any other country [19]. From Hippocrates on, many people have gained wealth trough the health problems of others, but this in Italy is well known, as an ever clear and present danger.

5 Research in Progress

The previous lines constitute the theoretical, historical and methodological framework for research that is still in progress. In future publications the empirical part of our research will be widely illustrated and shared. The point from which we start is that medicine saves human lives: we need efficient medicine and high ethical standards. The overwhelming majority of doctors and practitioners are animated by the best of good intentions.

The problem of medical errors must be dealt with in a objective way, without sensationalism and without hysteria. We can learn a lot from all previous experiences, both regarding the medical sciences and the criminal investigation, including the methodological principles that have been overcoming in many sectors through the reference to "evidence-based" practices.

In this perspective, working together within a group of researchers, we have collected a series of interviews, narrations, testimonies with health workers. Those stories, confessions, experiences are told by individuals operating in the sector and can statistically be considered representative of the current interpretation regarding the existing situation in Italian hospitals, in terms of the relationship between patients and their environment.

Our study paints a picture that is very much characterized by the real existence, within the Italian health system, of practices that are often dangerous, intentionally or unintentionally. One may think that it is a typical Italian deformation: for some years now, in Italy there has been a public discourse characterized by frequent analysis in terms of corruption and incompetence. According to an interpretation, however, the Italian case does not describe a particularly criminal social and institutional world, rather by a social and institutional world that is characterized by a particular sensitivity and attention towards criminal danger: Italy is the country of Machiavelli and Lombroso, of Beccaria and *Clean Hands*. Mafia, corruption and crime are more visible than in other countries (Italy is also the country of melodrama), where they can flourish undisturbed. Mafia, corruption and crime are investigated and prosecuted more than in other countries, making it seem like there is more than elsewhere.

In the international literature we can find support for our interpretation, according to which the Italian case does not stand out for errors committed within the health structures. The relationship between legality and medicine is felt to be complicated in many countries. In the most recent modern times, the narratives on medical care are often ruthless, for example in the cinema, from *The Barbarian Invasions*, in 2004 (a sad description of public health care in Canada, with particular emphasis on illegality, but described as "A triumph" by *The Boston Globe*) to *Sicko*, in 2007 (a sad description of private health care in United States, with particular emphasis on illegality, but described as "A very strong and very honest film about a health system that's totally corrupt and that is without any care for its patients" by *The Boston Globe*). Often, it is precisely in the narration of health systems that we find a mixed belch of anti-Americanism and anti-Westernism, seasoned with a taste of unthinkingly reactionary approach. In the good old days, previous to contemporary health systems, people died for minimal health problems and average life expectancies were very short.

6 Conclusion

The word legality in medicine is used as a label for nearly all of the clinical incidents that harm patients. Legality is not limited to the observation of error: it is a search for responsibility (the guilty one). In the international bibliography, long-term health is often quoted as a typical case of malpractice. In our empirical research we try to distinguish misunderstandings, exaggerated expectations, medical errors and clinical errors. In medicine, the mechanisms underlying errors are the same ones that influence people's incorrect behavior. Error can become an important learning opportunity [14] and it is not just negligence or incompetence. It can be inherent in cognitive processes that usually give good results and it can be remarkably connected to a social and institutional context that produces problems, tensions, misunderstandings, abuses.

References

1. Bell DA (2015) The China model. Princeton University Press, Princeton
2. Berggruen N, Gardels N (2012) Intelligent governance for the 21st century: a middle way between West and East. Polity, London
3. Canguilhem G (2002) Écrits sur la médecine. Éd. du Seuil, Paris
4. Canguilhem G (2011) Euvres complètes, tome I: Écrits philosophiques et politiques (1926–1939). Librairie Philosophique Vrin, Paris
5. Canguilhem G (2015) Le normal et le pathologique. Presses Universitaires de France, Paris
6. Chimisso C (2008) Writing the History of the Mind: Philosophy and Science in France 1900–1960s. Ashgate, London
7. Deneen PJ (2018) Why liberalism failed. Yale University Press, Yale
8. Gammone M (2017) One village, many tribes, countless wolves. Dangerousness and education in western thought. In: Gammone M, Icbay MA, Arslan MA (eds) Recent developments in education. E-Bwn, Bialystok, pp 17–21
9. Gawande A (2009) The cost conundrum. The New Yorker, vol 65, June 24–29

10. Levitsky S, Ziblatt D (2018) How democracies die. What history reveals about our future. Viking, New York
11. Luhmann N (1997) Das Recht des Gesellschaft. Suhrkamp, Frankfurt
12. Miller J (2002) The Passion of Michel Foucault. Anchor Books, New York
13. Nader R (2002) The Ralph Nader Reader. Seven Stories Press, New York
14. Popper K (1963) Conjectures and refutations: the growth of scientific knowledge. Routledge, London
15. Reason JT (1990) The contribution of latent human failures to the breakdown of complex systems. Philos Trans R Soc Ser B, London 327:475–484
16. Reason JT (1997) Managing the risks of organizational accidents. Ashgate, London
17. Rosenthal E (2017) An American sickness: how healthcare became big business and how you can take it back. Penguin Press, New York
18. Sartori G (2011) Cómo hacer ciencia política. Taurus, Madrid
19. Sidoti F (2012) Il crimine all'italiana. Una tradizione realista, garantista, mite. Milano: Guerini
20. Singh S, Edzard E (2008) Trick or treatment? Alternative medicine on trial. Bantam, London
21. Svarcova I, Hoskova-Mayerova S, Navratil J (2016) Crisis management and education in health. Eur Proc Soc Behav Sci EpSBS XVI:255–261. http://dx.doi.org/10.15405/epsbs.2016.11.26
22. Tomes N (2016) Remaking the American patient how madison avenue and modern medicine turned patients into consumers. North Carolina Press, Chapel Hill

Lifelong Investments for a Healthy Ageing

Daniela-Tatiana Şoitu ⓘ

Abstract Romania has one of the lowest healthy life expectancies in Europe. At the same time, it is facing an accelerated process of population aging. Today's generations of children, young people and adults may be aware of the factors and mechanisms that can ensure their healthy ageing. Options for preventing a sick, vulnerable aging can be educated. Lifelong well-being investments can be done individually, or in the family, but also at the level of the educational, economic, environmental, cultural system. Starting from concrete, current data, this chapter presents and analyses the endogenous and exogenous factors of healthy aging, measures and actions taken in Romania. The chapter makes significant contributions through the proposals it makes for three national programs: education for lifelong investments for a healthy ageing and awareness of health determinants throughout life; regular assessment and monitor the health status of adult citizens; financing services that support lifelong investments for well-being and a healthy ageing.

Keywords Healthy ageing · National health programmes · Prevention · Lifelong well-being · Education for health · Social determinants of health · Active and healthy life

1 Introduction

Recent research underlines the importance of all life stages, including on childhood, in the evolution of health status throughout life [15, 18]. The United Nations Development Programme, in assessing the Third sustainable development goal, *Good health and well-being*, found that infant mortality rate had dropped by half because of the programmes run over the past 10 years [23]. Also, through information an awareness programmes regarding determining factors in health, UNDP [23] and WHO [28] has identified a major potential of reducing the incidence of non-transmissible and

D.-T. Şoitu (✉)
Department of Sociology and Social Work, Alexandru Ioan Cuza University of Iaşi, Carol I bvd., 11, 700506 Iaşi, Romania
e-mail: danielag@uaic.ro

© The Author(s), under exclusive license to Springer Nature Switzerland AG 2021   41
D. Soitu et al. (eds.), *Decisions and Trends in Social Systems*,
Lecture Notes in Networks and Systems 189,
https://doi.org/10.1007/978-3-030-69094-6_4

chronic diseases, through primary and secondary prevention programmes seen, by us, as lifelong investments for a healthy life and ageing.

Europe has seen an increase in life expectancy, but not necessarily in "healthy life" expectancy. According to EUROSTAT, in the European Union, in 2016, almost six out of ten people over 65 years of age had suffered from chronic diseases [5]. The concern for general care and specially for long-term care adapted to needs is present among the European institutions. In 2015, the report concerning social protection and social inclusion adapted to the need for long-term care as well, made jointly by the European Union Social Protection Committee and the European Committee, underline the need for developing such services [4, 13]. A (semi)dependent state may occur in an individual irrespective of age, and society will have to be prepared with long-term care offers that have a systemic, integrative design [16, 20].

On the global arena, the central themes of the International Action Plan on Ageing, adopted by the United Nations General Assembly in Madrid (UN 2002), refers in a focused manner to "provision of health care, support and social protection for older persons, including preventive and rehabilitative health care", in close connection with the other ten themes:(1) the full realization of all human rights and fundamental freedoms of all older persons; (2) the achievement of secure ageing, which involves reaffirming the goal of eradicating poverty in old age and building on the specific United Nations Principles; (3) empowerment of older persons to fully and effectively participate in the economic, political and social lives of their societies, including through income-generating and voluntary work; (4) provision of opportunities for individual development, self-fulfilment and well-being throughout life as well as in late life: access to lifelong learning and participation in the community while recognizing that older persons are not one homogenous group; (5) ensuring the full enjoyment of economic, social and cultural rights, and civil and political rights of persons and the elimination of all forms of violence and discrimination against older persons; (6) commitment to gender equality among older persons through, inter alia, elimination of gender-based discrimination); (7) recognition of the crucial importance of families, intergenerational interdependence, solidarity and reciprocity for social development; (8) provision of health care, support and social protection for older persons, including preventive and rehabilitative health care; (9) facilitating partnership between all levels of government, civil society, the private sector and older persons themselves in translating the International Plan of Action into practical action; (10) harnessing of scientific research and expertise and realizing the potential of technology to focus on the individual, social and health implications of ageing; (11) recognition of the need to seek means to give ageing indigenous persons an effective voice in decisions directly affecting them (Romanian Government Decision 566, [12] regarding Strategy for a Healthy Ageing).

In recent years, The Global strategy and action plan on ageing and health [25–27] planned to contribute "to achieving the vision that all people can live long and healthy lives." By focusing on five strategic objectives: commitment to action on *Healthy Ageing* in every country; developing age-friendly environments; aligning health systems to the needs of older populations; developing sustainable and equi-table systems for providing long-term care (home, communities, institutions); and improving measurement, monitoring and research on Healthy Ageing. High expectations came from the partnerships and evidences brought during 2016–2020 from the Decade of Healthy Ageing (2020–2030).

In a recent report of WHO [28] it is stated that "the adoption of the 2030 Agenda for Sustainable Development and the Sustainable Development Goals have provided a framework within which to strengthen actions to improve health and well-being for all and ensure no one is left behind. Despite overall improvements in health and well-being in the WHO European Region, inequities within countries persist." The report identifies "five essential conditions needed to create and sustain a healthy life for all: good quality and accessible health services; income security and social protection; decent living conditions; social and human capital and decent work and employment conditions. Policy actions are needed to address all five conditions. The Health Equity Status Report also considers the drivers of health equity, namely the factors fundamental to creating more equitable societies: policy coherence, accountability, social participation and empowerment." [28]

Our analyse of Romanian institutional structures, decisions and actions, considering these principles and those of World Health Organisation referring to social determinants of health support adjustments and investments for lifelong well-being and healthy ageing.

2   The Complexity of the Determinants of Health Through the Live

Living standards, education, the quality of the physical and social environment, working conditions, nutrition, active life etc. are factors that influence health in old age. Synthetically, according to a general analysis of the determinant factors [24–26, 28, 29], an individual's health is influenced by: (1) child development at early stages of childhood (neural, immune and endocrine system, sensitive periods, care and attachment, stress); (2) biological factors (gender, age, genes), psycho-social (self-esteem, isolation, level of control, anxiety, stress, depression, anger), lifestyle (diet and nutrition, smoking, alcohol consumption, physical activity); (3) a sum of factors: access to services (of primary, secondary, tertiary care, social services, transport services, of employment, of leisure, of habitation, of recreation), socio-economic determinants (income, education, employment, work); social and community context (social networks, connection within the community, social capital), environment determinants (water, air and soil quality, noise level, built-up environment, habitation, system of transport to the workplace), culture and ethnicity (system of beliefs and perceptions, attitudes, values and norms); global forces (world economics, market conditions, trade environment; e.g.: global warming, natural or man-made disasters) and governmental policies (economic, habitation, taxing, welfare policies, local, national, regional policies) [22].

Starting from these considerations, even if we accept the definitions of the phrase of social determinants of health proposed in the literature or by the international bodies operating in the domain of health, we shall attempt to narrow down the areas of interest and the domains that we could target. One way of finding orientation in specialised literature and among the issues related to the social determinants of health status is to make a distinction between two groups of determinants: (1) factors related to system problems and to socio-economic inequalities, on the one hand and (2) factors related to attitudes and practices, values and culture, on the other [11, 18].

3   Romanian Context of Lifelong Investments for a Healthy Ageing

Health risk factors are identified as determinant in the latest National Report regarding the Health Status of Romania's Population [9]: body mass index, blood pressure, smoking and overall alcohol consumption, consumption of fruits and vegetables, exposure to particles (PM10) and poverty risk (ibidem, pp. 153–164). It is necessary to have a more complex approach, one that not only measures, monitors and identifies risk factors, but also provides genuine primary prevention, consisting in information and increased awareness of the complexity of the determining factors in health [14]. Of the 15 national programmes that impact public health, run and funded by the Ministry of Health starting with 2015, just one, the National programme for the assessment and promotion of health and of health education, has been aimed at educating the population in order to develop a healthy lifestyle, with long-term impact on the pressure on the long-term care (LTC) system.

The results of the Survey of Health, Ageing and Retirement in Europe (SHARE), wave 7 [1] regarding the subjective health status of older people, the yearly reports of the Ministry of Health (2019) and the Ministry of Health Report concerning the population health status, and morbidity and the incidence of degenerative diseases, respectively (2018), Eurostat data [5] and data from other institutions monitoring health status [10] underline the need for investment in primary health care and in primary and secondary prevention [19].

Romania's previous health assessment programme (2007–2008) gave the opportunity for a complex screening and for the subsequent development of national prevention programmes, as well as of healthcare programmes. The Service packages and the Framework contract regulating the provision of healthcare, of medication and medical devices within the social health insurance system for the years 2018–2019 (HG 140/[6]) provide the reimbursement of health check services once every three years for individuals younger than 45 and once a year for those over 45. The decision to choose the 45 years of age threshold is an informed one, but there are risk categories below this threshold, and this is the reason why the action should have as beneficiaries all adult individuals. Results from our research, accessing databases (SHARE 2019; [10] indicate that among death causes in Romania there are some that could be prevented or controlled, as they are influenced not just by genetic factors, but also by exogenous factors. These include diseases of the circulatory system (56.7%), tumours (19.6%), diseases of the respiratory system (6.5%), diseases of the digestive system (5.9%), traumatic lesions, poisoning and other consequences of external causes (3.8%).

According to Art. 3 from Law 292/[7] regarding Social work system [17], the responsibility for a dignified and autonomous life, "for developing one's own social integration abilities and the active involvement in solving difficult situations belongs to each individual and to the individual's family, the state authorities intervening by creating equal opportunities and, in subsidiary, by granting welfare payments and adequate social services" (Art. 3).

The employment rate among older individuals is very low. In 2014, Romania [30] was 10 percentage points below the European average, at a level comparable to that of neighbouring countries and to Southern Europe. The trend is towards a decrease of the presence in the job market of people over the age of 55. In 2009, 40% of individuals in this age group had opted for early retirement, this resulting in a population of pensioners 7.5 years younger than the legal retirement age. Individuals with low or no training lose their jobs as early as at the age of 50–55, and the general trend in this population segment is to have decreased employment rates. In 1997 we had an employment rate of 52% among the elderly, while in 2013 this became 41% [20, 30].

An analysis of the SHARE data (2019) reveals a worrying situation in terms of the decrease and limitation of work capacity after an adult age far from the standard retirement age for most occupation: the age of 50. In Romanian's case, the situation is, at first sight, paradoxical: although the state of health is perceived as being rather poor, and the limitation of current activities due to a health problem reaches the highest levels, the country records one of the smaller percentages of limitation of paid employment due to health issues or to disability. One explanation of this situation could be, in fact, the need to continue lucrative activities irrespective of the perceived state of health, because of the very low income [15, 20].

Although healthy ageing is, naturally, an increasingly important problem for the elderly, the public's focus on the scale and the cost of medical care, of pensions and other services oftentimes enhances the negative image of ageing. Recent European documents (Together for Health: A Strategic Approach for the EU 2008–2013 (COM[2] 630 final), European Union Strategy 2020 (COM [3] 2020), 2030 Agenda for Sustainable Development of United Nations) emphasize the importance of prevention in the shape of a healthy lifestyle, active and healthy ageing, as these are also financially effective solutions.

An activity that has involved the collaboration of the local Public Health Centres with community nurses and with health mediators has aimed to organise and implement interventions in order to promote health in vulnerable communities and groups through 1266 interventions in vulnerable communities, the number of beneficiaries being 173,143 [8].

Considering the above mentioned data, our main proposals for neccessary action-investments through the life are:

Action 1. Implement a national programme for increasing awareness about the factors that determine health status throughout life.

Purpose of the action: education the population of all ages in order to become aware and understand the impact of all these factors on health.

Action 2. Establish a national programme for the regular evaluation of the health status of adult citizens.

Purpose of the action: ill-preventive activities, increased individual responsibility, the prevention and control of risk factors, early interventions/corrections from the institutions in charge of public health, and provides data for the sustainability of the healthcare system.

Action 3. Funding services for supporting active life and healthy ageing.

Purpose of the action: supporting and funding services needed for an active life and a healthy ageing [21].

4 Institutional Involvement and Benefits of Investments-Action

For the Ministry of Labour and Social Protection, the national programme for increasing awareness about the determining factors in health status throughout life (proposed action 1) is a way of reducing risk factors, that the vulnerabilities, supporting, in the medium and long term, the design, development and adjustment of public policies regarding the improvement of long-term care (LTC) services.

Among the citizens, information is deficient, and so is awareness of the effects that health determining factors have in the long term. This programme will help citizens understand, become aware, have a level of personal competence that will help them adapt to the environment, change the environment, and thus have a healthy life for a longer period of time. Another category of beneficiaries of this programme includes the central and local public authorities, which can design sustainable development policies with improved results, lower costs in healthcare, better results in the economic domain, through the national programme for increasing awareness about the determining factors in health status throughout life. For this purpose, it is necessary to organise information campaigns (media and social media campaigns) for adults and the elderly.

For action 2, the main responsibility for organising and running such a national programme would go to the Ministry of Health, and to the INSP, through its four national centres (the National Centre for the Evaluation and Promotion of Health Status—CNEPSS and the National Centre for Monitoring Risks in the Community Environment—CNMRMC, the National Centre for the Oversight and Control of Transmissible Diseases—CNSCBT, the National Centre for Statistics and Information Technology in Public Health—CNSISP), as well as through the six regional public health centres (Bucureşti, Cluj, Iaşi, Timişoara, Târgu Mureş and Sibiu). With a secondary role in the implementation of this action, the Ministry of Public Works,

Development and Administration, the Ministry of Youth and Sports, together with deconcentrated bodies and the local public authorities can include the theme of this programme in the yearly sustainable development plan, and in local strategies, respectively. Civil society organisations and social and socio-medical service provides could also join in the effort. The direct beneficiaries of such a programme (action 2) would be adult individuals and individuals at risk of contracting diseases due to endogenous and exogenous causes. The Ministry of Labour and Social Protection would be an indirect beneficiary, as the yearly health status assessment programme would be a means of reducing risk factors and vulnerabilities, respectively, supporting, in the medium and long term the design, development and adjustment of public policies regarding the improvement of the quality of LTC services. Another category of beneficiaries of this programme includes the central and local public authorities, which can design sustainable development policies that would yield improved results, reduced healthcare costs and better results in the economic domain. The main responsibility for organising and running such a national programme would go to the Ministry of Health, through the National Centre for the Evaluation and Promotion of Health Status. It would have as partners the local public authorities, as well as bodies that manage health insurance in both the public and the private system. The Ministry of Labour and Social Protection would support the access to this programme of vulnerable individuals, groups, and communities (beneficiaries for whom the contribution to the National Health Insurance Fund (NHIF) is paid, co-insured individuals, uninsured beneficiaries, other vulnerable categories).

The direct beneficiaries of the services for an active life and healthy ageing would be all the citizens of Romania. Ministry of Labour and Social Protection would be an indirect beneficiary, such as services reduce and mitigate some of the social vulnerabilities and risks. Ministry of Health is an indirect beneficiary, as the funds from the public state budget and from the insurance fund could also be allocated for national programmes for the prevention of disease and for the promotion of a healthy society. The central and local public authorities would be able to design active policies and budget chapters that would complement those for education, employment, sport and health, investing, in the short, medium and long term, in local and regional development.

The main responsibility for organising and running such a national programme could go to an inter-ministry entity (Commission, Committee) for an active life and healthy ageing. The Ministry of Labour and Social Protection would become involved through the design, development and adjustment of public policies regarding the prevention of risk situations for vulnerable individuals and groups, encouraging active actions also together with the Ministry of Health: programmes for healthy communities. It would have as partners the local public authorities, as well as bodies that manage health insurance in both the public and the private system, trade unions, business owners associations, businesses, other associations from the civil society, social and educational services providers, professional associations [21].

5 Critical Future Risks and Funding Sources

The most targeted impact would be reached through a coherent three-year programme, reprised and adapted for as long as necessary. Based on the current institutional structure, the costs of such a programme should be included in the budget of the Ministry of Health and of the other ministries involved, as well as a percentage (3–5% initially, growing up to 10% in three years) of the budget of the National Health Insurance fund (NHIF). The new educational and preventive programme could continue the financing line of the national public health programmes, from the Ministry of Health budget, from the state budget and from its own income, from the NHIF budget, as well as from other sources, including donations and sponsorships, based on the current law. The best impact will be reached through a coherent programme run for a minimum of three years, reprised, and adapted for as long a necessary. Based on the current institutional structure, the costs of such a programme should be included in the budget of the Ministry of Health and of the budget of the National Health Insurance fund (as a percentage of 5–10% aimed at primary prevention).

The new health' evaluation programme could continue the financing line of the national public health programmes, from the Ministry of Health budget, from the state budget and from its own income, from the NHIF budget, as well as from other sources, including donations and sponsorships, based on the current law. Potential funding sources for services supporting active life and healthy ageing are public state budget for education, employment, sport and health; operational programmes; private sources and sponsorships; taxes. The best impact will be reached through a programme with a rapid start, but with long-term objectives and plans, run for as long a necessary. The costs of managing the insurance should be included in the budgets of the Ministry of Health and of the Ministry of Public Finance, as well as of the Ministry of Labour and Social Protection, of the Ministry of Education and Research, Ministry of Public Works, Development and Administration. The most rapid impact will be reached by recognizing or equating the private insurance already contracted by citizens and private insurers for LTC or for diseases with incapacitation potential.

6 Conclusions

The actions for the healthy development of children and youth through the Physical Education and Sports subject-matter in the school and university curricula, through the healthy diet programmes in schools, through investments in sports infrastructure and the promotion of performance in various sports, are steps that Romania has already taken. Such actions will have the results characteristic for an active lifestyle through a sustainable approach, rethinking investment in medium and long-term objectives for the younger generations: achieving and supporting today the actions and the programmes that create long-term behaviours and attitudes.

The programmes of the central and local authorities for supporting an active life and healthy ageing are successful among the elderly people in the urban environment. Actions for highlighting and stimulating a healthy lifestyle are needed especially in the rural environment.

The action supporting preventive-health activities are increasing individual responsibility, the prevention and control of risk factors, early interventions/corrections from the institutions in charge of public health, and provides data for the sustainability of the healthcare system. The programme would complement the National programme for the health of mother and child (Ministry of Health), as well as other programmes in the domain. Through the county Public Health Offices, the Bucharest Public Health Office, through the Regional Public Health Centres, the Ministry of Health monitors the risk factors from the environment (water quality, radiations), food and nutrition risk factors, as well as work environment risk factors. When correlated with the results of the population's health status screening, such data can support real improvements of the quality of life, the reduction of risk and social vulnerability on the short, medium and long term, with direct impact on reducing the pressure on the long-term care system.

Lifelong investments for a healthy life and ageing should be provided through a coordination of the health, educational and social systems. The Ministry of Labour and Social Protection, together with the Ministry of Education and Research can contribute as partners in the design, running and co-funding of the preventive educational programme. The Ministry of Education and Research would facilitate the inclusion on the general curricula themes related to a healthy lifestyle and to health determining factors (by generalising the facultative/optional subject-matter Education for health). The Ministry of Labour and Social Protection would support the access to this programme of vulnerable individuals, groups and communities.

References

1. Börsch-Supan A (2019) Survey of health, ageing and retirement in Europe (SHARE) Wave 7. Release version: 7.1.0. SHARE-ERIC. Data set. https://doi.org/10.6103/share.w7.710
2. Commission of European Communnities (2007) White paper together for health: a strategic approach for the EU 2008–2013. Brussels, 23.10.2007 COM(2007) 630 final
3. European Commission (2010) Europe 2020 strategy. A European Strategy for smart, sustainable ad inclusive growth. COM (2010) 2020, 03.03.2010. Brussel
4. Commission European, Spasova S, Baeten R, Coster S, Ghailani D, Peña-Casas R, Vanhercke B (2018) Challenges in long-term care in Europe. A study of national policies, European Social Policy Network (ESPN). European Commission, Brussels
5. EUROSTAT (2018) Available data: https://ec.europa.eu/eurostat/data/database
6. Governmental Decision 140 (2018) Framework contract regulating the provision of healthcare, of medication and medical devices within the social health insurance system for the years 2018–2019
7. Law 292 (2011) regarding national social work system
8. Ministry of Health (2018) Yearly report

9. Ministry of Health, National Public Health Institute, National Centre for the Evaluation and Promotion of Health Status. Cucu A (ed) (2018) National Report regarding Romanian Population Health Status. Available at: http://insp.gov.ro/sites/cnepss/wp-content/uploads/2018/11/SSPR-2017.pdf

10. National Institute for Public Health—National Centre for Statistics and Informatics in Health (2019) Newsletter. Main indicators of health awareness for the first quarter of 2019 compared to the first quarter of 2018. Available at: https://cnsisp.insp.gov.ro/wp-content/uploads/2019/10/Principalii-indicatori-trim-I-2019.pdf

11. Rebeleanu A, Şoitu D (2016) Perceptions of the family receiving social benefits regarding access to healthcare. In: Maturo A, Hošková-Mayerová Š, Soitu D-T, Kacprzyk J (eds) (2016) Recent Trends in social systems: quantitative theories and quantitative models. Springer International Publishing Switzerland, ISBN: 978-3-319-40583-4

12. Romanian Government Decision 566 (2015) The National Strategy for the Promotion of Active Aging and the Protection of the Elderly for 2015–2020 of the Operational Action Plan for 2016–2020 and of the Monitoring and integrated evaluation Mechanism

13. Social Protection Committee and European Commision (2014) Adequate social protection for long-term care needs in an ageing society. Report jointly prepared by the Social Protection Committee and the European Commission

14. Şoitu D (2014) Being healthy means being educated and acting accordingly. Procedia Soc Behav Sci 142(2014):557–563

15. Şoitu D (2015) Resilience and vulnerability—competing social paradigms? Analele Ştiinţifice ale Universităţii „Alexandru Ioan Cuza" din Iaşi (serie nouă). Sociologie şi asistenţă socială, tom VIII no.1, Editura Universităţii "Alexandru Ioan Cuza", Iaşi, pp 7–14. ISSN print 2065-3131. Available at: http://www.phil.uaic.ro/index.php/asas/article/view/364/313

16. Şoitu D (2018) The integrated social design of future long-term care. Analele Ştiinţifice ale Universităţii "Alexandru Ioan Cuza" din Iaşi (serie nouă). Sociologie şi asistenţă socială, tom XI no.1, Editura Universităţii „Alexandru Ioan Cuza", Iaşi, pp 5–15. ISSN print 2065-3131. Available at: https://anale.fssp.uaic.ro/index.php/asas/article/view/508

17. Şoitu D (2020a) Social work system and social innovations in Romania. Challenges and opportunities. In: Sarasola Sanchez Serano J et al (eds) qualitative and quantitative models in socio-economic systems and social work. Springer Nature Switzerland AG 2020, pp 17–24. https://doi.org/10.1007/978-3-030-18593-0_2; ISBN 978-3-030-18593-0

18. Şoitu D (2020b) Researching ageing by the life course perspective. In: Sarasola Sanchez Serano JL et al (eds) Qualitative and quantitative models in socio-economic systems and social work. Springer Nature Switzerland AG 2020, pp 83–88. https://doi.org/10.1007/978-3-030-18593-0_7, ISBN 978-3-030-18593-0

19. Şoitu D, Johansen K-J (2017) The space of innovation and practice on welfare, health and social care education and practice in Romania and Norway. Analele Ştiinţifice ale Universităţii „Alexandru Ioan Cuza" din Iaşi (serie nouă). Sociologie şi asistenţă socială, tom X no.1, Editura Universităţii „Alexandru Ioan Cuza", Iaşi, ISSN print 2065-3131, pp 5–17. Available at: https://anale.fssp.uaic.ro/index.php/asas/article/view/456

20. Şoitu D, Şoitu CT (2020) Romania. In: Ní Léime Á et al (eds) Extended Working Life Policies. International Gender and Health Perspectives, pp. 385-394, https://doi.org/10.1007/978-3-030-40985-2, ISBN 978-3-030-40984-5

21. Şoitu D, Matei A (eds) (2020) Îngrijirea de lungă durată. Cercetări. Politici şi Practici [Long-term care. Researches, policies and practices] Publishing House of Alexandru Ioan Cuza University of Iaşi

22. Svarcova I, Hoskova-Mayerova S, Navratil J (2016) Crisis management and education in health. Eur Proc Soc Behav Sci EpSBS XVI: 255–261. http://dx.doi.org/10.15405/epsbs.2016.11.26

23. UNDP (2020) Sustainable development goals. Goal 3: Good health and well-being. Available at: https://www.undp.org/content/undp/en/home/sustainable-development-goals/goal-3-good-health-and-well-being.html. Accessed June 2020

24. WHO (2010) The world health care report—Health systems financing: the path to universal coverage, http://www.who.int/whr/2010/en/index.html

25. WHO (2016a) Multisectoral action for a life course approach to healthy ageing: draft global strategy and plan of action on ageing and health, SIXTY-NINTH WORLD HEALTH ASSEMBLY A69/17 Provisional agenda item 13.4 22 April 2016. Available at: https://apps. who.int/gb/ebwha/pdf_files/WHA69/A69_17-en.pdf?ua=1
26. WHO (2016b) The Global strategy and action plan on ageing and health 2016–2020: towards a world in which everyone can live a long and healthy life. SIXTY-NINTH WORLD HEALTH ASSEMBLY WHA69.3 Agenda item 13.4 28 May 2016. Available at: https://apps.who.int/gb/ ebwha/pdf_files/WHA69/A69_R3-en.pdf?ua=1
27. WHO (2017) Global strategy and action plan on ageing and health. Available at: https://www. who.int/ageing/WHO-GSAP-2017.pdf?ua=1
28. WHO (2019) Healthy, prosperous lives for all: the European Health equity status report. Available at: https://www.euro.who.int/en/health-topics/health-determinants/social-determina nts/health-equity-status-report-initiative/health-equity-status-report-2019
29. WHO, Commission on Social Determinants of Health (2008) Closing the gap in a generation. Health equity through action on the social determinants of health, http://www.who.int/social_ determinants/thecommission/finalreport/en/index.html
30. World Bank, Human Development Network, Europe and Central Asia Region (2014) Long, active and forceful life. Promoting active ageing in Romania. Document of the World Bank

Welfare System, Aging Population and Long-Term Care Services in Italy

Vincenzo Corsi ⓘ

Abstract The chapter analyses the development of the Italian welfare system and focuses on the recent transformations of social and welfare needs. These changes concern the governance of the system and the forms of social and health assistance, in response to emerging needs. The aging of the population in Italy is the aspect of welfare policies with greater social impact; the aging of the population is a constantly increasing trend. The Italian local welfare system is built on the analysis of the social needs of the population and the planning of social services in the area. The planning of services for the elderly is becoming more focused on emerging welfare needs, and welfare systems are increasingly being required to give answers even for long-term care policies.

Keywords Sociology · Welfare · Aging of the population · Social services · Long-term care

1 Aging Population in Italy

The study of the Italian population has revealed an important aging phenomenon since the 1980s. Italy shows a greater accentuation of the phenomenon than other western countries. From ISTAT 2018 data, life expectancy at birth is 80.8 years for men (+0.2 on 2017) and 85.2 years for women (+0.3 on 2017) (Table 1). In Italy there are 168.9 people aged 65 and over for a hundred people under the age of 15.

These data show a positive situation for the Italian population, as we live more and better; this effect is determined by improvements in the quality of life and access to health care. As a negative consequence of the phenomenon, there is a relative decrease in the number of people aged 0–14; those aged 65 and over increase. The effect that is produced in the welfare system is the different distribution of resources and services over time in favour of the elderly population.

V. Corsi (✉)
Department of Management and Business Administration, "G. D'Annunzio" University—Chieti-Pescara, Viale Pindaro, 65127 Pescara, Italy
e-mail: vincenzo.corsi@unich.it

© The Author(s), under exclusive license to Springer Nature Switzerland AG 2021
D. Soitu et al. (eds.), *Decisions and Trends in Social Systems*,
Lecture Notes in Networks and Systems 189,
https://doi.org/10.1007/978-3-030-69094-6_5

Table 1 Life expectancy at birth and at the age of 65—Italy 1 January 2018

Years	Life expectancy at birth		Life expectancy at the age of 65	
	Men	Women	Men	Women
2014	80.3	85.0	18.9	22.3
2015	80.1	84.6	18.7	21.9
2016	80.6	85.0	19.1	22.3
2017	80.6	84.9	19.0	22.2

Source Istat

Today in Italy a 65-year-old man has a 19 years of life expectancy whilst a woman has 22.2 years (Table 1). The shift of resources to be allocated to welfare for the third and fourth age becomes a necessary and increasingly decisive condition in contemporary societies. Italy is the second oldest demographic country in the world after Japan. The consequences are not only demographic but also social because of the sustainability of health services and pensions within the welfare system.

The main demographic indicators describe a country with a significantly aged demographic structure, with an elderly dependency rate of 35.2 and an elderly index of 168.9 (Table 2). The population pyramid in 2018 shows an internal imbalance, with an average age of 45.2 and with 22.6% of the population over sixty-five (Fig. 1).

Aging has changed the structure of the Italian population; in the last few decades this process has undergone a particular acceleration and occurred simultaneously with the social and cultural changes that have affected the Italian family.

As the population ages, not only the demographic structure of a society changes, but also its organization with effects on the labour market, on the family structure, on the consumption of goods and services, on the organization of health systems and social assistance. It seems that there is a new society compared to the past, in which childhood and adolescence lose relevance for the benefit of the aging population and the related social dynamics; there is an increase in men and women who require greater participation in social life outside the constraints and responsibilities of work and childcare. This is particularly important for all those elderly people who maintain good personal health and autonomy; on the other hand, we have more and more

Table 2 Structure of the population—Italy 2015–2018

Years	Percentage			Index			
	Aged 0–14	Aged 15–64	Aged 65 and over	Elderly	Dependency rate	Elderly dependency rate	Average age
2015	13.8	64.5	21.7	157.7	55.1	33.7	44.4
2016	13.7	64.3	22.0	161.4	55.5	34.3	44.7
2017	13.5	64.2	22.3	165.3	55.8	34.8	44.9
2018	13.4	64.1	22.6	168.9	56.1	35.2	45.2

Source Istat

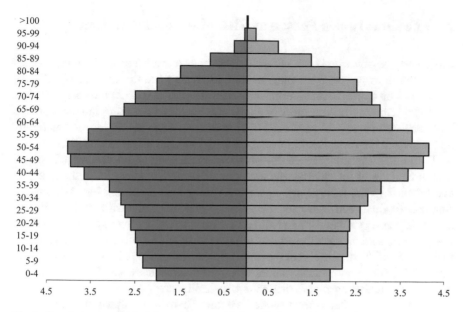

Fig. 1 Population pyramid of the resident population—Italy 2018—percentage. *Source* Istat

elderly people, often over seventy-five, who are not self-sufficient and require long-term care interventions.

The increase in life expectancy and the increasing presence of older people in contemporary societies have directed social policies on two main issues: the quality of life of these people and the social, health and welfare services to be put in place in their favour. The increase in the population aged seventy-five or over raises the question of a progressive transfer of economic resources to services for people who, over time, may find themselves in conditions of non-self-sufficiency.

Aging can under certain conditions lead to a partial or total, temporary or permanent loss of physical or mental autonomy in the person. The causes can be related to biological and psychological states, but also to the loss of loved ones in the family unit or to economic difficulties.

Aging does not necessarily mean non-self-sufficiency of the person, whether physical or mental. However, it is true that a percentage of the elderly aged seventy-five and over present some reduction in their autonomy. In Italy in 2018 four million families had a non-autonomous family member. In addition, in 2018, one million and two hundred thousand elderly people, aged sixty-five and over, described themselves as being alone, isolated and without friends.

Aging should not necessarily be seen as the source of health and social problems, but it is true that physical aging causes an increase in the economic, social and health conditions of risk and a potential increase in the use of social and health services.

2   The Elderly as a Person at Risk of Non-self-Sufficiency

The elderly person who is not self-sufficient needs a careful analysis of their needs and multiple responses from families and social and health services. Usually the elderly person who is not self-sufficient is a fragile person with whom social services must build an individual project of social and health care, taking into account the biological, psychological and social characteristics of the person, as well as family resources and the formal and informal network of social relationships [9].

The family, especially in the Italian social-welfare situation, has a fundamental role in the care of the elderly who are not self-sufficient. Changes in the structure and function of the Italian family pose new problems concerning the assistance of fragile individuals and require the intervention of social and health services. First of all, in Italy emerged the need to build a new local social-welfare system capable of integrating social and health services in order to provide a more effective response to the needs of the person and of the family. Local welfare in Italy is established with law 328/2000 and immediately arises as a system capable of building a network of services by integrating social assistance with healthcare.

In the mid-seventies of the twentieth century Charles F. Faheyil and the Federal Council on Aging (FCA) in the United States introduced the concept of fragile elderly to define a person who, usually over the age of seventy-five, due to an accumulation of various deficits, requires one or more help services to cope with everyday situations. The fragile elderly condition can be defined as a biological state that depends on age and is characterized by reduced resistance to stress, the cumulative decline of multiple physiological systems and is related to multiple health conditions, disabilities, institutionalization risk and mortality.

The FCA has made it possible to identify some important aspects of the fragility of the elderly, such as advanced age, the need for support and social and welfare services, but also the need of social policy interventions necessary to respond to the condition of need of the person [4]. Over the years, the social and relational aspects that can favour or hinder the daily life of non-self-sufficient elderly have increasingly been defined.

In 2001 the World Health Organization published the International Classification of Functioning, Disability and Health, known more commonly as ICF; it provides a standard language and framework for the description of health and health-related states. In the ICF, the term functioning refers to all body functions, activities and participation, while disability is an umbrella term for impairments, activity limitations and participation restrictions.

In this version, non-self-sufficiency refers to the person's health and age; it concerns the total or partial inability, temporary or permanent, to perform the normal actions of daily life. However, non-self-sufficiency also depends on economic and social factors such as, for example, having an adequate income, an accessible home, a social network of family relationships capable of protecting the individual and ensuring their mental and physical well-being [11]. These factors define a condition of socio-welfare need that concerns both medical and social spheres. They take into account the socio-relational and economic aspects that define a condition of non-self-sufficiency, including the social relationships in which the elderly person is involved.

The situation becomes more problematic in the presence of chronic and disabling diseases. The elderly person with these characteristics of health and limited psychological and social well-being is also a vulnerable person who, over the years, has an increased risk of developing diseases, disabilities and complex needs that require long-term care interventions and services.

Assistance within the family plays an important role in the Italian social and welfare state. The transformations that have taken place in the structure of the Italian family over the past decades have changed its role in assisting people with social and health difficulties [7]. The decrease in births, the increase in childless couples [1] and the trend towards a nuclear unit have reduced the family's ability to take care of the elderly in a comprehensive way (Table 3).

An analysis of the data on the Italian family structure shows an increase over the years in the percentage of elderly people living alone. The difference, in terms of socio-welfare needs, is considerable when the elderly also presents difficulties of economic nature. In these cases, the possibility of keeping the non-self-sufficient elderly at home is reduced. The condition of need is linked to isolation risks with reduced inclusion in the network of social relations.

Currently in Italy the majority of the elderly who are not self-sufficient continue to be assisted within their own family network, however it is becoming more common that elderly people are isolated and do not find among their family members the support and care they need [6].

Table 3 Population according to age group and family context—average 2017–2018, for 100 of the same age group

Age	Child in a family unit made by a couple with children	Child in a single parent family	In a family with two or more nuclei, isolated members, or without nucleus	In a couple without children	Parent in a family unit made by a couple with children	Parent in a single parent family	Alone
17 and below	83.2	12.1	4.7	0.1			
18–34	47.0	13.6	7.1	7.9	15.2	1.2	8.1
35–54	4.3	3.9	5.6	10.5	56.3	6.2	13.2
55–64	0.3	1.6	6.4	26.1	43.5	6.4	15.7
65 and over	0.0	0.1	8.2	43.2	13.0	5.9	29.6
Total	23.4	5.8	6.4	17.7	28.4	4.2	14.1

Source Istat

Istat data show an increase in the elderly who live within a fragmented network of social relationships. Over the years, these people can face problems, more or less serious, of not complete autonomy, without family or informal relational networks as support. This difficult condition is especially noticeable among the elderly who live alone, especially if they are unmarried, divorced or widowed, without the support of their children.

3 Long-Term Care Policies in Italy

Sociological analysis highlights how aging increases the risk of disability with the simultaneous increase in the need for care and social assistance. The transformations within society, the new role of women in the labour market and the structural transformations of the family have determined, especially for the elderly aged eighty and over, a growing need for assistance. More particularly, multigenerational families, those in which the elderly live with one or more of their children, have significantly decreased in Italy (Table 4). This structural modification of the family has effects on the living conditions of dependent people.

In almost all European countries, 10% of the adult population usually takes care of other family members. The role of informal caregiver, especially in Italy, is taken by women. The transformations of the family in a nuclear unit, the distance from children, divorces or widowhood have posed the problem of caring for the elderly as a growing social condition. The progressive increase in the society of people in a condition of serious non-self-sufficiency require, as the figure of the family caregiver weakens, specialized social-assistance interventions to be carried out on the territory or in residential social-assistance services for short or long-term periods [8, 12].

While in the sixties and seventies of the twentieth century the assistance services were aimed at elderly who were alone, without a family support network, today the situation in Italy has changed [10]; although Italian welfare still remains based on the family, there is a significant number of elderly people who need social services

Table 4 Families according to the number of members—average 2017–2018 (a), for 100 families

Year	Families (number in thousands)	Percentage					
		One member	Two members	Three members	Four members	Five members	Total
2013–2014	25.017	30.6	27.1	20.0	16.9	5.4	100.0
2014–2015	25.266	31.1	27.1	20.1	16.2	5.4	100.0
2015–2016	25.386	31.6	27.3	19.8	16.0	5.4	100.0
2016–2017	25.494	31.9	27.5	19.6	15.7	5.3	100.0

Source Istat, multi-purpose enquiry "aspects of every-day life"
The indicator of the years 2011–2014 have been updated following the recounting of the population after the 2011 census

outside the family [3]. Sociological analysis today highlights how care, previously performed almost entirely by the family, is today increasingly provided from outside with a growing demand for territorial social-welfare and socio-health services given by welfare services.

Care services have become real market goods. In Italy there is a situation in which, while current elderly aged seventy-five or over have had more stable marriages with many children, the next generations of elderly will be conditioned, especially in northern urban contexts, by less stable marriages and by a reduced presence of children. For these elderly people there will be less support and care from the family and an increase in the use of territorial welfare services, also of a residential nature, for situations of non-self-sufficiency. For these reasons, the analysis of the social needs of the population becomes important in the construction of local welfare services in order to plan activities according to the care needs of the territory.

The Area Plan (*Piano di Zona*) is the instrument for the construction of local welfare in Italy [5]. The Area Plan is a planning document in the Italian welfare system, which contains a separate section called 'Local Social Profile' that analyses the social needs of the area. The Local Social Profile describes the social, economic and cultural rights of the area [2].

In the construction of local welfare, in Italy there are four levels of government defined in the Law n. 328/2000: the municipalities have administrative functions and prepare the welfare services of the Local Plan; the provinces contribute to the planning of the system according to the procedures defined by the regions; the regions are responsible for the planning, coordination, management and verification of social interventions and prepares the Regional Social Plan; the State defines essential levels of care.

The Law n. 328/00 on the article 16 includes 'the provision of help and domestic support, also with economic benefits, with special regard to families that host and care for physical, psychical and sensorial disabled people and other people in difficulties, for children in care and for the elderly.' Regions and municipalities have introduced benefits for dependent people in different ways: economic benefits for families that care for non-self-sufficient elderly; vouchers for health services; a fund to support non-self-sufficient people.

When discussing the topic of self-sufficiency, it is important to take into account the following aspects: health (the presence of chronic diseases); functionality (autonomy of the people in every-day life and in social life); social context (formal and informal care, from the family or from external agencies). The increase in the number of non-self-sufficient elderly in Italy has transformed the system of interventions and services. The government, the regions and the municipalities realised that it was necessary to extend the offer of health and social services to meet the needs of old and new users.

The law 8th November 2000, n. 328 has restated the importance of domestic assistance for the non-self-sufficient elderly. This service must be present in each area. Moreover, the law outlines a new system of services where municipalities, citizens and the third sector are equally involved. This law underlines the importance of differentiating interventions according to the needs and the vulnerability of the old person, guaranteeing the opportunity to remain in a domestic environment, with the family, and only when domestic assistance is not an option, the lodging in residential structures.

In Italy there are different social services for non-self-sufficient elderly; the most common are:

- Aggregated Domestic Assistance (*Assistenza domiciliare integrata* ADI): medical and social assistance given at home, according to a personalised plan of care.
- Domestic Assistance: the assistance provided at home for the self-sufficient or partially self-sufficient elderly consisting of self-care, personal hygiene, provision of meals.
- Sheltered Accommodation (*Casa Protetta*): a residential structure that provides health services for the elderly in a condition of physical or psychological dependence and that cannot be assisted at home or within the family. The aim is to prevent further loss of independence and preserve mental, physical and social abilities.
- Health Residence for the assistance (*Residenza sanitaria assistenziale* RSA): it is a service that involves the non-self-sufficient elderly that cannot be assisted at home, affected by chronic and degenerative illnesses that require specific health services in a hospital environment.
- Other residential solutions are lodging houses, residential structures where the aim is to recreate a familiar and social environment, protected apartments that offers to the elderly the opportunity to remain in their social context facing only limited difficulties.

4 Conclusions

Italy is one of the countries in the world with the greatest presence of elderly people. The significance of non-self-sufficient people is a matter of concern in the field of social and social-welfare services. In particular, the integration between social and health services is necessary in the planning and organization of services in the context of local welfare.

Older people, especially if aged seventy-five and over, are more vulnerable and sensitive due to natural physiological decline. Non-self-sufficiency is often associated with psychological and physical health problems, as well as relational, family and economic problems.

The family member is often called to assist the elderly who are not self-sufficient, but the family's social-assistance intervention is not always sufficient. In Italy, the territorial organization of social and health services is the responsibility of the regions, local health authorities and municipalities. The latter are required to organise through Area Plans the social-welfare services to be guaranteed on the territory by looking specifically at the social-health interventions, in order to provide effective local welfare.

References

1. Barbagli M, Castiglioni M, Della Zuanna G (2004) Fare famiglia in Italia. Un secolo di cambiamenti, il Mulino, Bologna. ISBN: 9788815099389
2. Battistella A (2004) Costruire e ricostruire i Piani di Zona. In: Battistella A, De Ambrogio U, Ranci Ortigosa E (eds) Il piano di zona. Costruzione, gestione, valutazione, Carocci Faber, Roma. ISBN: 9788874660681
3. Caruso E, Gori C, Pelliccia L, Pesaresi F (2013) Le politiche nazionali per gli anziani non autosufficienti. In: Gori C (ed) L'assistenza agli anziani non autosufficienti in Italia. 4° Rapporto. Tra crisi e ripartenza. Rapporto promosso dall'IRCCS—INRCA per il Network nazionale per l'invecchiamento. Maggioli Editore, Santarcangelo di Romagna (RN). ISBN: 9788838735646
4. Casale G (2010) La persona "fragile" e i percorsi di cura: un enigma per il "SOSIA". In: Casazza S (ed) Continuum for care. Continuità e discontinuità nella cura dell'anziano fragile. FrancoAngeli, Milano. ISBN: 978-8856815313
5. Corsi V (2017) Methods and Models of Social Planning in Italy. In: Maturo A, Hošková-Mayerová S, Soitu D-T, Kacprzyk J (eds) Recent trends in social systems: quantitative theories and quantitative models. Studies in systems, decision and control, vol 66. Springer, Switzerland. ISBN 978-3-319-40583-4
6. Facchini C (2012) Les personnes âgées seules: conditions de vie, réseaux familiaux et vécu individuell. In: Pitaud P (ed) Soilitude et isolement des personnes âgées. Érès éditions, Toulouse. ISBN: 9782749229515/ISBN: 9782749202556
7. Facchini C (2010) Il sistema familiare nella cura: evoluzione e scenari futuri. In: Casazza S (ed) Continuum for care. Continuità e discontinuità nella cura dell'anziano fragile. FrancoAngeli, Milano. ISBN: 9788856815313
8. Gori C (ed) (2008) Le riforme regionali per non autosufficienti. Gli interventi realizzati e i rapport con lo Stato. Carocci, Roma. ISBN: 9788843049479
9. Gori C (ed) (2013) L'assistenza agli anziani non autosufficienti in Italia. 4° Rapporto. Tra crisi e ripartenza. Rapporto promosso dall'IRCCS—INRCA per il Network nazionale per l'invecchiamento. Maggioli Editore, Santarcangelo di Romagna (RN). ISBN: 9788838735646
10. Gori C, Ghetti V, Rusmini G, Tidoli R (2014) Il welfare sociale in Italia. Realtà e prospettive. Carocci, Roma. ISBN: 9788843073252
11. Ranci C, Pavolini E (2015) Le politiche di welfare. il Mulino, Bologna. ISBN: 9788815252319
12. Vallejo A, Hoskova-Mayerova S, Krahulec J, Sarasola JL (2017) Risks associated with reality: how society views the current wave of migration: one common problem—two different solutions. Stud Syst Decis Control 104:283–305. https://doi.org/10.1007/978-3-319-54819-7_19

Mediterranean Diet Patterns in the Italian Population: A Functional Data Analysis of Google Trends

G. Caruso and F. Fortuna

Abstract Internet search engines have become a popular and readily accessible source of information. Google Trends, by means of analyzing the popularity of search queries in Google Search, allows to provide deep insights into population behavior. Interestingly, Google is increasingly being used also to obtain health-related information, as well as to self-prescribe one's dietary intake. In particular, we analysed the search traffic related to the keywords Mediterranean diet since it has always been very popular. More specifically, we propose to use Google Trends data as proxies for the interest in Mediterranean diet and to analyze them through the functional data analysis (FDA) approach.

Keywords Google Trends · Functional clustering · Mediterranean diet · Health information seeking behavior

1 Introduction

Internet search engines have become a popular and readily accessible source of information. Google Trends is a free tool that allows to provide deep insights into population behavior, by means of analyzing the popularity of search queries in Google Search, across various regions and languages, allowing to identify and describe both seasonal and geographic patterns. Interestingly, Google is increasingly been used to obtain health-related information, too [35]. Nowadays, indeed, when noticing a new ache, it is very common to check symptoms online to self-diagnose, instead of scheduling a doctor appointment. The act of self-researching one's perceived medical

G. Caruso (✉)
University G. d'Annunzio of Chieti-Pescara, Pescara, Italy
e-mail: giulia.caruso@unich.it

F. Fortuna
Roma Tre University, Rome, Italy
e-mail: francesca.fortuna@uniroma3.it

© The Author(s), under exclusive license to Springer Nature Switzerland AG 2021
D. Soitu et al. (eds.), *Decisions and Trends in Social Systems*,
Lecture Notes in Networks and Systems 189,
https://doi.org/10.1007/978-3-030-69094-6_6

issues, named "Health Information Seeking Behavior" has become really pervasive [21, 27].

Moreover, people increasingly self-prescribe their dietary intake, too [17, 20]. Thus, and since the Mediterranean diet, hereinafter referred as MD, is the best-known nutrition regime, even recognized as "Intangible Heritage of Humanity" by UNESCO in 2010 [2], we decided to explore the functionalities of Google Trends using MD as keywords. The Mediterranean diet popularity is mainly due to the fact that it has always been recommended by many doctors and dietitians to prevent disease and keep people healthy for longer [8]. It helps, indeed, to prevent, amongst others, strokes and heart attacks, the metabolic syndrome, lung diseases, asthma, allergies, Parkinson, Alzheimer, and decalcification [10]. Furthermore, it keeps the bone mass in elderly people and it is linked to low incidences of many types of cancer. Since the Mediterranean diet owes its name to the traditional healthy living habits of people from countries bordering the Mediterranean Sea, including Italy, France, Greece and Spain, we decided to restrict our analysis to one of them, specifically to Italy. To implement our analysis, we used Google Trends, a free tool allowing users to interact with Internet search data, providing deep insights into population behavior and health-related phenomena [26].

In this chapter, Google Trends data have been analyzed in a functional framework [24, 28], providing several advantages. The most important of them relies on the fact that the functional data analysis approach enables an efficacious statistical analysis when the number of variables is higher than the number of observations, as in the case of search queries [1, 16, 23]. In this context, an unsupervised classification of the functional queries has been implemented, to identify specific common patterns among the Italian curves.

The remainder of this chapter is organized as follows. Section 2 deals with Google Trends data in a functional framework, whereas in Sect. 3 we expose the two functional unsupervised clustering methods employed, namely the k-means algorithm and the hierarchical technique. In Sect. 4, instead, we argue about the implemented application on a real dataset, concerning search queries for the keyword "Mediterranean diet" across the Italian regions. Finally, in Sect. 5, we draw some conclusions, summarizing our key findings deriving from the implementation of the applied methods.

2   Google Trends Data in a Functional Framework

Google Trends (http://www.google.com/trends) is a keyword research tool that provides real time trend data, regarding interest as operationalised by Internet search volume. It shows how often search terms are entered in Google, with regards to the total search volume over time, since 2004, and across different geographical locations. The search query index does not represent the raw levels of search queries, but rather a relative search volume index, RSV, that is:

$$RSV(q, r, t) = \frac{s(q, r, t)}{\sum_{q \in Q(r,t)} s(q, r, t)} \tag{1}$$

where $s(q, r, t)$ denotes the number of search queries for the keyword q in a specific geographical area r at time t, and $Q(r, t)$ is the set of all search queries from location r at time t. RSV is then normalized by the highest query share of that term, over the time series, as follows [7]:

$$GRSV(q, r, t) = \frac{RSV(q, r, t)}{max_{t \in [1,...,T]} RSV(q, r, t)} \times 100 \in [0, 100] \tag{2}$$

Since Google Trends data continuously flow from the server of a web site, they can be seen as functions in a continuous domain, rather than scalar vectors [16]. Despite the continuous nature of functional data, in real applications, sample curves are observed with error in a discrete set of sampling points, $t_1 < t_2 < \cdots < t_L$ of $\mathcal{T}$. Specifically, let $y_j(t) = \left\{ y_j(t_{jl}) \right\}_{l=1}^{L}$, $j = 1, 2, \ldots, n$, be a functional variable observed in a discrete set of sampling points, $l = 1, 2, \ldots, L$, in the temporal domain $\mathcal{T}$. Let us also assume that $y(t) \in L^2(\mathcal{T})$, where $L^2(\mathcal{T})$ is the Hilbert space of square integrable functions with the usual inner product $< y, g >= \int_{\mathcal{T}} y(t)g(t)\, dt$, $\forall y, g \in L^2(\mathcal{T})$ and the L^2-norm $||y|| =< y, y >^{1/2} < \infty$. Thus, the observed data satisfy the following statistical model:

$$y_{jl} = y_j(t_{jl}) + \epsilon_{jl} \quad l = 1, \ldots, L; \ j = 1, \ldots, n \tag{3}$$

with y_{jl} being the observed value for the jth sample path at the sampling point t_l.

One usual solution to reconstruct the functional form of the n samples, starting from the discrete observations, is to assume that sample paths belong to a finite-dimension space, spanned by a basis $\{\phi_1(t), \phi_2(t), \ldots, \phi_K(t)\}$, so that they can be expressed as follows:

$$y_j(t) = \sum_{b=1}^{B} a_{jb}\phi_b(t), \quad j = 1, \ldots, n \tag{4}$$

where (a_{jb}) is the bth basis coefficient for the jth functional observation $y_j(t)$, and $\phi_b(t)$ represents the bth basis functions.

Under the model in (3), the basis coefficients can be fitted by least squares approximation with B-splines basis [9].

3 Unsupervised Classification of Functional Queries

Functional queries may present a variety of distinctive patterns corresponding to different shapes and variation, which can be identified by clustering the functions

[16, 30, 34]. Specifically, a set of homogeneous clusters in L^2 can be identified by determining a partition of the space according to the minimal distance. To this end, an L^2 metric in function space has been applied, combined with both a k-means algorithm and hierarchical methods for finite dimensional data [11, 13].

In particular, the k-means algorithm [15, 22] consists in an iterative procedure initialized by setting K clusters, $C_k, k = 1, 2, \ldots, K$, and by randomly selecting in the function space a set of arbitrary initial centroids, $\{\phi_1^{(q)}(x), \ldots, \phi_K^{(q)}(x)\}$, one for each cluster. The curves, at each qth iteration, $q = 0, 1, \ldots, Q$, are allocated to the cluster whose centroid is nearest, on the basis of a specific distance. According to the assignment of curves to groups, the cluster means are updated and the algorithm reiterate until curves stop to be reassigned to clusters [29]. Specifically, the k-means algorithm finds a partition of the functional space into K clusters, by minimizing the sum of squared error criterion between the cluster center and the functions belonging to the cluster, as follows [3, 5, 18, 33]:

$$
J(C) = \min \sum_{k=1}^{K} \sum_{y_j(t) \in C_k} d^2\left(y_j(t), \phi_k(t)\right) = \min \sum_{k=1}^{K} \sum_{y_j(t) \in C_k} ||y_j(t) - \phi_k(t)||^2
$$

$$
= \min \sum_{k=1}^{K} \sum_{y_j(t) \in C_k} \left(\int |y_j(t) - \phi_k(t)|^2 d\,t\right)^{\frac{1}{2}} \tag{5}
$$

where the centroid is computed by averaging the functions across the replications [28]:

$$
\phi(t) = \sum_{y_j(t) \in C_k} \frac{y_j(t)}{n_k} \tag{6}
$$

where n_k is the number of functions in the kth cluster, with $\sum_{k=1}^{K} n_k = n$.

Hierarchical clustering techniques may also be adopted for functional data [14]. In this context, the classification strategy contains a series of partitions: from a single group including all the functions (divisive methods), to K groups, each containing a single function (agglomerative methods). To detect which group has to be merged (in the agglomerative approach) or split (in the divisive approach), it is possible to use different metrics and linkage methods. With regards to functional data, the L^2 metric is an appropriate choice for setting out dissimilarities among functions. Concerning the linkage methods, the most common are: the single linkage [31], the complete linkage [25], the average linkage [32], or the Ward's minimum variance method [36]. More specifically, in the context of the Ward's minimum variance method for agglomerative hierarchical clustering, the total within-cluster error sum of squares is minimized. Consequently, the algorithm finds, at each step, the pair of clusters generating a minimum increase in the total within-cluster variance after merging. This increase consists in a weighted squared distance among the cluster mean functions. The latter can be suitably computed through Eq. (6). Ward's method presents some similarities with the k-means algorithm, as it is the only one, among the agglomerative hierarchical clustering methods, to be based on a classical sum-of-squares criterion, producing groups minimizing the within-group dispersion at each binary fusion. However, Ward's method merges sub-clusters in order to achieve this goal, as opposed to k-means algorithm, which employs an iterative reassignment of points [4, 11].

4 Functional Clustering of Google Queries for the Keywords "Mediterranean Diet" in Italy

In this section, Google Trends data have been used to evaluate the public interest in Mediterranean diet argumentation among different Italian regions. Weekly search volumes for the terms "Mediterranean diet", conducted in the corresponding national language (that is "dieta mediterranea"), have been obtained from the website Google Trends (trends.google.com) by considering searches carried out in Italy. Data have been collected across a 5-year period, starting from January 2005 until May 2019 (for a total of 230 weeks) by considering the 20 Italian regions. The region of Aosta Valley is the only one with zero frequencies for the search term under consideration, hence it has been removed from the study. Figure 1 shows the Google Trends series for the Italian regions.

These data can be considered as raw functions observed in a discrete set of sampling points, that is 230 weekly observations in the period under consideration.

To reconstruct the functional form underlying the data starting from pointwise evaluations of it, a functional basis expansion with 20 cubic B-splines has been adopted (see Eq. 4). Figure 2 shows the reconstructed functional queries. In particular, we aim to identify specific common patterns among the functional queries, applying

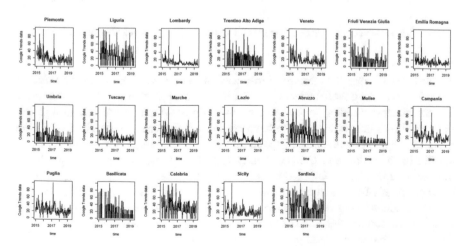

Fig. 1 Raw Google Trends data for the keyword "dieta mediterranea" across Italian regions

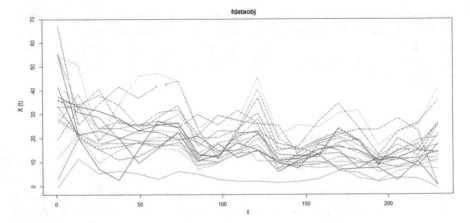

Fig. 2 Functional queries for "dieta mediterranea" in the Italian regions

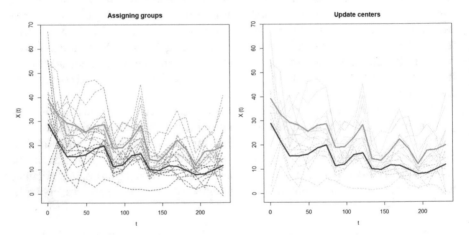

Fig. 3 Classification of the functional queries under the functional k-means algorithm

the functional k-means algorithm introduced in Sect. 3. We initialized the k-means procedure by fixing two clusters ($K = 2$), then, each function has been assigned to its nearest cluster center, by minimizing the sum of squared error criterion in (5). Figure 3 shows the partition of the function space into two clusters. The first cluster (in red in Fig. 3) is composed of functional queries of 9 regions, which are mainly located in the North and Centre of Italy (Lombardy, Trentino Alto Adige, Friuli Venezia Giulia, Emilia Romagna, Tuscany, Umbria, Lazio), with the exception of the regions of Molise and Basilicata, which are located in the South. The second cluster (in green in Fig. 3) is composed of functional queries of 10 regions which are principally located in the South of Italy (Abruzzo, Campania, Puglia, Calabria, Sicily, Sardinia) with the exception of three Northern regions (Piemonte, Liguria, Veneto) and one central region (Marche). Clusters features may be captured in an immediate

Fig. 4 Cluster dendrogram
for functional hierarchical
Ward's method applied to the
functional queries

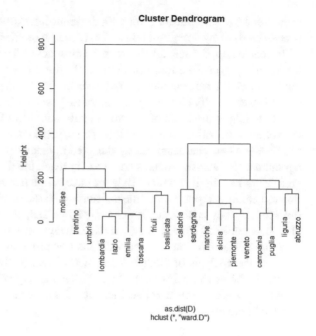

way by analyzing the shape of the functional centroids in the right panel of Fig. 3. The functional centroids present the same trend, but with different intensities. The second cluster (in green), indeed, is characterized by a major interest in "Mediterranean diet" search queries.

The same dataset has been analyzed through the functional agglomerative hierarchical Ward's method. As shown by the dendrogram in Fig. 4, we obtained the same results of the k-means procedure.

5 Concluding Remarks

The aim of this chapter is to evaluate the public interest in the Mediterranean diet, on the basis of weekly Italian search volumes for the terms MD on Google, in the period ranging from January 2005 until May 2019. To this aim, we implemented two functional unsupervised methods, namely the k-means algorithm and the functional agglomerative hierarchical Ward's technique, in order to classify the search queries regarding MD into two homogeneous groups and to finally compare their results. It resulted that both methods led to the same conclusions. The first identified cluster includes 9 regions, mainly located in the North-Centre of Italy, whereas the second one concerns 10 regions, principally located in the South of Italy. Furthermore, through both methods, it appears that the functional centroids present the same trend for both clusters, but with different intensities, since the second cluster is characterized by a major interest in MD search queries.

In conclusion, the second cluster is the most active in the search of MD in Google for several presumable reasons. First of all, the main reason is historical: the origin of the Mediterranean dietary traditions has his roots precisely in Mediterranean countries, especially Greece and southern Italy, since the early 1960s [37]. Then, the raw materials at the basis of the MD mainly come from the South of Italy, suffice is to think to the olive oil and to citrus fruits, just to name a few. Besides, the South has always been characterized by the predominance of activities as fisheries and agriculture [3], whose products could easily reach the table of the inhabitants of the area. Moreover, in the South of Italy the pace is much slower, favouring the preparation of healthy recipes, more time consuming, to the detriment of fast foods. Lastly, the South of Italy is historically economically disadvantaged compared to the North [19], thus poor but wholesome food have always been integral part of the diet. The final aim of this research is to provide practitioners with additional methodological tools for the analysis of Google trends data. Specifically, the use of the functional approach allows to deal with the high dimensionality of these data and to identify functional clusters, able to reflect the dynamics of the search process over the whole temporal domain.

References

1. Aguilera AM, Fortuna F, Escabias M, Di Battista T (2019) Assessing social interest in burnout using Google trends data. Ind Res Soc. https://doi.org/10.1007/s11205-019-02250-5
2. Bach-Faig A, Berry E, Lairon D, Reguant J, Trichopoulou A, Dernini S, Medina F, Battino M, Belahsen R, Miranda G, Serra-Majem L (2011) Mediterranean diet pyramid today. Science and cultural updates. Public Health Nutr 14(12A):2274–2284. https://doi.org/10.1017/S1368980011002515
3. Caruso G, Di Battista T, Gattone SA (2019) A micro-level analysis of regional economic activity through a PCA approach. In: Bucciarelli E, Chen S, Corchado JM (eds) Decisions economics: complexity of decisions and decisions for complexity. Advances in intelligent systems and computing (in print)
4. Caruso G, Gattone SA (2019) Waste management analysis in developing countries through unsupervised classification of mixed data. Soc Sci 8(6)
5. Caruso G, Gattone SA, Balzanella A, Di Battista T (2019) Cluster analysis: an application to a real mixed-type data set. In: Flaut C, Hoskova-Mayerova S, Flaut D (eds) Models and theories in social systems. Studies in systems, decision and control, vol 179. Springer, pp 525–533
6. Caruso G, Gattone SA, Fortuna F, Di Battista T (2018) Cluster analysis as a decision-making tool: a methodological review. In: Bucciarelli E, Chen S, Corchado, JM (eds) Decision economics: in the tradition of Herbert A. Simon's heritage. Advances in intelligent systems and computing, vol 618. Springer, pp 48–55
7. Choi HY, Varian H (2012) Predicting the present with Google Trends. Econ Rec 88:2–9
8. D'Adamo I, Falcone PM, Gastaldi M (2019) Price analysis of extra virgin olive oil. Brit Food J. https://doi.org/10.1108/BFJ-03-2019-0186
9. De Boor C (2001) A practical guide to splines. Springer, New York
10. Demarin V, Lisak M, Morović S (2011) Mediterranean diet in healthy lifestyle and prevention of stroke. Acta Clin Croat 50(1):67–76

11. Di Battista T, Fortuna F (2016) Clustering dichotomously scored items through functional k-means algorithm. EJASA 9(2):433–450
12. Di Battista T, Fortuna F (2017) Functional confidence bands for lichen biodiversity profiles: a case study in Tuscany region (central Italy). Stat Anal Data Min 10(1):21–28
13. Di Battista T, Gattone SA, De Sanctis A (2010) Dealing with FDA estimation methods. In: Ingrassia S, Rocci R, Vichi M (eds) New perspectives in statistical modeling in data analysis. Advances in intelligent systems and computing. Springer, Cham, pp 357–365
14. Ferreira L, Hitchcock D (2009) A comparison of hierarchical methods for clustering functional data. Commun Stat-Simul C 38(9):1925–1949
15. Forgy E (1965) Cluster analysis of multivariate data: efficiency vs interpretability of classifications. Biometrics 21:768–769
16. Fortuna F, Maturo F, Di Battista T (2018) Clustering functional data streams: unsupervised classification of soccer top players based on Google trends. Qual Reliab Eng Int 34(7):1448–1460
17. Golley S, Corsini N, Mohr P (2017) Managing symptoms and health through self-prescribed restrictive diets: what can general practitioners learn from the phenomenon of wheat avoidance? Aust Fam Phys 46:603–608
18. Jain A, Dubes R (1988) Algorithms for clustering data. Prentice Hall, Englewood Cliffs, NY
19. King R (1971) The "Questione Meridionale" in Southern Italy, vol 11. University of Durham, Department of Geography
20. Klement R, Frobel T, Albers T, Fikenzer S, Prinzhausen J, Kämmerer U (2013) A pilot case study on the impact of a self-prescribed ketogenic diet on biochemical parameters and running performance in healthy and physically active individuals. Nutr Med 1
21. Lambert SD, Loiselle CG (2007) Health information-seeking behavior. Qual Health Res 17(8):1006–1019. https://doi.org/10.1177/1049732307305199
22. MacQueen J (1967) Some methods for classification and analysis of multivariate observations. In: Proceedings of 5th Berkeley symposium on mathematical statistics and probability, 1967. University of California Press
23. Maturo F, Hoskova-Mayerova S (2016) Fuzzy regression models and alternative operations for economic and social sciences. In: Recent trends in social systems: quantitative theories and quantitative models. Springer, Switzerland, pp 235–247. https://doi.org/10.1007/978-3-319-40585-8_21
24. Maturo F, Hoskova-Mayerova S (2018) Analyzing research impact via functional data analysis: a powerful tool for scholars, insiders, and research organizations. In: Proceedings of the 31st international business information management association conference innovation management and education excellence through vision 2020, pp 1832–1842. ISBN 978-0-9998551-0-2
25. McQuitty L (1966) Similarity analysis by reciprocal pairs for discrete and continuous data. Educ Psychol Meas 27:21–46
26. Nuti SV, Wayda B, Ranasinghe I, Wang S, Dreyer RP, Chen SI et al (2014) The use of Google Trends in health care research: a systematic review. PLoS ONE 9(10):e109583. https://doi.org/10.1371/journal.pone.0109583
27. Pandey A, Hasan S, Dubey D, Sarangi S (2013) Smartphone apps as a source of cancer information: changing trends in health information-seeking behavior. J Cancer Educ 28(1):138–142
28. Ramsay JO, Silverman BW (2005) Functional data analysis, 2nd edn. Springer
29. Sangalli L, Secchi P, Vantini S, Vitelli V (2010) Functional clustering and alignment methods with applications. CAIM 1:205–224
30. Sangalli L, Secchi P, Vantini S, Vitelli V (2010) K-mean alignment for curve clustering. Comput Stat Data Anal 54:1219–1233
31. Sneath P (1957) The application of computers to taxonomy. J Gen Microbiol 17:201–226
32. Sokal R, Michener C (1958) A statistical method for evaluating systematic relationships. Univ Kansas Sci Bull 38:1409–1438
33. Tan K, Witten D (2015) Statistical properties of convex clustering. Electron J Stat 9:2324–2347
34. Tarpey T (2007) Linear transformations and the k-means clustering algorithm: applications to clustering curves. J Am Stat Assoc 61(1):34–40

35. Vance K, Howe W, Dellavalle RP (2009) Social internet sites as a source of public health information. Dermatol Clin 27(2):133–136
36. Ward J (1963) Hierarchical grouping to optimize an objective function. J Am Stat Assoc 58:236–244
37. Willett WC, Sacks F, Trichopoulou A, Drescher G, Ferro-Luzzi A, Helsing E, Trichopoulos D (1995) Mediterranean diet pyramid: a cultural model for healthy eating. Am J Clin Nutr 61(6):1402–1406

Healthy and Active Aging. Social Determinants and Policies

Mihaela Rădoi, Gabriela Irimescu, and Adrian Lucian Lupu

Abstract Romania faces a series of profound socioeconomic transformations due to demographic change, among which we mention the population aging process seen as a result of constant improvement in terms of life expectancy, but also the decrease in fertility rates and an increase in emigration among younger age groups. According to Eurostat [7], Romania will experience one of the most accelerated declines in the working age population in the European Union, so the working age population from 20 to 64 years old in Romania is expected to decline by 30% by 2060. In the absence of policies for healthy and active aging, all these processes will result in an increased demand for long-term health and care services, higher poverty rates, greater social exclusion, and addiction among the elderly [30, 31]. The paper aims to bring into focus active policies to involve the elderly on the labour market, to valorise seniors as an active part of civil society more after retirement, and analyze social determinants to prevent premature physical aging.

1 Demographic Trends

Romania has been facing deep socioeconomic transformations due first of all to the demographic changes. The population aging process was the result of constant improvements in the quality of life reflected in higher life expectancy, decreasing fertility rates and increasing emigration among the younger age groups. The projections of the United Nations Population Division [33] shows that in the subsequent decades, the number of persons aged over 65 will increase significantly, whereas the active population (between 15 and 64) will decrease considerably. Romania, according to Eurostat [7], will face one of the most rapid decreases in the active population within the European Union.

In retrospect, not only in perspective, it is worth noting for the last ten years, pursuant to the data provided by the National Institute of Statistics (I.N.S.) a decrease

M. Rădoi · G. Irimescu · A. L. Lupu (✉)
Faculty of Philosophy and Social-Political Sciences, Department of Sociology and Social Work, Alexandru I. Cuza University in Iaşi, Bd. Carol, Nr. 11, Iaşi, Romania

© The Author(s), under exclusive license to Springer Nature Switzerland AG 2021   73
D. Soitu et al. (eds.), *Decisions and Trends in Social Systems*,
Lecture Notes in Networks and Systems 189,
https://doi.org/10.1007/978-3-030-69094-6_7

in the number of persons from 21,130 thousand persons in the year 2007 to 19,644 thousand persons in the year 2017.

Concerning the social security system, the decreasing trend is manifested for all types of social security benefits. Hence, while in 2007, the number of pensioners was 5745 thousand persons in 2017, their number was 5228 thousand persons, namely 517 thousand persons fewer. Across the interval under analysis (2007–2017) (Fig. 1.), partial early retirements decreased from 117 thousand persons to 79 thousand persons, while invalidity retirements (first, second and third degree) from 882 thousand persons to 587 thousand persons, survivor pensions from 630 thousand persons to 531 thousand persons. The significant decreases recorded among partial early retirements and invalidity retirements may be explained on one hand by demographic factors (mortality rate) and by legislative modifications, on the other hand. We refer to the Pension law no. 263/2010 that modified the allowance calculation formula and Government's Decision 257/2011 that included the methodological norms for the application of this law and the entry into force on July 17th, 2016 of Law no. 142/2016, which modifies Article 65 of the Pensions Law. Starting from this period, the decrease in the amount for early retirement no longer uses a fixed percentage of 0.75 for each month of not working, but in a differentiated manner, by the duration of the additional contribution period besides the complete period.

The same modifications of the legislative framework resulted in more early retirements, from 12 thousand persons to 23 thousand persons in 2017, and age-limit retirements from 3153 thousand persons to 3629 thousand persons. Upon calculating the percentage of pensioners from the value of the total population, it is worth noting a slight increase in the analyzed interval (0.8%) until the entry into force of the law 263/2010 followed by a decrease of 1.4% until 2017.

Search results - Yearly average number of pensioners by system of retirement

Pension system	Year 2007	Year 2008	Year 2009	Year 2010	Year 2011	Year 2012	Year 2013	Year 2014	Year 2015	Year 2016	Year 2017
					MU: Thousands persons						
Number of persons	21130	20635	20440	20294	20199	20095	20020	19953	19875	19760	19644
Social insurance pensioners - total	5745	5701	5689	5675	5589	5487	5410	5357	5306	5257	5228
Social insurance pensioners - total (percent)	27.2	27.6	27.8	28.0	27.7	27.3	27.0	26.8	26.7	26.6	26.6
Social insurance pensioners (excluding farmers)	4794	4819	4877	4927	4903	4861	4840	4839	4838	4835	4849
State social insurance pensioners	4643	4664	4718	4767	4744	4702	4681	4682	4683	4678	4678
For age limit	3153	3186	3239	3309	3350	3383	3418	3453	3500	3566	3629
With due complete stage	2161	2099	2030	2282	2607	2579	2516	2537	2602	2675	2753
With due incomplete stage	992	1087	1209	1027	743	804	902	916	898	891	876
Anticipated pension	12	10	9	9	9	11	14	19	23	23	23
Partial anticipated pension	117	113	112	123	125	114	101	96	86	75	79
Invalidity	882	892	909	887	834	778	737	708	675	629	587
Survivor pension	630	618	608	599	585	575	570	563	554	542	531
Social insurance pensioners - farmers	932	866	799	737	677	619	564	513	464	419	376
Social benefit - pension type	3	2	2	2	1	1	1	1	1	1	1
IOVR	16	14	11	9	8	6	5	4	3	2	2
Invalidity	5	4	3	3	2	2	1	1	1	1	1
Survivor pension	11	10	8	6	6	4	4	3	2	1	1

© 1998 - 2018 INSTITUTUL NATIONAL DE STATISTICA

Fig. 1 Yearly average number of pensioners by system of retirement. *Source* National Institute of Statistics

In retrospect, it is worth highlighting for the last ten years, because the legislative framework in the field changed, a decrease in the number of partial early retirements, of invalidity retirements and of survivor pensions and an increase in age-limit retirements and in early retirements.

The higher retirement age correlated with the decrease in the total population imposes the elaboration of active demographic policies and social policies with the role of valorising-mobilizing the elderly in various activities on the labour force market. We will analyze the involvement of senior citizens on the labour force market in correlation with other aspects, such as health status and healthy life expectancy.

2 The Concept of Active Aging in the European Union Policies

According to the UN [32], a person is "labelled" older after the age of 60. Papalia et al. [19], WHO [40] define an older person as a person who turned 65. Abrams et al. [1], WHO [40] consider the age of 55 as a possible age for the onset of the third age. A criterion of defining the third age is related to the retirement age or to the retirement from professional activity [40]. The retirement age is variable at the level of various countries, depending on the economic development level of the countries and on the proportion of the elderly within the general population [14], but it is different even within the same country, throughout time. In Romania, the third age can be considered 63 years old for women and 65 years old for men (pursuant to Law 263/2010), but a variety of criteria may be taken into account: retirement, subjective perception, social perception, chronological age, health status, physical appearance [14].

As it has been defined by World Health Organization [38], active aging represents the process of optimizing opportunities for health, participation and security in order to enhance quality of life as people age, both within labour force, by postponing retirement, and within the society, by participating in social, economic, civic or cultural activities. The European Commission defines active aging as "helping people stay in charge of their own lives for as long as possible as they age and, where possible, to contribute to the economy and society" [29, 30].

The Europe 2020 strategy set up objectives in fields such as employment, innovation, education, poverty reduction and climate change. In order to measure the progress recorded in the attaining of these objectives, five main targets were settled at the EU level, two of which relevant for the concept of active aging. They are an employment rate of 75% for the population aged between 20 and 64 years old (70% for Romania) and at least 20 million persons who no longer have to face poverty or social exclusion or the risk of poverty or social exclusion.

In this context, the concept of active aging implies for the elderly to have the possibility of leading a life as healthy, productive, participative and independent life. The purpose of the policies elaborated for the implementation of this concept is to reduce premature physical aging, to encourage the elderly population to work for a longer time, to keep on contributing to the society through civic and political activities long after the retirement and to have an independent life even at older ages.

The Special Eurobarometer 378 (2012) referring to active aging represents an instrument developed for understanding the opinions and attitudes of the European citizens concerning older persons. The results of the survey show differences between Romania and the rest of the EU27 countries regarding the perceptions about age and the elderly. An almost double number of respondents in Romania indicated a negative perception of the people aged 55 years old and more, compared to the EU27 mean. In Romania, a person is considered "old" at the age of 61.3, compared to the EU27 mean of 64 years old. Romanians also indicate that they will have to retire from their current workplace four years earlier on average than their counterparts in EU27 (EU27—61.7, RO—57.7). A slightly smaller number of persons in Romania would want to continue to work after reaching the retirement age, compared to the EU27 mean (EU27—33%, RO—27%). Almost twice as many Romanian respondents do not fully agree with the need to increase the retirement age until the year 2030 related to the EU27 mean (EU27—36, RO—69). Due to the decrease in active population, Romania will have an employment rate 12% smaller until 2030 and a reduction by 32% until 2050, if the current employment models remain stable [43]. The survey also presents the answers to the questions referring to the useful methods to be applied by the government in order to help the persons who take care of older family members, the most common answers being as follows: (1) receiving a financial benefit for their caretaking services; (2) receiving pension credits for the caretaking period; (3) the possibility of working part-time; (4) he possibility of working on a flexible schedule; and (5) the possibility of temporarily leaving their job, with the right of resuming it later.

As we have previously shown, the concept of active aging entails for the elderly to lead as healthy a life as possible. The factors within the social environment that may contribute to the alteration of health status, the social and economic determinants influencing it will be featured in the following pages.

3   Health as a Multidimensional Concept of Individual Welfare

Health is a concept hard to define; it has been understood under various aspects throughout time. Researches in the field have explored the way individuals define and relate to health the most accepted and commonly used definition is the one provided by World Health Organization (WHO) [36, 37, 39], which defines health in positive terms (physical, mental and social wellbeing), not only in negative ones (absence of a disease). The concept of positive health [5] defines wellbeing from a physical viewpoint (an individual works, eats, conducts various activities in the family and community, in good conditions), a mental viewpoint (a person lives in an adequate climate and is able to face daily stress) and social viewpoint (an individual has to ability of establishing and maintaining interhuman relations in the society; they have a comfortable living standard; they integrate normally in the family and community life, in the professional and social activity).

Starting from the holistic concept on health, Ewles and Simnett [9] have redefined the concept of individual health, by describing the following dimensions: biological (optimal body function), psychological (optimal mental health), emotional, social (optimal family and social integration), spiritual and ecological (optimal integration of an individual in the surrounding environment). Consequently, health status is the expression of a person's adaptation to environmental conditions, in such a way that between the psycho-physiological structures, the body resources and the environment balance will be established. This balance allows people to conduct their activities, to attain their goals, to have a complete life and to be active members of the society [15].

Bircher [4] defines health as a dynamic wellbeing, characterized by a physical and mental status that meets the life demands of an individual taking into account the age, culture and personal responsibility. A good health status is an essential element of human welfare [2]; it represents a value in itself, an important component of human capital, based of the fundamental rights of the individuals [23].

The 1986 Ottawa charter of WHO, for Health Promotion, brings to light health not only as a state, but also as a "resource" for daily life. The charter established five fundamental strategies for a successful health promotion: elaborating public health policies, creating a favourable environment, intensifying community action, developing individual skills and reorienting health services.

The question: *How many years of our lives do we live in good health?* is answered by the Healthy Life Expectancy index. I.N.S. calculates the indicator of healthy life expectancy starting with 2007 (Fig. 2), using self-statement regarding the health status. For Romania, although the life expectancy of men is lower than that of women (71 years compared to 78.2 years), healthy life expectancy of men is higher than that of women (57.5 years compared to 57.1 years). At the level of the European Union, the highest healthy life expectancy is recorded in Malta (70.7 years for women, while 70.3 years for men) and Sweden (70.2 and 71.1 years), while the lowest in Slovakia (52 years) and Slovenia (54 years) [17].

For the interval analyzed in this paper, we notice not only a decrease in the number of population, but also a reduction of healthy life expectancy among the Romanian population. Therefore, the inclusion of this indicator under the age of 63

Healthy life expectancy, by gender, in Romania and at the level of EU-2007, 2007-2011

		EU-27	ROMANIA
2007	men	61,7 years	60,6 years
	women	62.6 years	62.6 years
2011	men	61.8 years	57.5 years
	women	62.2 years	57.1 years

Fig. 2 Healthy life expectancy, by gender, in Romania and at the level of EU. *Source I.N.S. 2013*

and of 65 years old, respectively when a person may retire, according to the national legislation in effect, highlights the risk of premature physical aging.

The decrease in healthy life expectancy and premature physical aging must be contrasted with active aging and valorised aging in the social and cultural space through active social policies.

4 Social Determinants of Health

The Healthy People 2020 strategy (Office of Disease Prevention and Health Promotion) features five categories of factors that determine health status. They were grouped as follows: economic stability, education, social and community context, health and health system and vicinity and environment. This new definition included determinants targeting access to higher education and to child education and development programs, social cohesion and civic participation, perceived discrimination and equity, institutionalization, access to medical care services [24]. The aim is for them to be materialized in action plans and strategies for a healthy population. In order to operationalize the phrase *determinants of health*, it is also useful to make the distinction between individual health and population health. Whereas, at individual level, the determinants are more likely characteristics with an important role, but usually unalterable (gender, age, genetic factors), the social determinants of population health are placed at the macrosocial level and they are out of the social control of the individuals ([25], p. 19).

The social determinants of health are represented by the conditions in which people are born, grow up, live, work, become old, as well as by the systems applied in order to cope with the disease ([34], p. 1641). These circumstances are determined by the distribution of many, of power and of resources at global, national and local level. The authors make the difference between two main categories of social determinants: structural determinants—(fundamental structures of the nation-state that create social stratification) such as the welfare of the country, income inequality, education status, gender of ethnic standards—and proximal or intermediary determinants—living standards, from the quality of family environment and the relationships with the peers, access to food, housing and leisure activities to access to education. The authors also highlight that the proximal determinants are generated by the social stratification created by the structural determinants, but also by cultural, religious and community factors. The authors state that the proximal determinants also establish individual differences concerning exposure and vulnerability to the factors compromising the health of a person. Numerous studies have underlined the close connection between socioeconomic factors (place of residence, incomes, social status, education, access to medical services, etc.) and health status [13, 28].

The health status of people is also determined by factors acting at macrosocial level (community, region, society), as well as by the quality of healthcare services, the quality of the environment [20]. Romania spends less than 5% of the GDP on health, which is a low figure compared to the European average of 8.7% [43]. The health status of a population within a society is not determined only by economic welfare, but also by the social inequality of incomes and by social cohesion. A higher level of income inequality determines a higher health inequality [42].

The socioeconomic determinants of health may be measured using indicators such as: GDP, the net income/capita, the GDP percentage allocated to health, instruction level (the relation between the educational level and mortality/morbidity, including the infant one; the relation between the educational level and diet, obesity, smoking, alcoholism, etc.), employment or unemployment level and professional stress [8].

The residence background also influences health status, through the lack of drinking water and sewerage system, the lack of electrical power in certain localities, through poor housing standards, the bas state of the roads, the limited access to information [6]. Background is often cited [26, 27] as an important factor that influences the health status of persons. It includes characteristics of the natural environment, of the created environment and the social environment, such as clean water and air, proper houses, as well as safe communities and roads—all of the above contribute to good health. On average, Romanians report that their local environment is less favourable to the elderly compared to the European average (EU27—65%, RO—47%). The most important improvement for their local environment to become elderly-friendly were reportedly as follows: (1) more facilities for the elderly to keep their good shape and health; (2) better public transportation; (3) better roads and road safety rules; and (4) more public areas, like parks.

The social and community networks where family is also included play a considerable role in the health of individuals. Often, it ensures, through its local structures, information regarding health and health services. Thus, people receive the necessary support in order to play an active role in the improvement of their own health. Reaching one's health potential depends not only on providing health services, but also on many other factors that must work together in an effort of increasing health status and of reaching the health potential of a nation (Report of the Presidential Commission for the analysis and elaboration of policies in the field of public health in Romania [21]. Almost half of the Romanian population lives in rural localities (46% according to the 2011 census). Both rural population aging and the migration of many inhabitants from villages, mostly of young people and of women (looking for a job abroad) accentuate the general financial shortcomings within the rural areas.

A quantitative study [12] on a representative sample ($N = 884$) for the persons aged over 60 years old who had a migration experience, conducted at the level of the northeastern region of Romania, shows the social isolation degree of the elderly and the identification of the (institutional an noninstitutional) elements of the support network (Fig. 3).

The social support network is centred on family (53.6%), on proximity community (neighbours 25.8%) and medical institutions (11.4%). The noninstitutional support model surpassed by far the institutional support model.

There is a pronounced urban-rural division in Romania regarding the physical availability of basic infrastructure and of services, given that rural communities are very much disadvantaged. The main fields requiring investments are as follows: (1) the road network (the density of rural roads accounts for less than a half of the

How would you characterize the relationship between yourself and ... (Percent)								
	Family	Relatives	Friends	Neighbours	Church	City -hall	Medical institutions	Police	Institutions providing support
Nonexistent	0.3	1.0	2.0	1.6	3.6	21.9	3.4	38.7	56.8
Unsatisfactory	1.6	3.2	1.5	2.5	3.5	10.2	0.0	6.2	3.2
Satisfactory	6.0	12.7	14.4	17.0	20.4	26.7	29.3	16.0	7.4
Good	31.0	42.4	46.6	45.8	34.7	23.5	39.4	21.7	10.4
Very good	53.6	33.6	27.8	25.8	30.0	8.1	11.4	6.8	3.3
DN/DWTA	7.5	7.1	7.7	7.4	7.8	9.5	8.5	10.6	19.0

Fig. 3 The relationship between persons aged over 60 and elements of the support network. *Source* [12]

national density); (2) basic utilities (water supply, sewerage network and natural gas network); (3) preschool and school establishments (agricultural high-schools included); (4) healthcare (the number of inhabitants per physician in the rural areas is seven times higher than in the urban areas); and (5) information and communication technology. The current difference between the rural and the urban provides a solid justification for the rapid increase in the provision of basic services and infrastructure in the rural areas.

Health and poverty are closely interconnected and interdependent. Poverty has a significant contribution to poor health, whereas in its turn poor health may have a major contribution to poverty, by reducing the work capacity of a person and by leading to high costs of treatment and healthcare. Poor and vulnerable people become sick rapidly and they die faster than the general population. Poverty creates poor health through various social determinants such as faulty diet and improper living standards (the absence of a decent home, clean water and/or adequate cleanliness).

Incomes and material living standards are important for health and they vary considerably between the social groups. The social protection and general welfare policies of the State can reduce the consequences of income losses; they are thus important regarding the reduction of health inequalities [11].

Income represents a universal factor in determining health inequalities, at the level of both the individual and the society [3]. There is a close connection between educational capital, health status and the economic resources. Health represents an important resource for individual development, thus allowing the participation on the labour force market and ensuring the income necessary for satisfying needs [35].

The study regarding the analysis of the unemployment benefit programs in Europe has shown that higher replacement rates (social benefits) are correlated with better health and that this correlation is stronger among those with lower instruction level. These benefits with positive effects are not limited to the unemployed because their existence seems to improve even the quality of life of people who do not need them [10]. Taking into account the considerable influence of poverty on poor health, the benefits of minimum income represent another important component of national social protection policies. An analysis based on OECD and on other data has shown that the countries providing higher rates of minimum income also record lower mortality rates [18].

According to the study called "The national strategy on social inclusion and poverty reduction" (2014–2020), conducted by the Ministry of Labour and Social Justice, the elderly recorded a significant decrease in poverty in the period 2008–2012. In 2008, the elderly (aged 65 and older) recorded the second poverty rate in value, after children (25%). In 2010, due to the great increase in the level of contributory and social pensions, poverty rates among the elderly dropped below the level of any other age group. Whereas the relative welfare of the elderly as a group has improved over the years, within this group, there is a higher gender gap. The difference in terms of poverty between the men and women aged 65 and older is 10 percentage points (19.3% for women, compared to 9.2% for men), and the difference is even higher for persons aged 80 and older. The elderly who are alone are more prone to poverty than other types of persons. Approximately 1.2 million persons aged 65 or older live alone (among whom three quarters are women). Whereas 25.8% of the older persons who are alone live in poverty, only 5.8% of the elderly couples are in this situation [16].

Income redistribution. The studies conducted on the association between work and health inequality show that the GDP percentage spent by a country for policies targeting the integration of disadvantaged population groups on the labour market and the decrease in the average level of stressful work was not improved [41]. The older persons who are alone are far more prone to poverty than other types of persons. Approximately 1.2 million persons aged 65 or older live alone (three quarters of who are women). Whereas 25.8% of the older persons who are alone live in poverty, only 5.8% of elderly couples are in this situation. The closest poverty rate to the one of older persons who are alone is the one of households without elderly members (22.7%). As we have mentioned previously, there is a significant gap between women who are alone and men who are alone—30.2%, compared to only 13.8% [43].

Education and professional status are in close connection with both relative poverty in terms of income and measured poverty. More than a third of the persons who only managed to graduate from middle school are exposed to poverty risk. The percentage drops significantly to 15% among those who managed to graduate from high school and/or post-secondary school and it decreases even more, to 6%, among those who obtained a higher education degree. By professional status (persons between 15 and 64 years old), the groups recording the lowest poverty rates are the persons employed and the pensioners (5.6% and 8.4%, respectively). The persons recording the highest poverty rates are persons who work independently in agriculture (60.6% live in poverty), followed by the unemployed (their poverty risk being 52.1%) [16].

5   Promoting Active Aging and the Protection of the Elderly in Romania

Taking into account the importance of the population aging topic for the Romanian society and the wide range of issues covered, the Ministry of Labour and Social Justice elaborated "The national strategy for the promotion of active aging and the protection of the elderly 2015–2020", in collaboration with the experts Of the World Bank, with public and private entities, including with the Ministries of Health, Public Finances, Regional Development and Public Administration, Transports and Economy, etc.

The national objectives of the Strategy for the promotion of active aging and the protection of the elderly comprise:

1.   The extension and improvement of the quality of active life of the elderly.
2.   The consolidation of the public pension system reform, which involves modifications of human resources policies for a better integration of older workers and for the creation of favourable jobs that do not affect negatively health, as well as the improvement of abilities, of employment capacity and of the independence of the older population.
3.   The promotion of active and dignified social participation of the elderly by highlighting the social image of the older population and the promotion of social participation, the improvement of accessibility/accessibilization of public space infrastructure, the prevention of abuses of older population and of exclusion.
4.   Obtaining independence and higher safety for the elderly with long-term care needs by creating a unified healthcare system and by ensuring enough financial, human and physical resources.

By approaching these needs, Romania should take into account policy aspects in the following fields: (1) reviewing eligibility criteria for retirement, by focusing on reaching higher actual retirement ages (2) human resource policies regarding the elderly, (3) modernizing the benefit programs for invalidity and disabilities (4) promoting working environments that do not affect health; and improving lifelong learning.

Upon analyzing the studied data, we notice that the elderly population represents an insufficiently used resource of the Romanian economy, which makes it difficult to attain the employment rate of 70% established by Europe 2020 Strategy. Around 70% up to 80% of the population support a mandatory retirement age and they disagree with continuing to work. Nonetheless, when asked to think about their own circumstances, 40% up to 50% of the population would like to work more, a strong support especially from older women, well-educated people and persons who struggle to pay their bills (B.M., 2014). The difference in opinions can be ascribed to the social standards referring to the value that older workers can provide to economy and to a fear that older persons may occupy the jobs of younger workers on the labour markets. The ensuring of healthy working environments adapted to the elderly, which prevents the onset of diseases and at the same time promote and encourage a healthy lifestyle will become increasingly important as the workforce in

Romania grows older. Employers and decision factors must monitor the health of the employees, through regular medical checkups, and they must intervene from time to time through counselling and necessary adjustments, in order to restore health. In addition, people must be aware of the importance of making jobs available for elderly employees because these adjustments are to the benefit of both employees and the productivity of companies.

Concerning lifelong learning, in the broadest sense, the concept comprises all forms of learning—formal education, non-formal education and informal education. A national strategy for lifelong learning (IPV) 2015–2020 was adopted by the Romanian Government through the Government's Decision 418 of June 23rd 2015, in order to increase participation to lifelong learning and to improve the relevance of education and professional training systems for the labour market. The strategy aims at developing programs for the demand targeting both the employees and the employers, through which investments and partnerships are developed for professional training.

Lifelong learning involves several components, besides the formal one. Non-formal education provided by both institutions and private providers, NGOs may provide information with the help of which the elderly may assimilate new life skills, new professional competences and general knowledge information that may help them conduct activities on the labour market.

Social participation refers to the activities people get involved in through their formal and informal social networks. A connection was noticed between civic implication, through both formal and informal social networks, and the increase in welfare level, in satisfaction concerning life, self-esteem, the feeling of control, of physical health and of the duration of life. Social participation as civic implication and volunteering contributes to the welfare of all persons involved—volunteers and beneficiaries, but they also provide a significant economic benefit for the communities that function on a low budget and that have high social needs. The current degree of social participation among the older population in Romania, compared to the EU28 States, using the components corresponding to social participation within the index regarding active aging shows that the persons aged 75 and older rank the 27th in activities with voluntary participation, compared to persons of the same age within the other EU28 countries (only Hungary has a lower level). By contrast, Romanians aged 45–59 years old rank the 8th in social connection, compared to the persons within the same group of the other Member States of the EU28 [43]. In the year 2014, the Law no. 78/2014 entered into force. This law concerns the regulation of the volunteer activity in Romania through which volunteering activity is acknowledged as an important factor in the creation of a competitive European labour market and, at the same time, in the development of education and professional training, as well as in increased social solidarity. However, Romania ranks among the last EU28 States regarding the volunteer hours completed by the elderly population. Financial difficulties, poor health status and low instruction level, as well as residence in the rural areas represent essential impediments regarding social participation among the adults [43].

6 Conclusions

The active population in Romania, aged between 20 and 64 years old, is expected to decrease by 30% until the year 2060; the number of the elderly will increase, which will entail increased demand in health and long-term care services.

A great part of the improvements of life duration and healthy life duration within the past decades has been determined by the evolution of medical systems, by higher instruction levels and by better socioeconomic standards [22]. Aging with better health status will help compensate the predicted decrease in active population, thus allowing more seniors to remain on the labour market for a longer period; it will also contribute to expense control corroborated with a development of health services.

The standard retirement ages in Romania have increased gradually to 65 years old for men and 63 years old for women, from 64.8 years and 59.8 years in 2013; the condition regarding seniority has increased to 35 years for both genders. These harsher conditions regarding age-limit retirement are a reflection of a longer and a healthier life, of the expected decrease in the tax-paying workforce and of the need for longer careers in the context of slower accumulation of pension rights. Therefore, these are necessary steps on the path towards the required adjustments in the social security system, even if they have been implemented quite recently, in order to attain their objective as late as 2030.

For Romania, the amendments to the legislative framework determined modifications at the level of each category of pensioners. The decrease in the total population that our country has been facing imposes the development of active demographic policies and social policies with the role of valorising-mobilizing the elderly in various activities on the labour force market. The transition from a full-time job to a complete pension in Romania has no intermediary paths. Such intermediary paths should be introduced, and they would consist in fewer responsibilities, easier work, fewer work hours, individual and/or group counselling before retirement. This transition could be included in the active social policies for the promotion of employments for the elderly. At the same time, the gradual passage from an intensely active life in the labour field to adapted activities would bring an extra benefit, by focusing on the health status of the older person, on healthy life expectancy and it would maintain seniors as an active source on the labour force market.

Currently, the decrease in healthy life expectancy, the onset of premature physical aging can be accompanied by vulnerability, dependence, disability, disease, due to the inefficiency of social policies for the promotion of active and healthy aging. Population aging will contribute to an increase in the pressure upon health-related expenses. Romania should invest in the prevention, early detection and treatment of chronic diseases. A special attention must be paid to the reduction of tobacco and alcohol consumption and to the promotion of healthy diet and sport, including among the elderly, through a better integration of these preventative measures in the primary services and in the community life. Early detection and the management of cardiovascular diseases, of diabetes and depression will become ever more important, as the population gets older. To this end, the following measures should be taken into account:

- Obtaining a decrease in risk behavioural factors, including tobacco and alcohol consumption, reducing obesity, as well as promoting a healthy lifestyle through a better diet and by exercising,
- Consolidating the policies regarding the prevention, early detection and treatment of chronic diseases,
- Creating a health system centred on the elderly, focusing on geriatric and family medicine,
- Activating the older population and ensuring a healthier, more productive, participative and independent life.

In order to satisfy the needs of an increasing percentage of aging population through a fiscally sustainable method, Romania will have to increase the workforce in the field of geriatric health services. The next generation of workers in the sector of health will have to understand better the specific health needs of elderly patients. Given that the number of persons involved on the labour force market has decreased and that its financial resources are limited, Romania will have to introduce a transition from hospital care to consolidated primary healthcare and prevention services. Furthermore, healthcare will have to be provided more and more at the level of the community.

The social participation of the elderly, the implication in volunteer activities, civic activities, activities for the group of peers, all of these are elements of active social policies. For Romania, in order to involve seniors in such programs, the traditional cultural model should be taken into account, a model that focuses on participation in philanthropic and charitable activities. The valorisation of older persons as an active resource in the community life may determine on a medium- and long-term basis a higher healthy life expectancy, a positive perceived self-esteem and a contribution of the economy of the community.

References

1. Abrams D, Russell PS, Vauclair CM, Swift H (2011) Ageism in Europe: findings from the European Social Survey. https://www.ageuk.org.uk/documents/en-gb/id10704%20ageism%20across%20europe%20report%20interactive.pdf?dtrk=true
2. Alber J, Köhler U (2004) Health and care in an enlarged Europe. Office for Official Publications of the European Commission, Luxembourg. http://www.eurofound.europa.eu/pubdocs/2003/107/en/1/ef03107en.pdf
3. Anderson R, Mikulič B, Vermeylen G, Lyly-Yrjanainen M, Zigante V (2009) Second European quality of life survey: overview. Office for Official Publications of the European Commission, Luxembourg. http://www.eurofound.europa.eu/pubdocs/2009/02/en/1/EF0902EN.pdf
4. Bircher J (2005) Towards a dynamic definition of health and disease. Med Health Care Philos 8:335–341

5. Bowling A (1997) Research methods in health: investigating health and health services. OUP, Buckingham
6. Doboş C (2003) Accesul populaţiei la serviciile publice de sănătate. The quality of life, nr. 3–4, pp 3–14
7. EUROSTAT (2018) Population structure and aging. http://ec.europa.eu/eurostat/statistics-exp lained/index.php?title=Population_structure_and_ageing/ro
8. Evans RG, Barer ML, Marmor TR (eds) (1994) Why are some people healthy and others not? The determinants of health of populations. Aldine de Gruyter, New York
9. Ewles L, Simnett I (1999) Promoting health: a practical guide, 4th edn. Bailliere Tindall, London
10. Ferrarini T, Nelson K, Sjöberg O (2014) Decomposing the effect of social policies on population health and inequalities: an empirical example of unemployment benefits. Scand J Public Health 42(7):635–642
11. Lundberg O, Åberg Yngwe M, Bergqvist K, Sjöberg O (2014) DRIVERS final scientific report: the role of income and social protection for inequalities in health, evidence and policy implications. Report produced as part of the 'DRIVERS for Health Equity' project. Centre for Health Equity Studies, Stockholm. http://health-gradient.eu/
12. Lupu AL (2015) Migraţia internaţională a persoanelor în vârstă din Regiunea de Dezvoltare Nord-Est. Pro Universitaria, Bucharest
13. MacDonald G (2000) Redesigning the evidence base for health promotion. Internet J Public Health Health Educ 2(B):9–17
14. Man GM (2017) Psihologia vârstei a treia şi a patra. Editura Trei, Bucharest
15. Mărginean I, Precupeţu I, Tsanov V, Preoteasa AM, Voicu B (2006) First European quality of life survey: quality of life in Bulgaria and Romania. Office for Official Publications European Communities, Luxembourg. http://www.eurofound.europa.eu/pubdocs/2006/67/en/1/ef0 667en.pdf
16. Ministry of Labour and Social Justice (2014) Strategia naţională privind incluziunea socială şi reducerea sărăciei (2014–2020). http://www.mmuncii.ro/j33/images/Documente/Familie/ 2015-DPS/2015-sn-is-rs.pdf
17. National Institute of Statistics (2013) Life expectancy. http://www.insse.ro/cms/files/publicatii/ pliante%20statistice/07Speranta%20de%20viata_n.pdf
18. Nelson K, Fritzell J (2014) Welfare states and public health: the role of minimum income benefits for mortality. Soc Sci Med 2014(112):63–71
19. Papalia DE, Olds SW, Feldman RD (2010) Dezvoltarea umană. Editura Trei, Bucharest
20. Precupeţu I (2008) Evaluări ale protecţiei sociale şi îngrijirii sănătăţii. In: Mărginean I, Precupeţu I (eds) Calitatea vieţii şi dezvoltarea durabilă. Politici de întărire a coeziunii sociale. Editura Expert—CIDE, Bucharest, pp 137–146
21. Raportul Comisiei prezidenţiale pentru analiza şi elaborarea politicilor din domeniul sănătăţii publice din România (2008) Un sistem sanitar centrat pe nevoile cetăţeanului. http://www.pre sidency.ro/static/ordine/COMISIASANATATE/UN_SISTEM_SANITAR_CENTRAT_PE_ NEVOILE_CETATEANULUI.pdf
22. Rechel B, Doyle Y et al. (2009) World Health Organization. Regional Office for Europe, Health Evidence Network, European Observatory on Health Systems and Policies. How can health systems respond to population ageing? https://apps.who.int/iris/handle/10665/107941
23. Saracci R (1997) The World Health Organization needs to reconsider its definition of Health. BMJ 314:1409–1410
24. Şoitu D, Rebeleanu A (2012) Echitatea în sănătate. Studia UBB Bioethica, LVII 2:19–27
25. Şoitu D, Rebeleanu A (ed) (2016) Noi perspective asupra cursului vieţii. Editura Universităţii "Alexandru Ioan Cuza", Iaşi
26. Solar O, Irwin A (2006) Social determinants, political contexts and civil society action: a historical perspective on the Commission on Social Determinants of Health. Health Promot J Austr 17(3):180–185
27. Solar O, Irwin A (2007) A conceptual framework for action on the social determinants of health. Discussion paper for the Commission on Social Determinants of Health, World Health Organization, Geneva

28. Svarcova I, Hoskova-Mayerova S, Navratil J (2016) Crisis management and education in health. Eur Proc Soc Behav Sci EpSBS XVI:255–261. http://dx.doi.org/10.15405/epsbs.2016.11.26

29. The European Commission (2012) EU contribution to active aging and to solidarity between generations

30. The European Commission (2012) Special Eurobarometer 378 "Active ageing". http://ec.eur opa.eu/public_opinion/archives/ebs/ebs_378_fact_ro_ro.pdf

31. The European Commission (2015) The ageing report. Underlying assumptions and projection methodologies. http://ec.europa.eu/economy_finance/publications/european_economy/2014/pdf/ee8_en.pdf

32. United Nations (2013) World populations prospects. Revised edition 2012. ONU, NY

33. United Nations, Department of Economic and Social Affairs, Population Division (2013). World populations prospects. http://esa.un.org/unpd/wpp/

34. Viner RM, Ozer EM, Denny S, Marmot M, Resnick M, Fatusi A, Curri C (2012) Adolescence and the social determinants of health. Lancet 379:1641. http://download.thelancet.com/pdfs/journals/lancet/PIIS0140673612601494.pdf?id=3d35b1b5aa0ec416:58d1b5c9:13be60 db275:-3781356778794886

35. Voicu B (2005). Penuria Pseudo-Modernă a Postcomunismului Românesc. Volumul II: Resursele. Expert Projects, Iași

36. WHO (1998) The world health report. Life in the 21st century. A vision for all. Report of the Director-General, WHO, Geneva. http://www.who.int/whr/1998/en/whr98_en.pdf

37. WHO (2000) The world health report 2000—health systems: improving performance. WHO, Geneva. http://www.who.int/whr/2000/en/

38. WHO (2002) Active ageing: a policy framework (active aging: cadru de politici). WHO, Geneva. http://www.who.int/ageing/publications/active_ageing/en/

39. WHO (2006) Preventing disease through healthy environments. Towards an estimate of the environmental burden of disease. WHO, Geneva. http://www.who.int/quantifying_ehimpacts/publications/preventingdisease.pdf?ua=1

40. WHO (2009) Definition of an older or elderly person. WHO, Geneva. http://www.who.int/hea lthinfo/survey/ageingdefnolder/en/

41. Wahrendorf M, Siegrist J (2014) Proximal and distal determinants of stressful work: framework and analysis of retrospective European data. BMC Public Health 14:849

42. Wilkinson RG (1996) Unhealthy societies: the afflictions of inequality. Routledge, London

43. World Bank (2014) ECA regional flagship on ageing. In print World Bank. Viață lungă, activă și în forță. Promovarea îmbătrânirii active în România. http://www.seniorinet.ro/library/files/raport_banca_mondiala_viata_lunga,_activa_si_in_forta.pdf

Approaches of Long Term Care: The Geriatric Syndromes Among Older Adults with Comorbidity in Moscow Population

N. V. Sharashkina⬤, N. K. Runikhina⬤, K. A. Eruslanova⬤, and O. N. Tkacheva⬤

Abstract The aim of this chapter is to analyze the influence of chronic non-infectious diseases and geriatric syndromes on the progression of autonomy loss and needs of the elderly for medical care and social assistance. *Materials and methods* We have conducted a cross-sectional study in several Moscow polyclinics, for 1730 patients, 60 years and older. These people were divided into subgroups according to a dependency level in activities of daily living (ADL) and their need for assistance. The average age of the participants was 74.9 ± 6.1 years. Among them, a Group I of patients with comorbidity who underwent a comprehensive geriatric assessment included 315 older adults. In the focus of our attention was patients' past with comorbidity and its associations with other geriatric syndromes. The associations with other geriatric syndromes were analyzed with chi-square test. Results with $p < 0.05$ were treated as statistically significant. *Results* 32% (619 people) of older people needed no social or medical help (subgroup 0), 21% (people needed minimal help (subgroup 1). Subgroups 2, 3, and 4—39% of patients need social assistance and medical care at home. The prevalence of chronic non-infectious diseases in the study group was high. Most participants of the study (95.5%) had no autonomy loss and dependency in the basic functional activity, and only two-thirds of participants (65.2%) were independent in instrumental activities of daily living (IADLs). *Conclusion* With increasing age, there was a rise in the prevalence of such geriatric syndromes as autonomy loss, reduced mobility, cognitive impairment, sensory deficits, and urinary incontinence,

N. V. Sharashkina (✉) · N. K. Runikhina · K. A. Eruslanova · O. N. Tkacheva
Russian Clinical Research Center for Gerontology, Pirogov Russian National Research Medical University, 1 Leonova, 16, 129226 Moscow, Russia
e-mail: sharashkina@inbox.ru

N. K. Runikhina
e-mail: nkrunihina@rgnkc.ru

K. A. Eruslanova
e-mail: kae.07@mail.ru

O. N. Tkacheva
e-mail: tkacheva@rgnkc.ru

© The Author(s), under exclusive license to Springer Nature Switzerland AG 2021
D. Soitu et al. (eds.), *Decisions and Trends in Social Systems*,
Lecture Notes in Networks and Systems 189,
https://doi.org/10.1007/978-3-030-69094-6_8

but there was no significant increase in frailty, risks of malnutrition, symptoms of depression and falls.

Keywords Geriatrics · Autonomy loss · Geriatric syndrome · Frailty · Comorbidity

1 Introduction

The number of older adults is steadily rising in Russia nowadays, and problems of aging are vital because aging is connected with increased needs in medical care and social assistance and increases the socioeconomic burden. According to data from the Federal State Statistic Service, the rise in the number of older adults (persons over working age) significantly outstrips the growth of the entire population. In recent ten years, the population growth in Russia stood at 2.8%, whereas the share of the elderly increased by 23.4%. The number of people over working age in Russia rose from 29,732 people, or 20.8% of the population as of January 1, 2007, to 36,685 people or 25% of the population by 2017 [7]. An increase in the number of older adults is caused by an increase in life expectancy, a tendency which is present not only in Russia but in the world as a whole. Figure 1 shows the differences in life

Countries	Men, years	Women, years
Norway	19.0	22.2
The United Kingdom	18.6	22.1
France	18.3	21.4
Finland	18.2	21.2
Lithuania	17.6	21.0
EU-27*	17.2	20.9
Germany	14.0	19.2
Russia	11.6	16.4

*EU-27 comprises 27 member-countries, including the United Kingdom, France, Finland, Lithuania and Germany

Fig. 1 Life expectancy at the age of 65 years (European countries)

expectancy at the age of 65 years in Russia and other European countries. The highest life expectancy is observed in Norway: 19.0 years for men and 22.2 years for women, which is by 7.4 and 5.8 years, respectively, more than in Russia. Such a noticeable gap between Russia and other European countries in life expectancy at 65 years is a combined effect of economic, social, and environmental factors.

Russian demographic changes happen following the accelerated aging model, which is based on the low birth rate and high mortality rate of the younger population, especially the male population. To respond to this challenge, we need measures to increase the quality of life of older people and the higher involvement of this population in social and economic affairs. A range of programs for older adults has been adopted in Russia recently.

In February 2016, authorities approved the strategy of measures to protect older people in Russia by 2025, which is the first step to develop the single overarching framework for the Russian policy regarding older people. The strategy is designed to promote active longevity, as well as the Plan of measures for 2016–2020 to implement the first stage of the strategy, Russia implements the "Demography" national project and the "The Older Generation" federal project. The federal project aims to promote active longevity and develop programs for comprehensive support and higher quality of life by establishing a system of long-term care for the elderly and disabled and improvement of social assistance in the federal subjects of the Russian Federation. To assess progress in the implementation of programs for the elderly, we need a system to collect statistical data regularly and assessment tools for quantitative evaluation of the current situation and potential.

Russian strengths include the achieved level of education of older people, their involvement in care for children and grandchildren, and physical safety. The level of formal education in Russia is one of the highest in Europe, as 87.5% of people aged 55–74 had education not lower than secondary general education as of 2017 (as of 2017, only Lithuania and the Czech Republic had higher education levels) [7, 8].

Less success has been achieved in the involvement of the older generation in family care for elderly patients: Russia took fifth place from the end in the ranking of EU countries plus Russia. On the one hand, this can be explained by short life expectancy in Russia, while a significant part of elderly parents, relatives, and friends die before their children and other relatives aged 55 and older can take care of them. On the other hand, there is no systemic support to these people, particularly regulatory and financial, as well as support for family care by both the state and business. With the aging of a large generation of the 1950s, this problem will become more acute. It is estimated that the total number of people who need long-term care in Russia will stand at 10.3 million by 2025, including 4.65 million people who will need home care (a threefold increase against 2017) and 0.52 million of disabled who will need hospitalization.

The main barriers limiting the quality of life of older people in Russia are connected with poor health in the elderly, low life expectancy at older ages, and problems with infrastructure for volunteer and social activities, lifelong education, and physical activity [3, 9]. Although the life expectancy in older ages increased in Russia, the country consistently takes the last place in a corresponding ranking of the

European countries plus Russia and has a large gender difference in life expectancy (the difference is higher only in Latvia, Lithuania, and Estonia). According to data from the UN for 2010–2015, life expectancy at the age of 60 in Russia amounted to 15.5 years for men and 21 years for women. The measures aimed at preventing and treating diseases, health promotion in the elderly and higher life expectancy should be based on a health care system that provides medical care, as well as the accessibility of pharmaceuticals, safe life and working conditions, environmental well-being and efforts of the elderly to maintain own health.

It should be stressed that the elderly are a heterogeneous population. Age is the main differentiating parameter, an increase in which is connected with worsening objective and subjective health characteristics that also influences the engagement of a person in various activities. According to studies, health indicators in Russia deteriorate faster than in other countries. Thus, according to the data of the Russian Monitoring of the Economic Situation and Public Health of the Higher School of Economics as of 2016, the share of people who assess their health as poor and very poor increased among men from 20.2% among 60–69-year-olds to 35.6% among 70–79-year-olds and 50.9% among people aged 80 years and older, and also rose among women from 23.3% to 39.4% and up to 60.7%, respectively. In addition to age, other factors affecting the quality of life in older people include the state of health, disability, level of education, marital status, and household composition. Living alone, with other things equal, impacts negatively on psychological well-being in the elderly.

2   The Influence of Chronic Diseases and Geriatric Syndromes on the Progression of Independence Loss and Their Connection with Needs for Medical Care and Social Assistance

Higher morbidity, which is present in the elderly, is connected with higher needs for medical care. At the same time, quality of life in the elderly and the prognosis for their life and health depend not only on chronic diseases but also on geriatric syndromes. Frailty or senile asthenia is the leading geriatric syndrome present in the most vulnerable older patients; it is characterized by general weakness, slowness, and unintentional weight loss. Frailty is accompanied by a functional and physical loss, as well as decreased adaptation and regeneration, development, and progression of which increases the risk of developing dependency and worsens the prognosis for a patient. Frailty is an age-related condition that usually occurs in people aged over 65 years, but in some cases, it can be present in younger patients, while frailty is characterized by high variability in manifestation and progression. Frailty is closely related to other widespread geriatric syndromes, such as sarcopenia, malnutrition, urinary incontinence, mobility loss, recurrent falls, cognitive impairment, anxiety and depression, which are often diagnosed too late, leading to functional dependency and lower quality of life, increasing the number of hospitalizations and risk of death [1, 4, 6].

In order to study the influence of chronic non-infectious diseases and geriatric syndromes on the progression of autonomy loss and needs in medical care and social assistance in the elderly, we conducted a simultaneous study in several Moscow polyclinics.

Group II consisted of 1730 patients living at home. These people were divided into subgroups according to a dependency level in activities of daily living (ADL) and their need for assistance. All participants in the study gave informed consent to participate in research. Participants unwilling to cooperate were excluded from the study. Among them, a Group I of patients with comorbidity who underwent a comprehensive geriatric assessment included 315 older adults. In the focus of our attention was patients' past with comorbidity and its associations with other geriatric syndromes. The associations with other geriatric syndromes were analyzed with chi-square test. Results with p < 0.05 were treated as statistically significant.

Patients who had higher education accounted for 54.8%; patients who still worked accounted for 9%. As for concomitant diseases, 55 patients (17.4%) had a past medical history of myocardial infarction, 42 patients of 13.3% acute cerebrovascular accident, 76 patients (24.1%) of type 2 diabetes mellitus. The most common form of systemic arterial hypertension was isolated systolic hypertension.

Frail patients accounted just for 8.9% of the group according to phenotypic model criteria and 4.2% using deficit accumulation model criteria, but every second patient had prefrailty (61.3% and 45.8%, respectively). Most participants of the study (95.5%) had no autonomy loss and dependency in the basic functional activity, and only two-thirds of participants (65.2%) were independent in instrumental activities of daily living (IADLs). Only one patient (0.3) had malnutrition, but every fourth patient had an increased risk of developing malnutrition (25.8%). Mobility parameters were satisfactory, as the average walking speed was one ±0.2 m/s, and the result of the Timed Up and Go Test (TUG) was 12.1 ± 4.5 s; 18.8% of patients had low walking speed (<0.8 m/s). Problems with balance were more widespread, as 41.0% of patients failed to perform a balance test. Falls during the last year were in 35.4% of patients, and 11.0% had repeated falls, 19.4% of patients had previous osteoporotic fractures. Symptoms of depression (>5 in the GDS-15) were identified in 36.0% of cases. The mean MMSE score was 27.3 ± 2.1, and the mean score of the clock drawing test was 8.1 ± 1.6. The number of patients who underwent neuropsychological testing amounted to 258 (average age 75.6 ± 5.9 years, women—77.0%). The share of patients without complaints of memory loss and signs of cognitive impairment was 6.6%. Subjective cognitive impairment was detected in 10.5% of cases, and mild cognitive impairment was in 25.2% of cases. Every second patient (50.0%) had mild cognitive impairment, and 7.8% of patients had dementia as it was diagnosed. Alzheimer's disease, the leading cause of severe cognitive impairment according to data obtained by researchers in other countries, was diagnosed less frequently (35.0%) than vascular dementia (40.0%). With increasing age, there was a rise in the prevalence of such geriatric syndromes as autonomy loss, reduced mobility, cognitive impairment, sensory deficits, and urinary incontinence, but there was no significant increase in frailty, risks of malnutrition, symptoms of depression and falls. The prevalence of geriatric syndromes, including functional loss, reduced mobility, falls, risks of malnutrition, symptoms of depression, and cognitive impairment was higher in women (1.7 ± 1.4 against 1.4 ± 1.2, $p < 0.05$) [2, 4, 10]. Moreover, the prevalence of prefrailty, autonomy loss, and cognitive impairment increased with age in the older women (Table 1).

The prevalence of chronic non-infectious diseases in the study group was high: 88.5% of patients were diagnosed with arterial hypertension, 55.3% had coronary heart disease, 39.9% had chronic heart failure, 17.4% of patients had an old (previous) myocardial infarction, 13.2% had a previous stroke, 24.2% had diabetes mellitus, 73.0% had joint pathology, 12.9% of patients had COPD, and 16.9% of patients had oncological diseases. The average number of non-infectious diseases in patients with frailty assessed by the questionnaire and using phenotypic and deficit accumulation models was higher than in patients without frailty (3.7 ± 1.5 against 3.0 ± 1.6,

Table 1 Comprehensive geriatric assessment (CGA) results depending on age and gender

Parameters	Group 1 (65–74 years)		Group 2 ($\geq$75 years)	
	Women (n = 134)	Men (n = 24)	Women (n = 152)	Men (n = 46)
The Barthel Scale $\leq$ 95, n (%)	2 (1.5)	2 (8.3)	9 (5.2)	3 (6.5)
IADL, scores, M $\pm$ SD	26.5 $\pm$ 1.2 A	26.1 $\pm$ 1.6 AB	26.0 $\pm$ 1.6 B	26.0 $\pm$ 1.7 AB
BMI, kg/m^2, M $\pm$ SD	30.0 $\pm$ 5	29.5 $\pm$ 5.2	28.4 $\pm$ 4.6	26.5 $\pm$ 4.3
MNA, scores, M $\pm$ SD	25.3 $\pm$ 2.3	25.9 $\pm$ 1.7	24.7 $\pm$ 2.5	25.1 $\pm$ 2.3
Speed of walking, m/s, M $\pm$ SD	1.1 $\pm$ 0.2 B	1.1 $\pm$ 0.2 B	0.9 $\pm$ 0.2 A	1.0 $\pm$ 0.3 AB
The timed up and go test, s, M $\pm$ SD	10.9 $\pm$ 3.9 B	10.3 $\pm$ 2.4 B	13.3 $\pm$ 4.7 A	12.7 $\pm$ 5.2 AB
Walking stick use, n (%)	13 (9.7) A	1 (4.2) AB	41 (27) B	4 (21.7) AB
Falls, n (%)	47 (35.1)	3 (12.5)	64 (42.1)	12 (26.1)
Repeated falls ($\geq$2), n (%)	11 (8.2)	0	24 (15.8)	4 (8.7)
Previous osteoporotic fractures n (%)	21 (15.7) AB	2 (8.3) AB	42 (28.3) B	4 (6.5) A
GDS-15, scores, M $\pm$ SD	4.7 $\pm$ 3.2 AB	3.1 $\pm$ 2.7 B	5.2 $\pm$ 3.6 A	4.0 $\pm$ 2.8 AB
GDS-15, >5 scores, n (%)	44 (32.8)	4 (16.7)	67 (44.1)	13 (28.3)
MMSE, scores, M $\pm$ SD	28.0 $\pm$ 1.6 A	27.5 $\pm$ 2 AB	26.8 $\pm$ 2.3 B	26.9 $\pm$ 2.4 B
The clock-drawing test, scores, M $\pm$ SD	8.4 $\pm$ 1.6 A	8.7 $\pm$ 1.4 A	7.7 $\pm$ 1.7 B	8.3 $\pm$ 1.4 AB

Note Groups without common letter differ significantly (p < 0.05) by results of the Holm multiple comparisons

$p < 0.001$, 4.0 ± 1.3 against 2.8 ± 1.4, $p < 0.001$, and 5.8 ± 0.9 against 2.5 ± 1.2, $p < 0.001$, respectively).

According to the results of questionnaire use, the risk of frailty was higher in patients with joint diseases (OR 2.2, $p < 0.005$) and oncological diseases (OR 2.0, $p < 0.05$). According to the phenotypic model, the risk of prefrailty was higher in patients with chronic heart failure (OR 2.7, $p < 0.005$), and according to the deficit accumulation model it was higher in participants with coronary heart disease (OR 5.6, $p < 0.001$), joint diseases (OR 4.2, $p < 0.001$), diabetes mellitus and oncological diseases (OR 3.1, $p < 0.005$) (Table 2).

We observed no relevant association between frailty or prefrailty and COPD or previous stroke. Patients with arterial hypertension or previous myocardial infarction had a lower risk of developing prefrailty according to the phenotypic model that may be associated with the administration of drugs of particular groups and requires further research. Thus, a strong relationship between frailty or prefrailty and the total number of non-infectious chronic diseases, as well as particular chronic non-infectious diseases, was observed. The study also identified the association between developing geriatric syndromes and cardiovascular diseases, as well as diabetes mellitus, chronic kidney disease, joint diseases, lower respiratory tract diseases, oncological diseases. Aging and chronic non-infectious diseases precipitate the progression of existing deficits according to the vicious circle principle. Of

Table 2 Results of a generalized linear model for assessment of the relationship between frailty/prefrailty and chronic non-infectious diseases taking into according to age of patients

Predictor variables (categorical)	Frailty according the questionnaire (≥ 3 scores)		Prefrailty according to the phenotypic model		Prefrailty according to the deficit accumulation model	
	OR	p	OR	p	OR	p
Age	1.1	<0.001	1.0	0.45	1.1	<0.01
Arterial hypertension	1.3	0.55	0.4	<0.01	0.7	0.334
Coronary heart disease	1.8	0.05	0.9	0.84	5.6	<0.001
Previous myocardial infarction	1.1	0.74	0.5	<0.05	1.0	0.98
Chronic heart failure	1.1	0.72	2.7	<0.005	1.7	0.11
Previous stroke	0.6	0.14	0.7	0.35	1.0	0.94
Joint diseases	2.2	<0.005	0.8	0.42	4.2	<0.001
Diabetes mellitus	1.0	0.93	0.9	0.62	3.7	<0.001
COPD	1.0	0.95	0.9	0.76	1.8	0.12
Oncological diseases	2.0	<0.05	1.2	0.53	3.1	<0.005

geriatric syndromes, depression and mobility loss can influence functional deficits. Data on the prevalence and structure of geriatric syndromes improves understanding of elderly needs and planning comprehensive medical care and social assistance for a better quality of life and extended period of functional independence [5].

These new challenges for society increase the need for care and assistance called the "long-term care system" in different countries. The amount of support needed to an individual depends on limitations to a person's life activities and support of other people and can be defined with the help of typification or dividing people into groups (groups of typification). The current procedure of getting social assistance in the form of an application has obvious shortcomings. A large number of people who need regular help and care do not apply to social protection authorities for various reasons, and, as a result, the system fails to provide assistance to people who have right to get it and is faced to problems with accurate calculations based on an assessment of needs.

Dr. Marjory Winsome Warren, one of the first geriatricians, who worked in the UK, was the one who proposed for the first time that people who need external assistance can be divided into several groups depending on the complexity of care, in 1937. Later this approach was applied to people who live in their own homes regardless of their age, including people with mental disorders. Nowadays, people in different countries use questionnaires and scales to assess dependency from external assistance, which are adjusted for national, cultural, legal, and other characteristics of each country. Some states, such as France, legally define capabilities (GIR) depending on particular diseases (Pathos) and self-care abilities. The long-term care system is a comprehensive social and medical care for people with persistent disabilities who need external assistance aimed to help people to regain quality of life and social participation; it includes activation, rehabilitation, habilitation, socialization, and education. The long-term care system includes care at home, at the hospital, and semi-hospital care, as well as accompanied living for those who need assistance. Russia should abandon the outdated system, according to which social workers visit people who need social assistance only two or three times a week, people who need long-term care do not receive proper medical care, and families take care for patients at their own risk without state support, and there is no accompanied living. A new system should be introduced to meet individual needs in a range of aspects, from comprehensive social and medical care to communication and socialization.

The obligatory assessment of self-care abilities, as well as independency and function loss, to determine the amount of care is being introduced on the federal level for the first time in the Russian Federation. Nevertheless, the federal subjects of Russia have the right to apply scales and questionnaires to assess needs according to the 442-FZ federal law.

How can we determine the amount of needed care? After all, long-term care comprises both medical care and social assistance. The most comprehensive and easy to use assessment tool is the Barthel Scale [4], which is applied by some federal subjects of Russia in the assessment of needs in social assistance. The Barthel Scale cannot be used to assess ADLs, needs to be looked after, or to get assistance during education, although these needs cause significant dependency in Russia. Currently, federal subjects of Russia are testing different ways to identify people who need assistance. The methodology of the assessment should include main aspects of human activity, including self-care ability, ability to move independently, to orientate, communicate, and control own behavior. The assessment of dependency (also assessment of "eligibility for benefits") is aimed to check the older adult's functional abilities and need for external assistance. A person who makes the assessment during a visit examines needs for assistance in several aspects, including self-service abilities, free movement at home, eating, dressing, washing, and controlling physiological functions, as well as overall functional, cognitive and emotional state. On the results of the assessment, five levels of dependency were distinguished. Thus the participants fell into five corresponding subgroups. The participants of the first subgroup needed only one-time assistance, while the participants of the fifth subgroup were totally dependent. Various limitations in activities (ability to self-care, move, orientate, communicate, control own behavior, educate and work) can be compensated in different ways depending on age, place of living, educational and social status, and also family.

The results of typification can help to create an individual program of social services/social assistance (IPS), as well as to develop an individual care plan (ICP) in the future. The typification of outpatients who can need long-term care was carried in order to assess the level of dependence in ADL, their need for assistance, and is being looked after. We received data on a dependency level, on the results of an interview. All patients who had undergone the typification were divided into these subgroups for effective organization of social assistance, medical and other care-related to long-term care, including economic and management aspects in the healthcare industry, for determining the number of staff and staffing, training, as well as for choosing services needed for each group and in each form of social assistance. We assessed the health status and performance of 1730 outpatients in their own homes in the study.

The distribution of outpatients by the level of social services and medical care needed is shown in Fig. 2. The participants of the study were divided into five subgroups, 32% of them (619 people) needed no social or medical help (subgroup 0), 21% or 405 people needed minimal help (subgroup 1). Subgroups 2, 3, and 4 are of particular interest because such patients need social assistance and medical care at home. According to different sources, the share of older people in need of social and medical help at home is between 20 and 50%. In our study, the share of such people was 39%. The outpatients who entered subgroup 5 had diseases in terminal stages and needed the constant presence of a caregiver; most such patients live in a hospice or a special care unit of a nursing home.

According to statistics, every fifth adult of advanced age cannot move independently and fully serve himself/herself. Over 70% of such people live in their own homes. Stress, which can be caused by hospitalization, even if its conditions are very comfortable, is harmful to patients' health, and the desire of frail elderly with medical disorders to stay at home can be well understood. It is also easier to create conditions for comfortable and safe living at home than in an institution. Moreover, patients willingly agree to receive care from marriage partners, relatives, and friends,

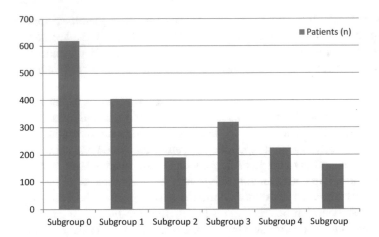

Fig. 2 The distribution of outpatients by needs in social assistance and medical care

who are in touch with older people, and it also has psychological benefits for the elderly.

In the last 5–7 years of life, older adults can lose independence due to developing and progressing chronic non-infectious diseases and frailty (senile asthenia). As a result, they will need external help and care. The share of older people with autonomy loss will increase in Russia and lead to a higher socioeconomic burden. Today, in order to solve the problem with different degree independence loss, comfortable living conditions at home should be provided together with appropriate adjustments and changes in social services, as well as outpatient (ambulatory care) settings, while the modern long-term care system must meet needs of the elderly and of the society as a whole.

3 Conclusion

According to our study, 39% of patients need social assistance and medical care at home. With increasing age, there was a rise in the prevalence of such geriatric syndromes as autonomy loss, reduced mobility, cognitive impairment, sensory deficits, and urinary incontinence, but there was no significant increase in frailty, risks of malnutrition, symptoms of depression and falls. The prevalence of geriatric syndromes, including functional loss, reduced mobility, falls, risks of malnutrition, symptoms of depression, and cognitive impairment was higher in women (1.7 ± 1.4 against 1.4 ± 1.2, $p < 0.05$) [4, 10]. Moreover, the prevalence of prefrailty, autonomy loss, and cognitive impairment increased with age in the older women.

References

1. Christensen KA, Holm NV, McGue M, Corder L, Vaupel JW (1999) Danish population-based twin study on general health in the elderly. J Aging Health 11:49–64
2. Karpman C (2014) Measuring gait speed in the out-patient clinic: methodology and feasibility. Respir Care 59(4):531–537
3. Lawton MP (1988) Scales to measure competence in everyday activities. Psychopharmacol Bull 24(4):609–614
4. Mahoney F (1965) Functional evaluation: the Barthel Index. Md State Med J 14:61–65
5. Noguchi N, Blyth FM, Waite LM (2016) Prevalence of the geriatric syndromes and frailty in older men living in the community: the Concord Health and Ageing in Men Project. Australas J Ageing 35(4):255–261
6. Podsiadlo D, Richardson S (1991) The timed 'Up & Go': a test of basic functional mobility for frail elderly persons. J Am Geriatr Soc 39(2):142–148
7. Russian Federation Federal State Statistic Service (2019) Population. Age. Available from: http://www.gks.ru/wps/wcm/connect/rosstat_main/rosstat/en/figures/population. Accessed 25 Sept 2019
8. Russian Federation Federal State Statistic Service (2019) Population. Education level. Available from: http://cbsd.gks.ru/#. Accessed 25 Sept 2019
9. Tkacheva ON, Runikhina NK, Ostapenko VS, Sharashkina NV, Mkhitaryan EA, Onuchina US, Lysenkov SN (2017) Validation of the questionnaire for screening frailty. Adv Gerontol 30(2):236–242
10. Tkacheva ON, Runikhina NK, Ostapenko VS, Sharashkina NV, Mkhitaryan EA, Onuchina US, Lysenkov SN, Yakhno NN, Press Y (2018) Prevalence of geriatric syndromes among people aged 65 years and older at four community clinics in Moscow. Clin Interv Aging 13:251–259

Physical Therapy for Increased Quality of Life Among Patients with Arthroses

Marius Neculăeş and Paul Lucaci

Abstract Nowadays, arthrosis is one of the major causes of discomfort and localized pain in the joints. The most commonly affected are the small joints of the hand, the knee and the hip joints. According to estimates of the "Ion Stoia" Rheumatism Center, over four million Romanians are diagnosed with osteoarthrosis. In medical practice, the recent years have recorded an increase in the incidence of this type of rheumatism, which is most likely due to the aging process of the population. The prospective study was conducted on a group of 24 patients with arthrosis, between the ages of 53 and 68, 16 females and 8 males, suffering from knee arthrosis. The aim of the chapter is to demonstrate the efficiency of physical therapy in improving the quality of life of patients diagnosed with arthrosis. The results of the research highlight the fact that the physical therapy methods which we used have led to pain relief, improved mobility and increased muscle strength, which have in their turn increased the well-being of the patients.

Keywords Mobility · Muscle strength · Pain · Medical gymnastics · Joints

1 Introduction

Arthrosis can be described as a group of symptoms with similar clinical manifestations and with common pathological and radiological modifications. This disorder represents the result of certain mechanical and biological factors that destabilize the normal circuit between the degeneration and the synthesis of chondrocytes within the extracellular matrix of the joint cartilage and of the subchondral bone. Arthrosis may be caused by several predisposing factors, such as genetics, environment, metabolism, traumas, bodyweight, lifestyle and nutritional deficits.

M. Neculăeş (✉) · P. Lucaci
Faculty of Physical Education and Sport, "Alexandru Ioan Cuza" University, Iaşi, Carol I Boulevard, No. 11, 700506 Iasi, Romania

© The Author(s), under exclusive license to Springer Nature Switzerland AG 2021 101
D. Soitu et al. (eds.), *Decisions and Trends in Social Systems*,
Lecture Notes in Networks and Systems 189,
https://doi.org/10.1007/978-3-030-69094-6_9

According to the American College of Rheumatology, arthrosis represents a heterogeneous group of disorders that evolve with articular manifestations due to the alteration in the integrity of the joint cartilage and to consecutive modifications of the subchondral bone.

The prevalence of this disorder increases with the age; between 30 and 50% of the adults over 65 suffer because of arthrosis and there are a series of differences related to gender. Until the age of 50, men are more prone to suffer from this impairment of the joint cartilage, but after the age of 50, more women suffer from digital arthrosis and gonarthrosis. Whereas joint degeneration is most common in the hands, involvement of the knees and hips is much more disabling [6].

The emergence of arthrosic pain, besides the functional deficit it provokes, also determines a series of negative modifications related to the psycho-emotional state of individuals.

Taking into account the degree of pain, invalidity, anxiety, depression and reduced wellbeing, it is no surprise that many musculoskeletal conditions have social consequences [5].

The risk factors of arthrosis are associated with distinct causes, depending on the impaired joint, on age, on the presence of osteoporosis, on genetics, on gender. Furthermore, this category also includes the local articular factors, such as traumas, anatomic joint deficiencies and deformities, the individual's occupation and professional demands, the cartilage loading, sporting activities and muscle weakening.

Deformity is the abnormal shape or attitude—permanent and imposibile to correct voluntarily—one part of the body (physical impairement of the musculoskeletal system) [1].

Regardless of the underlying cause leading to the onset of an arthrosis process, the fibroblasts of the synovial membrane secrete cytokines and inflammatory factors as a response [3].

Numerous studies demonstrate the existence of a close connection between obesity and knee arthrosis; at least three theories support this argument: increased pressure on the joint, the metabolic alterations induced by the emergence of excess body fat, the diet elements favouring obesity, all of these can damage the cartilage or other articular structures. These theories fail to justify, however, the onset of joint degeneration at the level of the hands.

Gonarthrosis affects 15% of the population around the Globe, causing significant pain and functional limitations [4]. Hand arthroses are more common among women and their onset concerns mostly the dominant hand [2]. Thus, in terms of the weight supported by the knee joints, obesity may favour the erosion of the cartilage through the pressure upon these joints. In what regards hand joints, there if no mechanical wear and tear due to obesity, but the cause is represented by certain elements within the diet that favour weight gain, which in its turn may damage the cartilage.

It has been proven that knee arthrosis is associated with physical activities and with the jobs involving repeated and prolonged flexions of this joint. Moreover, many sports expose the knee and hand joints to traumas or overloading, which favour the onset of degenerative processes over time.

Physical therapy has significant effects concerning the improvement of joint function by determining the secretion of synovial fluids and by intensifying the vasculotrophic phenomena at the level of soft tissues. Physical exercise programs also contribute to the mitigation of joint pain, but these routines must be performed constantly in order to maintain the results [7, 9, 10].

One of the important components of therapeutic success is represented by the trust that the patient invests in the physiotherapist, this element being named in a few words as a patient's expectation for the care provider to act in his/her interest [8].

2 Material and Method

The study was conducted on a sample of 24 patients aged between 53 and 68 years old, 16 females and 8 males, suffering from gonarthrosis. We performed the initial and final evaluation through the articular assessment, through the muscular assessment and using the Visual Analogue Scale (scale of pain intensity from 0—lack of pain to 10—maximum pain).

The patients followed targeted physical therapy programs for 12 weeks, with a frequency of three sessions per week. Within the medical gymnastics sessions, the subjects performed joint mobility unloading exercises and stretching exercise for the joints in question. In addition, the patients pedalled on the horizontal ergometer bike, they performed exercises from supine positions to unload the knee joint compression, in order to determine the secretion of synovial fluid, the limbering up of the periarticular muscles, and to get the capsulo-ligamentous system tense. After 14 days, the muscle toning exercises began, in open kinematic chain from decumbent positions, with low and average resistance in order for muscle and capsulo-ligamentous tensions to adapt, the purpose being to prepare the muscles for muscle toning and to stabilize the joints. From the 21st day, exercises in closed kinematic chain from sitting were introduced, the purpose being lower limb muscle toning and the indirect loading at the level of the knee joint. After the first month from the beginning of the rehabilitation program, we added the muscle toning exercises with full loading in order to stimulate joint stability; at the same time, the gait re-education exercises were introduced. The gait re-education stage included exercises for correcting step length and the support time for each lower limb. They consisted in making the patient move through specific tracks that marked the step distance, thus also determining the support time for each lower limb. Another method for correcting step length, support time and movement speed was represented by walking on the treadmill, thus also controlling and correcting the gait pace.

For knee stability, the intervention consisted in specific exercises after 6 weeks, namely walking tracks on unstable surfaces and muscle toning exercises on balance plates and other unstable surfaces (bossu, trampoline).

3 Findings

In order to highlight the findings, we illustrated the values obtained at the initial and at the final testing for joint mobility, for the flexion and extension movement, the muscle strength of knee flexors and extensors, pain evaluation through the Visual Analogue Scale, as well as gait testing through the distance covered in two minutes.

Within the initial evaluation, an important reduction of knee mobility was observed, with an average of 96.25° per flexion and an extension deficit of maximum 6.25°.

Figure 1 shows that the physical therapy programs led to joint mobility improvements, thus reaching an average flexion of 111.25° per flexion and an extension deficit of maximum 5°.

According to Fig. 2, it is worth noting an important reduction of muscle strength at the level of the impaired joints with an initial average strength of the flexor muscles of F3.08 and an improvement of muscle strength up to F3.91 at the final evaluation. The strength of the extensors had an initial average value of F3.16, with an improvement up to the value of F4.08 at the final evaluation.

Figure 2 highlights that muscle strength is low because of the presence of pain, which limits knee function and which implicitly leads to significant muscle hypotonias at the level of the impaired joints. The early debut of individualized programs for each case demonstrates a significant recovery of muscle strength for the impaired muscle groups.

Figure 3 underscores the presence of relatively great pain present at the level of the knee joints at the initial evaluation with an average intensity of 6.5 on the Visual Analogue Scale, it improved considerably at the final evaluation, reaching an average value of 4.

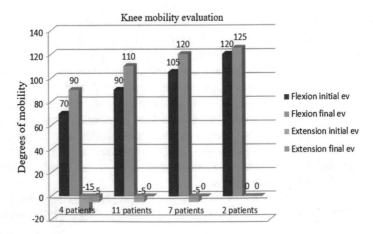

Fig. 1 Knee mobility evaluation

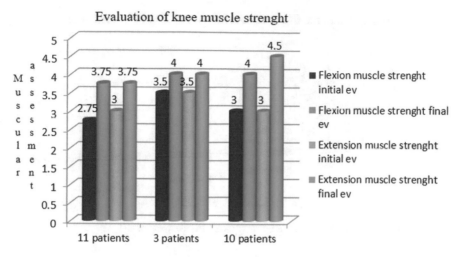

Fig. 2 Evaluation of knee muscle strength

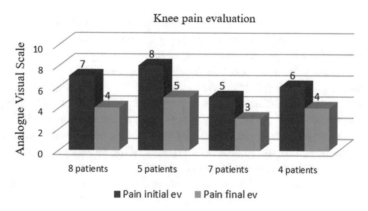

Fig. 3 Pain evaluation through the visual analogue scale

According to Fig. 3, the low level of joint mobility attracts pain intensification and ongoing inflammatory process, which reduces gradually, as the physical therapy sessions advance.

Figure 4 highlights that patients managed to cover a small distance in 2 min, with an average of 93.25 m, while at the end of the medical gymnastics sessions; the distance reached an average value of 103 m.

As illustrated in Fig. 4, the lack of muscle strength and the presence of pain limit the patients' capacity to move and the speed of their gait. Muscle toning and joint stabilization exercises accompany pedalling on the horizontal ergometer bike, on the treadmill and the walking tracks on unstable surfaces, the purpose being to restore an optimal movement capacity and to shorten the time necessary for covering certain distances.

Evaluation of the distance covered by walking in two minutes

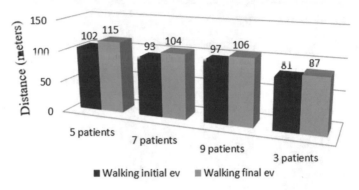

Fig. 4 Evaluation of the distance covered by walking in two minutes

4   Conclusions

The study conducted on the 24 patients suffering from gonarthrosis highlights that targeted physical therapy programs have a special importance concerning the increase in the quality of life for patients with arthroses.

The significant pain mitigation and the restoring of the range of motion contribute to the possibility of performing muscle toning and joint stabilization exercises; these elements finally enable the individuals to perform optimally their daily activities.

This chapter consolidated the theory according to which movement is the engine of life and an important factor in maintaining the wellbeing of the individuals. Another finding is that the elderly can improve their quality of life by using a physical exercise routine that they must perform constantly and correctly.

This work may represent a starting point for further studies on pathologies that elderly people face. Also with the help of this research in the present chapter, we have concluded that it is necessary and important to identify the shortcomings that the elderly face and to identify the most suitable strategies in order to increase the quality of life to this category of individuals.

References

1. Abalaşei B (2012) Psychomotricity and Psychomotor Re-education. Dumat Ofset, Ankara
2. Coskun N, Benlidayi I (2014) Non-pharmacological management of hand osteoarthritis: from a perspective of physiatry. J Arthritis 3(141):1–5. https://doi.org/10.4172/2167-7921.1000141
3. Figueroa R, Martinez Figueroa C, Rodriguez R, Figueroa PD (2015) Osteoartritis de rodilla. Chilean J Orthopedics Traumatol 56(3):45–51
4. Herman A, Mor A, Segal G, Shazar N, Beer Y et al (2018) Knee osteoarthritis functional classification scheme—validation of time dependent treatment effect. One Year Follow-Up of 518 Patients. J Arthritis 7(264):1–6. https://doi.org/10.4172/2167-7921.1000264

5. Hochberg M, Silman A, Smolen J, Weinblatt M, Weisman M (2011) Rheumatology, 5th edn. Mosby Elsevier, Philadelphia
6. Loeser R (2010) Age-related changes in the musculoskeletal system and the development of osteoarthritis. Clin Geriatr Med 26:371–386
7. Medeiros J, Rocklin T (2016) Manual therapy, therapeutic exercise, and HipTrac for patients with hip osteoarthritis: a case series. Physiother Rehabil 1(108):1–5. https://doi.org/10.4172/2573-0312.1000108
8. Rădoi M, Lupu A (2017) Understanding institutional trust. What does it mean to trust the health system, vol 66. Springer International Publishing Switzerland, pp 11–22. https://doi.org/10.1007/978-3-319-40585-8_2
9. Silvis M, Sylvester J, Hacken B, Wawrzyniak J, Kelly R et al (2016) Comparison of three exercise modalities on patient reported symptoms of knee osteoarthritis: a randomized prospective study. J Arthritis 5(220):1–5. https://doi.org/10.4172/2167-7921.1000220
10. Torbjørn M (2017) Osteoarthritis can be treated in several ways. Prim Health Care 7(252). https://doi.org/10.4172/2167-1079.1000252

Medical Rehabilitation of the Elderly with Chronic Lumbar Pain

Paul Lucaci, Marius Neculăeş, and Raluca Mihaela Onose

Abstract One of the most common causes of elderly presentation in medical recovery room is lumbar pain. It is known that the lumbar spine absorbs most of the shocks, this region supporting most of the weight of the body. Because of these particularities, elderly people can easily cause a back pain to themselves when they lift a weight, stretch for an object or twist their trunk. Also, the fact that age predisposes to the degeneration and dehydration of vertebral discs, low back pain has a higher prevalence in elderly people. The aim of the chapter is to highlight the role of medical rehabilitation in the treatment of low back pain. The study was conducted on a group of 43 patients with lumbar pain, 17 males and 26 females, aged between 60 and 73. The results of the research highlight the effects of medical rehabilitation in the improvement the lumbar spine pain for the patients included in this study.

Keywords Low back pain · Treatment · Super inductive system

1 Introduction

Low back pain is one of the most important causes for which patients come to the medical rehabilitation practices. It affects all categories of age, given that more than half of the population has had at least one back pain episode throughout their lives. This condition is manifested to the same extent among men and women, with increased prevalence among the elderly.

Chronic back pain has a high incidence. It is among the most prevalent musculoskeletal disorders that increase the economic costs in the field of health. The individuals with this type of pain have a low endurance of the muscles in the lumbar and hip region; at the same time, the low back spine flexibility is reduced due to extended faulty postures, which can also be associated to causing pains [8].

Chronic low back pain is usually defined as persistent or consistent when it goes on for more than 3 months or 12 weeks. The symptoms are diverse and they include

P. Lucaci (✉) · M. Neculăeş · R. M. Onose
Faculty of Physical Education and Sport, "Alexandru Ioan Cuza" University, Iaşi, Carol I Boulevard, No. 11, 700506 Iasi, Romania

© The Author(s), under exclusive license to Springer Nature Switzerland AG 2021   109
D. Soitu et al. (eds.), *Decisions and Trends in Social Systems*,
Lecture Notes in Networks and Systems 189,
https://doi.org/10.1007/978-3-030-69094-6_10

the impairment of physical functions by limiting the mobility of the spine, by causing pain, irritability, anxiety, depression [7].

Human body was designed for movement and it was endowed with motor capabilities to serve this action. Aging determines a series of degenerative processes at the level of the spine, they are the consequence of the dehydration of intervertebral discs and of the deterioration of regenerative biological processes.

Due to the aforementioned particularities, the seniors are more prone to injuries because of the aging process [3]. For this reason, the physical therapy programs are carefully selected and particularized for each patient.

Motor ability is a complex reaction to the environmental stimuli including, in a characteristic unit, several elemets: psychomotor skills, as a natural psychophysical endowment and the motor-athletic skills, influenced by a multi-level maturity of the functions and by practice, and accompanied by internal motivational factors [1].

Back pain becomes more difficult as its management becomes ever more limited, especially when sleep disturbances are associated: sleep disorders, short duration and, finally, dissatisfaction concerning the quality of sleep [5].

Until a couple of years ago, the most common therapy prescribed for acute low back pains was represented by bed rest and the administration of anti-inflammatory medication. Nowadays, numerous studies have proved that extended bed rest has no beneficial effect. Specific physical exercises, correctly and constantly executed, do not intensify pain. On the contrary, they have an important role in the immediate treating of low back pain, by retrieving the integrity of the musculoskeletal system [2].

In what regards the physical therapy treatment of the elderly, one must take into account their characteristics and the associated disorders that they display. In order to make sure that we do no harm and that we do not overload the patient, it will be necessary to also take into consideration the methodological principle of physical exercise, namely from easy to hard and from simple to complex [4].

According to medical and socio-medical services providers' perception, lack of income and health insurance are the main triggers of vulnerability in access to health care [6].

2 Material and Method

This paper was conducted on a sample of 43 patients aged between 60 and 73, of whom 17 male subjects and 26 female subjects. They were divided into two groups: group 1 comprised 21 patients (8 male and 13 female) and group 2 included 22 patients (9 male and 13 female).

The initial and final evaluation was conducted using the Shober test, the finger-floor distance, the lateral trunk mobility and the Visual Analogue Scale.

Both groups of subjects benefited from physical therapy, while the second group also benefited from electrotherapy procedures conducted using the Super Inductive System device.

Concerning the medical rehabilitation sessions, the subjects underwent personalized physical therapy programs for 4 weeks, with a frequency of 3 sessions per week. In case of both groups, the objective of the first week of medical gymnastics was to mitigate pain and to restore the mobility of the lumbar spine using specific exercises of spinal unloading and active elongation of the spine. For pelvic mobilization, the patients executed steps on the device called Stepper. After the first two weeks, patients began executing muscle-toning exercises, which targeted the toning of abdominal paravertebral and gluteal muscles, meant to stabilize the lumbar spine. Starting from the third week, we also introduced exercises for the re-education of the cardiorespiratory system and exercises for the limbering up of the spine. For the cardiorespiratory system, we introduced pedalling on the horizontal ergometer bike and walking on the treadmill. For the limbering up of the spine, we also introduced stretching and pulling exercises on the fixed ladder.

In case of the second group, besides the aforementioned physical therapy protocol, applications of electrotherapy with BTL 600 Super Inductive System were also performed. This device uses successively induced electromagnetic waves with myorelaxant effects, anti-inflammatory drugs and rapid biostimulators. Throughout the four weeks of treatment, the patients within the second group also benefited from two applications per week using the Super Inductive System, with the predefined programme of the device, called Vertebral Syndrome.

3 Findings

In order to highlight the findings, we have presented the values obtained at the initial and final tests for the two groups of subjects. Hence, we have displayed comparatively the arithmetic mean of the values obtained by each group for the Shober test, the finger-floor distance, the lateral mobility of the spine and the Visual Analogue Scale.

The purpose of the graphical representation is to underline the efficacy of combining the two types of therapy, namely physical therapy and electrotherapy, as well as the new-generation Super Inductive System.

Figure 1 features comparatively the evaluation of lumbar spine mobility within the two groups through the Shober test; the first group benefited only from the physical therapy programs, while the second benefited from both the physical therapy program and the electrotherapy sessions using the Super Inductive System device.

According to Fig. 1, the average initial value of the first group was 1.61 cm, and it finally reached the average value of 3.09 cm. It is worth noting that, for the second group, the average value obtained at the final evaluation was 4.5 cm, a significant evolution due to the therapy with Super Inductive System, which increased the myorelaxant effect of physical therapy, which in its turn facilitated the mobility of lumbar spine.

Figure 2 shows comparatively the results obtained by the two groups upon testing the finger-floor distance. The initial evaluation provided an average finger-floor distance of 25.6 cm for the first group and of 25.43 cm for the second group. At

Fig. 1 Comparative evaluation of lumbar spine mobility—Shober test

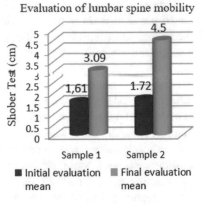

Fig. 2 Arithmetic mean fingertip-to-floor test

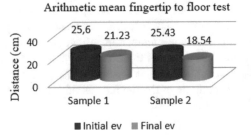

the end of the rehabilitation program, the average finger-floor distance was 21.23 cm and 18.54 cm, respectively.

Figure 2 underscores the effects of physical therapy concerning the improvement of lumbar spine mobility, but the therapy using the Super Inductive System provides more mobility, which is also reflected in the mitigation of the pain perceived by the patients.

According to Figs. 3 and 4, the lateral trunk mobility improved considerably for the first group, from an initial average of 11.61 cm for the left side and 11.73 cm for the right side, to a final average of 13.8 cm for left lateral tilt and 14.1 cm for the lateral tilt on the right side.

In what regards the second group, the mobility of the spine for lateral tilt increased from an initial average of 11.5 cm on the left side and 11.68 cm on the right side to a final arithmetic mean of 14.4 cm on the left side and 15.3 cm on the right side. These values highlight the beneficial and additional effects of the therapy with successively induced electromagnetic waves.

Figure 5 illustrates the comparative interpretation of pain evaluation for the two groups of subjects. The first group recorded at the initial evaluation an average pain of 7.19 on the Visual Analogue Scale, while in the end the value dropped considerably, to a value of 3.04. This decrease stands to show once again the beneficial effects of physical therapy for low back pain among the elderly. As for the second group, which reported at the initial evaluation an average pain level of de 6.9, the final value

Fig. 3 Evaluation of right lateral and right mobility of the trunk

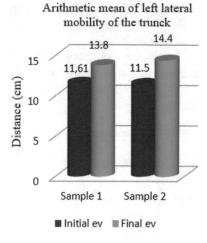

Fig. 4 Evaluation of left lateral and right mobility of the trunk

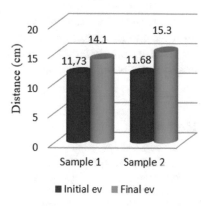

Fig. 5 Evaluation of low back pain through the visual analogue scale

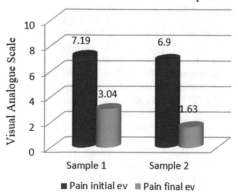

shows a significant and fortunate mitigation, given that the final average value was 1.63, which suggests a minimum pain level, described as a mere discomfort rather than pain per se.

The increased improvement degree at the level of the lumbar spine among the subjects of the second group is due to the painkiller and myorelaxant effects of the Super Inductive System therapy, combined with the effects of targeted physical therapy.

4 Conclusions

This research suggests that low back pain experienced by the elderly can be treated and mitigated using physical therapy programs designed and adapted to their particularities.

The individualized medical rehabilitation programs contribute to both pain mitigation and the restoration of spine mobility and muscle strength, which determine the stability of the rachis.

Furthermore, this paper highlights the efficiency of combining movement therapy with the modern means of electrotherapy. Hence, the association of the effects provided by physical therapy and by the Super Inductive System marked higher positive results and those obtained using medical gymnastics alone.

References

1. Abalașei B (2012) Psychomotricity and psychomotor re-education. Dumat Ofset, Ankara
2. Benedetto T (2008) Back School: un programma completo per prevenire e curare le algie vertebrali. Red edizioni, Milan
3. Bevers K, Hulla R, Rice O, Verdier G, Salas E et al (2017) The chronic low back pain epidemic in older adults in America. J Pain Relief 6(285):1–3. https://doi.org/10.4172/2167-0846.1000285
4. Neculăeș M, Lucaci P (2017) New therapeutic approaches in the treatment of low back pain. Timisoara Phys Educ Rehabil J 10(19):11–15. https://doi.org/10.1515/tperj-2017-0009
5. Olufemi O, Isaac O (2017) The effect of a six-week structured physiotherapeutic modality on the chronic low back pain sufferers' activities of daily living. J Spine 6(390):1–4. https://doi.org/10.4172/2165-7939.1000390
6. Rebeleanu A, Soitu T (2017) Perceptions of the family receiving social benefits regarding access to healthcare. Recent Trends Soc Syst Quantitat Theor Quantitat Models Stud Syst Decis Control 66:157–171. https://doi.org/10.1007/978-3-319-40585-8_14
7. Rofail D, Myers L, Froggatt D (2016) Treatment satisfaction and dissatisfaction in chronic low back pain: a systematic review. J Psychol Psychother 6(260):1–10. https://doi.org/10.4172/2161-0487.1000260
8. Sipaviciene S, Kliziene I, Pozeriene J, Zaicenkoviene K (2018) Effects of a twelve-week program of lumbar-stabilization exercises on multifidus muscles, isokinetic peak torque and pain for women with chronic low back pain. J Pain Relief 7(309):1–10. https://doi.org/10.4172/2167-0846.1000309

Top Management of Social Care Services for the Elderly in Romania

Conţiu Tiberiu Şoitu and Daniela Şoitu⑩

Abstract This chapter makes an analysis of social care services for the elderly in Romania, focused on the particularities of top management. These are highlighted both as self-declared training needs and from the perspective of legal and administrative regulations. The analysis correlates the current demographic and social data, with the forecast of their evolution and with the changes that have led, in the last decades, to a significant change of the traditional Romanian model of intergenerational family care. The chapter radiographs the current realities, but it is and can be a reference document for persons or institutions involved in (re) thinking the training offer. The conclusions converge towards the need to update both the procedures and criteria for recruiting management staff and the offer of its initial and continuous formation.

Keywords Social care services · Elderly · Top management · Education and training · Ageing · Models of care · Social and demographical changes

1 Introduction

Until half a century ago in Romania the traditional model of caring for children, the elderly and people with health issues, including disabilities, was an inter-generational one. It used to be quasi-generalised in the rural environment and widespread in the urban environment. During the past fifty years, several social, economic and geographical changes have changed the reality known two generations ago.

The accelerated urbanisation of the 1970s and 1980s separated children and grandchildren moving into cities from parents and grandparents staying behind in villages. The new habitation model (apartments with modest areas in buildings with hundreds

C. T. Şoitu (✉) · D. Şoitu
Department of Sociology and Social Work, Alexandru Ioan Cuza University of Iaşi, Carol I bvd., 11, 700506 Iaşi, Romania

D. Şoitu
e-mail: danielag@uaic.ro

© The Author(s), under exclusive license to Springer Nature Switzerland AG 2021   115
D. Soitu et al. (eds.), *Decisions and Trends in Social Systems*,
Lecture Notes in Networks and Systems 189,
https://doi.org/10.1007/978-3-030-69094-6_11

and sometimes thousands of inhabitants) has led to a loss of those habitats that were favourable to several generations of a traditional rural households living together.

The political, social and economic changes of the 1990s changed the nature of social relations and behaviours, including of intra-familial ones. The impact of these changes became obvious in the subsequent decades. The fact that the borders opened for emigrating in search of work literally broke the chain of inter-generational care. A significant part of the active population has chosen to work abroad, oftentimes even in the area of care for the elderly, the sick or of individuals with disabilities. The "export" of care capability has left those at home stranded.

From that moment on, caring for the elderly became a topic with important social and economic relevance. The topic has also been amplified by demographic developments and prognoses that were not conducive to a natural, automatic resolution: the growth of life expectancy and the decrease of birth-rates. In other words, in the predictable future a dwindling population will have to look after a growing one. This discrepancy between the need for assistance and the possibility of providing it is the most visible in the area of elderly care services [1].

2 Caring Market and Care Needs

As of January 1, 2018, in Romania the total volume of population resident in the country was 19,523,621 persons, of which 3,550,957 persons (i.e. 18.2%) aged 65 years and over. Their distribution differs by regions of the country and residential areas (rural/urban) reflecting several structural dimensions. The least aged area of Romania is the one in the capital Bucharest and its surroundings (15.9%, up to over 20% in counties in the south-east of the country). The care needs of these people are covered, again, differently: among the 41 counties, six have almost 10 licensed social services reported per 10 thousand inhabitants over 60 years of age, while three counties have only one each. Nine counties—mostly covering the central and north-western part of the country—have five licensed social services for 10,000 inhabitants over the age of 65, and more than half of the country's counties provide three services in the same proportion. The dynamics of these services is noticeable, and so is the correlation with the higher standard of living in the areas with more services, compared to the higher poverty rate in the counties with fewer licensed services for the elderly. We can conclude that the market of social services for the elderly in Romania reflects the social need, but it is more closely correlated with affordability, given that two thirds of services are provided by private entities.

The providers of social services in Romania may be, in accordance with the legislation in the field (Law 292/2011 art. 37): natural or legal persons, of public or private law [2]. The public providers of social services may be: the specialized structures within/subordinated to the local public administration authorities and the executive authorities from the administrative-territorial units organized at the level of commune, city, municipality and sectors of Bucharest; the central public administration authorities or other institutions subordinated to them or under their coordination which have, by law, attributions regarding the provision of social services for certain categories of beneficiaries; healthcare facilities, educational institutions and other public institutions that develop, at community level, integrated social services. The private providers of social services may be: non-governmental organizations, respectively associations and foundations; religious organisations recognized by the law; natural persons authorized under the law; subsidiaries and branches of international associations and foundations recognized in accordance with the current legislation; and since 2015 (Government Ordinance 31/2015) [3]: for-profit economic operators, for all categories of social services organized under the law, except for services for preventing the separation of the child from the family and those for children temporarily or permanently deprived of their parents, and except for social services aimed at preventing and combating domestic violence—victims and aggressors—and services for people with disabilities [4, 5].

In December 2004, there were 59 public residential centres for the elderly in Romania (of which 20 homes with approximately 100 places each and 39 former old people's homes transformed into care and assistance centres with approximately 120 places each), 17 private centres (with residential and day services, with 776 places, therefore with an average of 45 places/institution) and 39 home care services subsidized from the state budget [6], p. 197.

15–16 years later, the changes in the field of social services for the elderly are remarkable: the provider market is undergoing a sustained development and the types of services have diversified.

In terms of social policies, a law regarding quality in social services has been enacted (Government Emergency Ordinance GEO 68/2003, followed by Law 197 of 2012), the quality standards for each type of social services have been updated three times, the latest update taking place in January 2019 –, and procedural for accrediting and licensing the providers of social services and the services themselves have been published [7, 8].

In early 2019, according to the Ministry of Labour and Social Justice, 348 providers of residential social services (homes) for the elderly were accredited in Romania, providing a total of 15,460 places in licensed homes for the elderly—with an average of 45 places/institution. The development was made with differences from one area to another, most places in homes being available in the North-West Region (2,641 places).

In the first months of 2020, in Romania were accredited over ten times more providers (compared to 2004) for twice as many (eight) types of services for the elderly; thus: 568 homes for the elderly, 5 respite/crisis centres, 19 sheltered housing; to these are added the 57 medical-social residential centres and the 5 palliative care centres. The day centres have separated the assistance and recovery activity (53 accredited providers) from the socialization and leisure activity (95 accredited providers). Whereas in 2004 there were 39 home care services, in May 2020 they are almost six times more numerous (261). The 131 licensed social canteens supplement basic services for the elderly and other vulnerable groups. A novelty is the 128 integrated community assistance services, developed through projects of the Ministry of Labour and Social Protection, in partnership with the Ministry of Health and local authorities in disadvantaged areas. Along with another 408 social services intended for vulnerable categories, the market of licensed services in Romania, in May 2020, registers 1730 social service providers aimed solely at or especially designed for the elderly.

Analysing the data of the social services market for the elderly we find, once again, a sustained development and diversification [1, 9]. However, the services provided by the residential care and assistance centres (592) remain predominant, together with the residential care and medical-social assistance centres for the elderly, chronically ill and/or terminally ill (181), followed by the home care services for the elderly, people with disabilities, people with addictions (261) and day centres (148).

On the entire market of social services for all categories of vulnerable people in Romania, there were, in May 2020, 3018 accredited providers. It should be noted that a provider may have one or more licensed social services. A proportion of two thirds of private providers to one third of public providers is also maintained at this level. Of the accredited public providers (993), most are specialized structures within or subordinated to the local public administration authorities (793), central public administration authorities (38) and other public outfits—in health or education (73). In the category of private suppliers, non-governmental organizations predominate (1623), followed, at a distance, by business operators (203). Considered, traditionally, the main providers of social assistance, the religious organisations accredited to provide social services are now below 5% (128). There are even two authorized individuals who have accreditation to provide social services, but the branches of international organizations, present in large numbers in Romania in the early 1990s, disappeared. This points to an evolution of the capacity to manage local social needs and provide support to vulnerable people and categories, but also to the effects of decentralization processes (203).

3   Caring Market and Human Resources. Top Manager's Survey

As a response to the growing need, the offer began to develop. The shortage of places in residential centres has been gradually reduced through various means. The establishment of new centres and/or increasing the capacity of the existing ones has had a relatively low impact in terms of quantity. The remaining deficit has been covered by the significantly larger number of new private residential care centres. We find here both not-for-profit initiatives from nongovernmental (including confessional) associations and organisations, as well as the numerous, of late, businesses in the domain.

The proof of the commercial attractiveness of this type of services can be found in the number (estimated to be high, but actually unknown) of unaccredited private centres. The latter provide care services disguised as hotel/hospitality services, which allow them to avoid the expenses needed in order to meet the minimum quality standards for, but not limited to, the specialised staff. This is only one of the sensitive aspects that have caused both the accredited service provides and the institutions regulating their operation to become more concerned with a more careful regulation of the domain, so that the beneficiaries do not lose confidence in the services.

For this purpose, several initiatives of founding professional associations in the area of social services for the elderly have appeared in the past decade. One of the first such initiatives was the Association of Elderly Care Institution Directors (Asociaţia Directorilor de Instituţii pentru Vârstnici/ADIV-România). Responding to their inquiries and based on our interest for the field, in October-December 2016 we have collected answers to a questionnaire applied to those who fill coordination and management positions in (predominantly residential) care centres, both public and private, for-profit and not-for-profit. One of the aims of the questionnaire was to identify the names of the top management positions and the initial and in-job education of those filling them. The interest for such information was justified by the absence, at the time, of clear administrative regulations, especially in the case of the private providers of care services.

We collected over 50 responses, which have confirmed the suspected diversity of circumstances. The labels of the management positions was (in decreasing order of frequency): "preşedinte" ["president"] (especially in foundations and associations), "director/director general/director executiv/director administrativ" ["director"/"executive director"/"administrative director"] (in public and private institutions), "manager", without further details (especially in private service providers), "administrator", "administrator general" ["general administrator"], "coordonator general" ["general coordinator"], "coordonator centru/serviciu" ["centre/service coordinator"], "coordonator personal de specialitate" ["coordinator of specialised staff"]. All these, in conjunction—in some cases—with additional positions such as "vicepreşedinte" ["vicepresident"] and "director adjunct" ["deputy director"].

The surprise was the very wide margin of variation of the accomplished level of education. Whereas for the public institutions the minimum education level (graduate diploma/BA, BSc) was observed, the diversity appearing in terms of the domains of study, in the case of the NGO providers, any homogeneity disappeared. Around one-quarter of responses indicated high-school diplomas (including one case in which education had not been completed); another approximately 5% of post-high-school education; the rest—graduate and postgraduate education, including master degrees and PhD.

In terms of areas of initial education, among those with graduate education we found medicine, law, economics, social work, sociology, administrative sciences, psychology, kineto-therapy, human resources, as well as many others, further away from the profile of the activity, mainly engineering (approximately one-quarter of all tertiary education graduates). This reality explains the diversity of the continued training of managers. Access to such training was conditioned by the level of education completed [10, 11]. This explains the high-school graduates' predominant training at post-high-school level as nurses. University graduates had found it useful to have training in the management domain (as shown by the relatively high frequency of MBA-type programmes attended), in social work, economics, but also in communication, counselling etc.

In the responses to the same questionnaire, the managers indicated—without exception—their need for continued training. It was natural for the lack of homo- geneity of educational and professional background to result in the impossibility of recommending a single type of training programme (postgraduate, post-university continued training and so on) and a single type of contents. As a complement or as a replacement of initial training, all categories of training contents suggested in the questionnaire, deriving from medicine, gerontology, law and legislation, social work, economics and management, sociology, psychology as well as others, were indicated as useful and necessary.

In 2017 the Romanian Classification of Occupations introduced a new entry: "social services manager". This is the first national occupational standard for this field and level. As at the time the occupational standard there were no legally mandatory requirements in terms of education and training in order for an individual to access this occupation, the document does not mention them. All it indicates is the necessary competences, the professional experience in social work, and a certificate attesting the attendance of a study programme for "social services managers". This certificate was deemed necessary but not sufficient and recommended potential future training programmes. According to our information, several universities (including ours) are in different stages of preparatory activities for new life-long education study programs, adapted to this standard.

4   Analysis and Interpretation. Legislation Details

Two years after the responses described above were recorded, in 2018, references as to the type, level and domain of education, both initial and continued, of the individuals in management positions in regional and local coordination structures, as well as among the providers of social services for the elderly, can be found n the national legislation.

For the management of service providers, public and private, for-profit and not-for- profit, the landmark information can be found in Government Decree Nr. 867/2015 of 14 October 2015 for the approval of the Social Services Occupations List, as well as of the framework regulations for the organisation and operation pf social services (amended in 2019, Government Decree 476) [12, 13]. The document lists the positions that are assimilated to that of management staff: "director" (or "şef de centru" ["centre head"]); "coordonator/coordonatori personal de specialitate" ["coor- dinator/s of specialised personnel"]. The amendments of 2019 also added the label of top management position for the chief accountant in the case of self-standing centres that have legal personality (Art. 9.1). It is only for the social services provided within the community that the name "project manager" is included for the top management position, as an alternative to that of "director" or "şef de centru".

Following the revision of the 2019 normative act, access to management positions is conditioned by the following minimum requirements for training and professional experience: Candidates for management positions must be graduates of bachelor's degree, respectively long-term higher education., graduates with a bachelor's degree or equivalent in the field of psychology, social work or sociology, with a minimum of 2 years of experience in the field of social services, or graduates with a bachelor's degree in law, medical, economics or administrative sciences, with experience of at least 5 years in the field of social services. (Art. 9.4). The application instructions of the document establish the possibility of further updating the list of graduated study programs, in accordance with possible changes to be made by the Ministry of Education and recorded, annually, in the Nomenclature of fields and university study programs. At the same time, they specify the recognition as quantifiable seniority/experience the one acquired in another member state of the European Union and the one accumulated through voluntary activities (only in the field of graduated education).

At the local and county level, there are structures that, in addition to their strategic coordination responsibilities, can (and generally exercise this potential role) provide social services. These are the General Directorate of Social Assistance and Child Protection, at county or sector level in the Municipality of Bucharest, and the Directorate of Social Assistance organized under the subordination of the local councils of the municipalities and cities.

According to Decree No. 797/2017 of November 8, 2017 for the approval of the framework regulations for the organization and operation of public social assistance services and the indicative staff structure (Art. 1 of Annex 1 of the Framework Regulation for the organization and operation of the General Directorate of Social Assistance and child protection), the General Directorate of Social Assistance and Child Protection is the public institution with legal personality established under the subordination of county councils/local councils of the sectors of Bucharest in order to ensure the application of social policies in the field of child protection, family, elderly, disabled, as well as other persons, groups or communities in social need, with a role in the management and provision of social assistance benefits and of social services [14].

At local level, according to Annex 2 to the Framework Regulation for the organization and operation of the Directorate of Social Assistance established under the subordination of county councils/local councils in the same GD Nr. 797/2017, but amended through GD no. 417/2018, the Directorate for Social Assistance is the public institution specialized in the management and provision of social assistance benefits and social services, established under the local councils of municipalities and cities, as a social assistance directorate, in order to ensure the application of social policies in the field of protecting children, families, the elderly, people with disabilities, as well as other people, groups or communities in social need. As an exception to these provisions, the Directorate can be organized as a functional department of the mayor's specialized staff or as a general directorate, a public institution subordinated to the local councils of municipalities and cities, depending on the demographic structure and socio-economic indicators of the city/municipality [14, 15].

The main difference between the two types of structures, from the perspective of the direct provision of social care services, consists in the different typology of the offer.

The social services provided by the local directorates aim to prevent or limit situations of difficulty or vulnerability, which can lead to marginalization or social exclusion.

Specialized social services aim to maintain, restore or develop individual capacities to overcome a situation of social need. They are administered by the General Directorate of Social Assistance and Child Protection—a structure of the County Council.

According to the same normative act (GD 797/2017, art. 12), the management of local—if organized as a public institution—and county directorates is provided by the executive/general director, assisted by deputy executive directors/deputy generals [14]. All of them have the quality of civil servant or, as the case may be, of contractual staff. In the case of the municipal or city directorate organized as a functional compartment in the specialized staff of the mayor, its management is provided by the director.

In the case of both types of directorates, candidates for the positions of executive director/general or, as the case may be, deputy executive director/deputy general must have a seniority in the domain of at least 3 years and be graduates of undergraduate studies, with diploma, in one of the following fields: (a) social work or sociology; (b) psychology or educational sciences; (c) law; (d) administrative sciences; (e) health; (f) economics or management, finance, accounting. The law also allows exceptions, so that in the competition organized for the positions mentioned above can also participate graduates with a bachelor's degree of long-term university education in other fields than those mentioned, provided they have graduated master's or postgraduate studies in public administration, management or in the specialty of studies necessary for the exercise of public office.

5   Academic and Professional Initiatives

The idea of a consistent education for managers in institutions for the care of the elderly, discussed in this chapter, emerged in 1993 in Austria. The "Educational Framework for the Management of Salzburg Elderly Organizations" was adopted at the time and recommended for implementation in all Austrian federal states. Later on, E.D.E. (as of 1989, the European Association of Directors and Providers of Long Term Care Services for the Elderly, which has become, in April 2018, E.A.N.—European Ageing Network) recognized this framework, supporting at European level the education of managers of senior care institutions and offering a certificate with European recognition. The same type of recent concern and strategies are described for North American and United Kingdom [16], p. 10; [17], pp. 3–22; [18], pp. 29–36; [19], p. 6.

Following the Austrian example, several other European countries (Germany, the Netherlands, Italy, Slovenia) have implemented compulsory education for senior staff in care institutions for the elderly. Training is provided by universities and accredited agencies. The requirements of the labour market and the socio-demographic transformations mandate an innovative update of the content and structure of the educational program through which it will be possible to obtain the E.D.E./E.A.N.—certificate that will offer long-term care providers a guarantee of service quality and a European recognition.

An example of an academic and professional initiative is an ongoing partnership, in which the project coordinator is Asociace poskytovatelů sociálních služeb ČR, z.s. (Association of Social Service Providers, Czech Republic), and the partners are, apart from the Alexandru Ioan Cuza University of Iaşi (Romania): Akademie für Sozialmanagement (Austria), Bundesverband der Alten- und Pflegeheime Österreichs „Lebenswelt Heim" (Austria), Rath Wolfgang Consulting e.U. (Austria), IBG—Institut für Bildung im Gesundheitsdienst GmbH (Austria), E.D.E.—European Association for Directors and Providers of Long-Term Care Services for the Elderly (Luxemburg)/which was transformed, in April 2018, in Bucahrest, into the European Ageing Network (EAN). With the title "Modification of the educational module—E.D.E. certificate for the providers of long-term care for seniors", the project is financed through Erasmus +, Strategic Partnerships supporting innovation, Strategic partnership in vocational education. KA2—Cooperation on innovation and exchange of best practices, with a duration of 36 months (01.09.2018–31.08.2021).

Within the mentioned program, three focus groups were created with the topic *Adapting the training offer in order for the directors of services for the elderly to obtain the EAN/EDE certificate, recognized at European level.* The meetings took place in Iasi and Bucharest, in March, May and July 2019. More than 50 social service managers for the elderly from all over Romania participated in the survey. The objectives of this survey were to identify the specific areas and themes of training in services for the elderly and ways to implement training-specific content.

We summarise hereby part of the responses provided by the managers of residential and non-residential services of the elderly taking part in the survey.

The general considerations regarding the problems faced by the coordinated centres touched upon the following aspects: dysfunctional relations with the partner institutions, especially the medical ones; insufficient specialized staff; issues regarding the licensing and operation of services for the elderly; differences regarding the observance of the standards of services provided in public and private institutions; the need for state involvement in supporting NGOs that provide care services for the elderly; unstable and incoherent legislative framework.

The first question was what training and level of education should the manager in a nursing home or generally the management in social services have?

All participants in the survey supported the idea that the directors of residential centres for the elderly should have higher education. The need for experience in the field for people running residential centres was emphasized. The individuals holding management positions in residential centres must/or may have economic or legal education, but they need to have experience in the field of providing social assistance services. The need for complementary training of residential centre managers in the field of social services was supported, regardless of the nature of the initial training.

Subsequent questions focused on the knowledge and skills required in a manager. The most frequently found answers include: the manager's ability to understand the legal framework; the ability to make decisions in situations where they do not have sufficient resources; the ability to attract various types of resources; the ability to understand customer needs; communication with beneficiaries and legal relatives; willingness to learn; flexibility/adaptability.

There were discussions as to the types of providers of the training offer and how to achieve it. All survey participants argue that the training provider should be a higher education institution; it is desired that the individuals involved in the training/evaluation activity have experience, including practical experience, in the field of social services for the elderly; the professional experience of the participants must be considered on an equivalent scale; completing training hours in a flexible program and system that includes e-learning.

6 Conclusion

The evolution and the diversity of the market of social services for the elderly in Romania requires a significant number of institutions and professionals. We have already mentioned that the first educational programs for the broad category of social services managers are not yet available. The future training program will be more targeted and will benefit from the longer and broader experience of previous training programs and of several EU national systems of social services for elderly. The international dimension of the program will be useful in the cases that can already be found in practice, of professional mobility of managers from one EU country (and not only) to another. In this situation, the beneficiaries of training programs compatibility will be both the potential employee and the employers.

However, the accelerated change of social, demographic, economic, technological, and professional context will require a periodic resumption of efforts to reassess training needs and propose adapted content and training methods.

References

1. Șoitu D, Șoitu CT (2020) Romania. In: Ní Léime Á et al (eds) Extended working life policies. International Gender and Health Perspectives, pp 385–394. https://doi.org/10.1007/978-3-030-40985-2, ISBN 978-3-030-40984-5
2. The Parliament of Romania (2011) Law No. 292 of 20 Dec 2011; Law on social assistance: Monitorul Oficial NO. 905 of 20 Dec 2011
3. The Government of Romania (2015) Ordinance No. 31 of August 26, 2015 for the amendment of art. 37 paragraph (3) let. e) of the Social Assistance Law no. 292/2011, as well as for the

abrogation of art. 11 paragraph (5) of Government Ordinance no. 68/2003 on social services: MONITORUL OFICIAL No. 651 of 27 Aug 2015

4. Șoitu D, Șoitu CT (2011) Social action of civil society, Analele Științifice ale Universității Alexandru Ioan Cuza din Iași. Sociologie și Asistență Socială, 4/2011, pp 107–122, Print ISSN: 2066-8961; Online-ISSN: 2066-8961

5. Șoitu CT, Șoitu D (2010) Europeanization at the EU's external borders: the case of romanian-moldovan civil society cooperation. J European Integr 32(5):491–506. https://doi.org/10.1080/07036337.2010.498633

6. Gîrleanu-Șoitu D (2006) Vîrsta a treia [third age], Iași, Institutul European. ISBN 973-611-391-4

7. The Government of Romania (2003) Government Emergency Ordinance GEO 68/2003, on social services, Monitorul Oficial, Part I no. 619 of 30 Aug 2003

8. The Parliament of Romania (2012) Law 197/2012 on quality assurance in the field of social services

9. Șoitu CT (2020) Measurement of disability in Romania. In search for comparability. In: Sarasola Sanchez-Serrano JL et al (eds) Qualitative and quantitative models in socio-economic systems and social work, studies in systems, decision and control, vol 208, pp 73–82. ISBN 978-3-030-18592-3, https://doi.org/10.1007/978-3-030-18593-0_6

10. Hoskova-Mayerova S (2016) Education and training in crisis management. In: The European proceedings of social and behavioural sciences EpSBS, vol XVI, pp 849–856. ISSN 2357-1330. https://doi.org/10.15405/epsbs.2016.11.87

11. Svarcova I, Hoskova-Mayerova S, Navratil J (2016) Crisis management and education in health. In: The European proceedings of social and behavioural sciences EpSBS, vol XVI, pp 255–261. https://doi.org/10.15405/epsbs.2016.11.26

12. The Government of Romania (2015) Decision No. 867/October 14, 2015, for the approval of the Nomenclature of social services, as well as of the framework regulations for the organization and functioning of social services: Monitorul Oficial NO. 834 of 9 Nov 2015

13. The Government of Romania (2019) Decision No. 476/2019 of July 3, 2019 for the amendment and completion of the methodological norms for the application of the provisions of Law no. 197/2012 on quality assurance in the field of social services, approved by Government Decision no. 118/2014, and of the Government Decision no. 867/2015 for the approval of the Nomenclature of social services, as well as of the framework regulations for the organization and functioning of social services: Monitorul Oficial NO. 581 of 16 July 2019

14. The Government of Romania (2017) Decision No. 797/2017 of November 8, 2017 for the approval of the framework regulations for the organization and functioning of the public social assistance services and of the indicative personnel structure: Monitorul Oficial NO. 920 of 23 Nov 2017

15. The Government of Romania (2018) Decision no. 417/2018 regarding the amendment of the Government Decision no. 797/2017 for the approval of the framework regulations for the organization and functioning of the public social assistance services and of the indicative personnel structure: Monitorul Oficial, Part I no. 497 of 18 June 2018

16. Hafford-Letchfield T et al (2008) Leadership and management in social care. Sage. ISBN 978-1-4129-2960-8

17. Meerabeau L, Wright K (eds) (2011) Long-term conditions: nursing care and management. Wiley-Blackwell. ISBN 978-1-4051-8338-3

18. Powers JS (2017) Creating a value proposition for geriatric care the transformation of American healthcare. Springer Briefs in Health Care Management and Economics. ISBN 978-3-319-62270-5.https://doi.org/10.1007/978-3-319-62271-2

19. Sullivan-Marx E, Gray-Miceli D (eds) (2008) Leadership and management skills for long-term care. Springer. ISBN 978-0-8261-5993-9

Actors in Social Systems. Decisions and Trends of Care

The Home Care Sector on the Pathway of Neo-Liberalism: The Belgian Case

Nathalie Burnay

Abstract Traditionally, social protection in Belgium has adhered to the corporatist model. However, the transformations observed in recent years in the elder care sector truly show a paradigm shift. This contribution analyses the penetration of neo-liberal logics into this sector, in particular through an increasing presence of the private sector in retirement homes and a transformation of the public authorities' role in the management of old age. The consequences of these new policies affect elderly people themselves, their families and the professionals in the sector. Women caregivers are particularly under pressure in this context of change.

Keywords Elderly care sector · Welfare state · Neo-liberalism · Caregiver · Gender · Working-life balance

1 Introduction

Traditionally, social protection in Belgium has adhered to the corporatist model, as is the case in Germany, France, Austria, and the Netherlands. The Bismarckian model has been favoured since the Belgian government in exile established the modern social welfare system in December 1944. It is based on a community of references that centres around the worker, who must be protected from life's vicissitudes—and above all from temporary or permanent exclusion from the workforce. Thus, the Belgian model is based on a generous social welfare system that approaches the Scandinavian universalist model in some aspects. However, it is also characterised by a low level of decommodification and defamiliarisation [6]. It is undoubtedly generous, but also favours the model of the *male breadwinner* by providing assistance differentially

N. Burnay (✉)
Transitions Institute, University of Namur, Namur, Belgium
e-mail: nathalie.burnay@unamur.be

Institute for the Analysis of Change in Contemporary and Historical Societies (IACCHOS), Université Catholique de Louvain, Louvain-la-Neuve, Belgium

© The Author(s), under exclusive license to Springer Nature Switzerland AG 2021
D. Soitu et al. (eds.), *Decisions and Trends in Social Systems*,
Lecture Notes in Networks and Systems 189,
https://doi.org/10.1007/978-3-030-69094-6_12

according to gender: cohabitants, predominantly women, systematically receive less generous benefits than heads of household, predominantly men.

The Belgian social system is also based on complementarity amongst a multiplicity of associations that assist populations in need. In contrast to Austria or Germany, where social benefits are granted directly to the beneficiaries, Belgium is characterised by an important institutionalisation of social benefits, which are channelled through associations that are supported by the regional and federal authorities in a kind of regulatory guardianship. *"The notion of regulatory guardianship describes a mode of regulation by the public authorities in which the provision of services is financed and overseen by the public authorities, who thus act as 'guardians' for the beneficiaries* [10]. *It is defined through regulatory frameworks enacted by the public authorities, which prescribe standards for oversight, training requirements for employees, and so on. Funding is thus accorded to approved organisations, public or private of the associative type (called the 'non-profit sector' in some countries), based on predefined quality standards. These standards determine eligibility for financing by means of issuance of operating permits, prior approval of estimated budgets, annual review of costs and investments, and definition of target populations and of the services and benefits intended for these populations"* [11].

Nevertheless, for the past several years, the Belgian social welfare system has been undergoing profound modifications. Regulatory guardianship is being progressively replaced by a competitive framework of regulation in the context of *new public management* [14], i.e., through the introduction of market principles into the forefront of governance. There is severe criticism of a mode of state functioning said to suffer from excessive bureaucracy and lack of effective management, which allegedly leads to a kind of disempowerment amongst recipients of public assistance, if not to perpetual dependence on this assistance. This penetration of neo-liberal logic into the welfare system leads to the call for development of pseudo-markets, in which public—and private—sector service providers compete for public funding in a logic of *marketization*. *"The concept of marketization is a wide concept that covers a broad span of arrangements where private sector organizations contract with public sector bodies with the purpose of delivering a welfare service in exchange for public funds"* (Brown and Potoski [2] cited by [12], p. 8). The idea is simple and is based on the neo-liberal postulate of *less government and more market* [3]. It is in this spirit that the European Commission White Paper was issued in 2001, aiming to impose principles of management based on financial performance [5].

For Hall [8], any change in public policy can be interpreted according to its degree of importance. He thus defines three orders of change, with differing effects on the social welfare system. Whilst the first two orders concern only more or less important adjustments to the instruments of social policy, the third order corresponds to a true paradigm shift as defined by Kuhn [9]. *"First and second order change can be seen as cases of 'normal policymaking,' namely of a process that adjusts policy without challenging the overall terms of a given policy paradigm, much like 'normal science.' Third order change, by contrast, is likely to reflect a very different process, marked by the radical changes in the overarching terms of policy discourse associated with a 'paradigm shift'"* [8], p. 279]. This analysis allows us to go beyond the historical neo-institutionalist concept of path dependence [13]. This concept demonstrates the difficulty of public authorities making any fundamental change to public policies that are entrenched in routine, as change would inevitably lead to wasted energy and additional expense. The financial cost—but also the human cost—associated with change would thus render real change nearly impossible.

This article will demonstrate the extent to which this *marketization process* is destructuring the Belgian social welfare system through the dual process of privatisation of the home care sector and transformation of the benefits provided to recipients. This dual process is thus leading to a true paradigm shift through the introduction of a neo-liberal logic that undermines the principles of the Belgian welfare state while shifting the burden of assistance to the family, and particularly to women.

2 A Policy of Ageing in Place

Like many other developed countries, Belgium is characterised by an ageing population. People over 80 years old made up 5.3% of the Belgian population in 2014 and are expected to reach 6.5% in 2030 and 9.8% in 2060. In 2016, people aged 65 years and over (65+) accounted for 2 million of the 11 million total inhabitants. This tendency is structural, and the following figure illustrates the progressive increase in this population aged 65+ (Fig. 1).

Belgium is a federal state made up of three regions, each with its own competences. This first figure illustrates the pronounced differences between regions. The year 2001 represents a turning point at which the percentages of people aged 65+ were

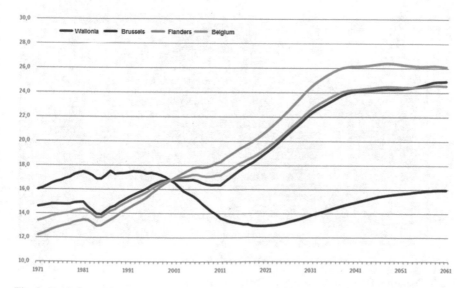

Fig. 1 Evolution of the percentage of the population aged 65 years and over by region

similar in the three regions of the country. Since then, the situation has changed substantially, and the central Brussels-Capital region (with approximately one million inhabitants) has experienced less ageing of its population than the other two regions, although its percentage of those aged 65+ is expected to increase beginning in 2020. This significant difference stems primarily from two factors: a higher proportion of populations of immigrant origin (which tend to be younger) in the Brussels-Capital region and higher birth rates in these groups than in the autochthonous population.

The situation also differs slightly between the northern region of Flanders (approximately 6 million inhabitants) and the southern region of Wallonia (approximately 4 million inhabitants). Flanders is experiencing a more pronounced ageing of its population than Wallonia. It is expected that in 2030, more than 26% of the population in Flanders will be aged 65+, whilst the corresponding percentage in Wallonia will be approximately 24%.

The age pyramid below confirms this ageing of the population and provides additional elements (Fig. 2).

This age pyramid clearly demonstrates an increase in the oldest age groups between 2001 and 2013, amongst both men and women. For both sexes, there is a clear decrease in the group aged 35–39 years in 2013 compared to 2001, and evidently an increase in the group aged 45–49 years: those 35–39 years old in 2001 have become the 45- to 49-year-olds of 2013. However, more surprisingly, the number of babies increases significantly between 2001 and 2013, the birth rate having increased slightly in recent years.

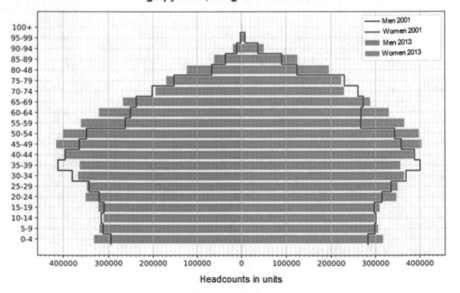

Fig. 2 Age pyramid for Belgium between 2001 and 2013 (number of inhabitants by age)

These factual data demonstrate what is at stake in the public authorities' response to this phenomenon of ageing, at the regional as well as at the federal level. If the payment of pensions remains a federal competence, the management of ageing is primarily a regional competence.

For the past several years, the regional authorities have developed policies encouraging ageing in place. The main reasons given are related to maintaining elderly people's autonomy, the well-being of the aged, and their desire not to leave their own homes. Thus, Wallonia's centre-right government, in its latest policy statement of 25 July 2017,[1] made ageing in place a key aspect of policy, according tax deductions and financial aid to encourage the elderly to remain in their homes.

This policy of ageing in place is not new; it was already promulgated by the previous government of Wallonia, which was socialist. Beginning in 1997, this socialist government advocated a policy of ageing in place using two approaches: firstly, the development of the home care sector, and secondly, the reinforcement of alternatives to nursing homes to provide respite for relatives and caregivers, such as day and overnight centres for those elderly people still enjoying relative independence.

However, behind these laudable goals of preserving elderly people's autonomy, there are other concerns at work. Thus, two further arguments have been introduced to legitimise the current policy. First and foremost, home care costs less than institutionalisation. As Wallonia's coffers are in chronic deficit, it is important to keep the budget allotted to the ageing population from exploding. Moreover, there is a drastic shortage of care facilities for the elderly in Belgium as a whole. Although the total number of beds has increased by 11.5% since the beginning of the 2000s, there are clearly not enough nursing homes. At the end of 2014, there were 1534 accredited facilities in Belgium, representing approximately 135,000 beds. Five years after, in 2018, 150,560 beds have been identified in Belgium, including more or less 50,000 in Wallonia.

In order to encourage ageing in place and in the Belgian tradition of the welfare state, the authorities have made efforts to support the home care sector. Thus, public or parapublic structures (subsidised by the region) have been able to develop around a very broad and open conception of home care. People in need can make use of this sector, with related costs assumed partly by the beneficiaries themselves (according to their means), but mainly by the public authorities.

Beneficiaries can call upon the home care sector in a number of different situations. Thus, four different professions are represented in this sector, offering a broad spectrum of services:

- *family assistants ("Aides familiales")* help the beneficiaries carry out the activities of daily life, whether during the day, in the evening, or on weekends or public holidays. Their role is to permit beneficiaries to continue living in their own homes while maintaining quality of life. Concretely, they may accompany beneficiaries to do their shopping, prepare their meals, help with administrative matters, organise

---

[1] https://www.lexalert.be/fr/article/d-claration-de-politique-r-gionale-wallonne-2017-2019-les-mesures-concernant-limmobilier (in French).

their budgets, or keep house. Their roles are also health-related (monitoring health, hygiene, and safety; assisting with personal hygiene and grooming), educational (advising and supporting families, promoting autonomy), and relational (listening and identifying difficulties, being present, conversing).

- *Housekeepers* (*"Aldes ménagères"*) ensure the cleanliness of the home; they perform domestic tasks such as tidying up, maintaining and improving household hygiene, and laundry.
- *In-home monitors* (*"Gardes à domicile"*) ensure a presence and an active surveillance, night and day. Concretely, they ensure that beneficiaries take their medication and eat their meals, and provide moral support through relational exchanges and taking the time to accompany beneficiaries in their daily tasks.
- *Handymen* (*"Ouvriers polyvalents"*) can intervene to adapt the home to the limitations of age and dependency. Concretely, they might clean a slippery terrace, repair a defective step, brighten the home by doing minor painting or upholstery work, change curtain rods, do minor plumbing or electrical repairs, or pack belongings before a move.

These different job descriptions illustrate how widely the nature of the services provided can vary in the global and holistic care of a diverse population.

3   The Penetration of a Neo-Liberal Logic

Faced with an increasing demand and an inadequacy of public finances, we are now witnessing a dual process, both of whose components are consistent with a neo-liberal logic—if not with a process of *marketization* in which public services give way progressively to the private sector.

First of all, the retirement home sector in Belgium has undergone a veritable transformation over the past few years, as in other European countries, with private sector involvement gradually replacing public investment. In Wallonia, the allocation formula that determined percentages of market share for the public and associative sectors is now being called into question. This formula guaranteed minimum market shares of 29% for the public and 21% for the associative sectors, and a maximum of 50% for the private sector. Its disappearance would seal the fate of the retirement home sector, which would see a substantial increase in private investment. Moreover, small family-style structures with only a few beds have totally disappeared from the institutional landscape, as a new regulation obliges retirement homes to have a capacity of between 50 and 150 beds. The "small independents," which still dominated the landscape ten years ago, now manage only about half of private retirement homes. Initially, Belgian institutions clustered together, hoping to attain a size that would permit them to confront the future. However, the sector has been opened progressively to foreign investors—huge conglomerates, predominantly French and sometimes listed on the stock exchange. For example, in the Brussels-Capital region, only six international groups now manage more than one bed in three. The French

Orpea group and the Belgian *Senior Living* group together account for more than 3500 of the total of 15,000 beds.

Thus, small retirement homes, which can no longer comply with regulatory standards and therefore can no longer receive public funding, are being progressively bought out by these large financial groups. They consequently lose their specific character, coming to resemble all of the group's other homes. These buyout conditions result in the development of a two-tiered sector in which the cost to residents increases progressively but may not be associated with corresponding improvement in their well-being. The nursing home becomes a business above all, functioning according to rationalisation of costs: grouping of purchases for multiple facilities, staff reductions, fragmentation of tasks, difficulty in pursuing continuing education for the personnel, and so on. The groups are also characterised by centralisation: decisions are made at the summit and no longer within the individual homes. Finally, the group's different facilities are managed according to standardised procedures that permit reduction of operating costs. For example, the residents at different sites eat the same thing at the same time; furniture is similar at the different sites; little opportunity is left for personalisation of a resident's private space; rooms are repainted in the same colour; kitchens are generally grouped into a single unit that distributes pre-cooked, pre-packaged dishes to the different homes. Along with this economic management comes human resources management that oversees personnel, work schedules, and distribution of tasks, often deliberately introducing competition between institutions.

In parallel to these structural transformations of the retirement home market, we are witnessing another phenomenon affecting the home care sector in Wallonia. A legislative proposal currently before Wallonia's parliament would redefine the eligibility requirements for public funding in this sector, limiting the assistance provided to beneficiaries through a radical change in the framework of care. Henceforward, only those beneficiaries recognised as dependent would be eligible for partial reimbursement for home care services. Their status would be evaluated using a dependency scale validated by the public authorities (the BelRAI Screener scale), which would permit assessment of the nature and degree of dependency and thus the admissibility of the request for assistance.

According to the home care sector, the use of the BelRAI Screener scale[2] can have nonsensical consequences. For example, an elderly woman who is quadriplegic due to multiple sclerosis might obtain a BelRAI Screener[3] score too low to qualify for public assistance if she has no behavioural or cognitive problems. Moreover, the BelRAI Screener scale evaluates people's dependency based on their ability to perform certain tasks, but does not consider factors such as their distance from care facilities or services or their family situations. Thus, the home care sector estimates that the new regulation would exclude between 30 and 40% of present beneficiaries. Those excluded from the system would then need to pay the total cost of the home

---

[2]The scientific validity of this scale, which was constructed and validated in Belgium, is not being questioned here—only its use in this highly political context.

[3]https://orbi.uliege.be/bitstream/2268/207752/1/BelRAI%20Screener%20IWEPS%2008122016. pdf.

care services received. This would represent a clear and substantial increase (from
a present maximum of 7 euros per hour of services to 35 euros if dependency is not
recognised), which would only exacerbate social inequalities.

Moreover, the new system would limit the services eligible for reimbursement
to delivery of health care—i.e., to the medical domain exclusively, social services
being excluded from the system henceforward. This redefinition of home care would
eliminate a wide range of services that are not medical in nature. For the home care
sector, this would lead to a redefinition of its mission and of the professional identity
of its personnel. Indeed, their entire core business would be modified, and thereby
the definition of the professions practiced. Comprehensive, holistic care would be
replaced by narrowly targeted, medicalised care.

4 Caregivers, Gender, and Social Roles

The development of the home care sector, together with the shortage of available
beds in nursing homes, clearly favours the maintenance of elderly people in their
homes. The combination of professional services and support provided by families
and friends has made it possible to envisage this solution serenely, particularly given
that most elderly people would prefer to remain in their own homes. However, the
profound changes taking place are likely to affect this fragile equilibrium and to
create tension amongst the different parties' investments. Indeed, the reduction of
public funding for the home care sector makes it less feasible for elderly people to
remain at home, but the shortage of available beds in institutions (and the resulting
increase in costs) forces them to do so. This paradox can only be resolved through an
increase in familial investment: thus, the burden of care is returned to the family when
financial constraints do not permit recourse to the services of private institutions.

A recent study conducted in Belgium[4] on the reduction of working hours amongst
workers aged 45 years and over (45+) was able to demonstrate the real tensions seen
in this population when subjected to the imperatives of care for family members,
particularly amongst women. Indeed, the gendered division of labour leads to the
expectation that women will play a central role in the care of the elderly, particularly
in a society in which social policies remain firmly anchored in the model of the *male
breadwinner*. For approximately 20% of workers aged 45+ wanting to reduce their
level of activity towards the end of their careers, the reasons given were family-
related, largely concerning the need to care for a family member. This percentage
was even higher (more than 30%) amongst women [4].

In this conception of *caregiving*, it is indeed the woman who is usually mobilised,
in an inegalitarian reshaping of gender relations in which the woman sees herself
assigned to traditional social roles, even in a family not initially characterised by
such inequalities [7], 151]. Thus, a breach is opened in the organisation of the family

---

[4]A research project concerned with the reduction of working hours at end of career [4], financed
by the CESI, an independent service concerned with occupational health and safety.

and in the familial investments of each member. The moral obligation to care for a relative necessarily imposes a reduction in working hours.

A family's negotiations concerning this decision therefore take place in a charged climate of competing values and normative models. They are distorted by this need for care, which disrupts the family's functioning and equilibrium. Arguments are debated, crystallising around a search for appeasement and resolution of the divergence, and there is a progressive imposition of the decision to reduce working hours through the elimination of realistic and feasible alternatives. Time will finally force the issue by necessitating action. The practical impossibility of finding satisfactory alternatives stems from a powerful moral imperative concerning caregiving. The moral weight of caregiving cannot be questioned, and guilt may strike in case of doubt: familial pressure intensifies around a kind of non-decision which reduces possibilities and locks the woman into the imposed routine of family caregiver. Caregiving seals destinies through a process of rationalisation which imposes itself as the unquestionable moral basis for assuming the role of caregiver. This is the case for Lucie, whose interview shows the extent to which she is trapped in her caregiving role.

Lucie, 52 Years, Middle Manager

What are the factors that influenced you, that made you reduce your working hours?

Let's say that, well, I had to take care of my mum. She's really not in good health. So, I made a few calculations, the salary that I had and what they offered me for half-time. Now, they did offer me a half-time, the boss was still offering me a fairly high salary range, okay, a high one. A little bit higher. So, with the half-time, from that point of view, all in all I'm not losing very much. That seemed to be a good solution, even if I would have preferred to stay at full-time. I have a brother who could also take care of Mum, but that seems impossible to him. He doesn't have time to do it. So, finally it's up to me to do it.

Did you discuss it with your husband?

Yes, because he could see clearly that it wasn't going well. At one point I cried almost every night. I'd come home and I was exhausted. I cried all the time. Family life got to be hellish too. I let things slide. You know, I've got a kid, I didn't take care of him during his school term when he was having a really hard time. I helped him afterwards, but, well, I didn't have time. It was crazy, or it was me who wanted to do too well, I don't know...

My husband helped me too. He's self-employed, so he doesn't have the same schedule either. He also comes home later.

Do you have leisure activities?

Not really. No, because I need to give my mum a lot of attention. She has ups and downs. There are days when she's really not okay. Sometimes I do a lot of cooking for nothing because it won't go down. They put a bit of intestine here, instead of the oesophagus. So, that means that food doesn't go down like for us, it's much more difficult. So, I need to grind all her meat, and still it barely goes down. I make a lot of soups, she really eats a lot of soup, so I make it at least...almost every other day, I need to make more soup. That's right, when I get home I make her some orange juice. I ask her if she's eaten her snacks, because, well, after all she's lost 30 kg. She should put on a little bit, but it's not so easy. So, she also has vasovagal attacks. I need to go to emergency, go quickly with her, because she's blocked [constipated], three days that she hasn't eaten, she vomits all day long. So I always have this stress from her sickness, it follows me. There's not a single day when I leave with my mind at peace. Because I'm afraid for her. Because I saw her, a fortnight ago, she had a vasovagal attack in the WC; it wasn't a pretty sight. So that frightened me a bit, because I said, "fortunately I was there." If I hadn't been there, what would have happened? Every day at 10 or 11 am I call her to know "are you okay?" I'm reassured, just a few words. That's hard, really hard.

This situation is typical of the conflict of values that can surround caregiving by family members. On one hand, Lucie likes her work, investing herself fully and finding personal fulfilment in it. On the other hand, she must assume her role as a woman by taking care of her mother single-handedly, her brother being only sporadically involved. The only solution for Lucie is to reduce her professional activity so that she can devote herself more fully to her mother—even if this means abandoning her job. Familial pressure, combined with the absence of public support structures, leaves Lucie very few options.

The current transformation of institutional resource allocation can only aggravate these difficult situations. Caught between the social roles of woman and of worker, these women perceive part-time work as a temporary solution that side-lines them from their careers, if not as a sacrifice.

5   Conclusions

The transformations observed in recent years in the elder care sector in Belgium testify to a true paradigm shift as defined by Peter Hall. The penetration of neoliberal logics into this sector can be analysed through different indicators, which reveal a kind of *marketization process*. Indeed, one observes at the same time an increasing presence of the private sector in retirement homes and a transformation of the public authorities' role in the management of old age. The costs associated with the ageing of the population are being progressively transferred from the public to the private sector, provoking an increase in social inequalities. Other countries are following the same pathway, including Scandinavian countries like Sweden [1, 15] that are known for their universalist model of social protection.

The consequences of these new policies are numerous, affecting elderly people themselves, their families and the professionals in the sector. The change in policy concerning home care leads to a reduction in the capacities of home care services through decreases in public funding, but also leads to a transformation of the kind of services that are funded, with holistic, global care being replaced by increasingly targeted and medicalised services. It is women, as family caregivers, who will probably need to compensate for the public authorities' disengagement by increasing their degree of involvement with the elderly. This greater involvement in caregiving may result in these women being obliged to renounce other kinds of symbolic investments, reducing their professional or leisure activities without being given a real choice in the matter.

References

1. Andersson K, Kvist E (2015) The neoliberal turn and the marketization of care: The transformation of eldercare in Sweden. Euro J Women's Stud 22(3):274–287
2. Brown T, Potoski M (2003) Transaction costs and institutional explanations for government service production decisions. J Publ Administr Res Theory 13(4):441–468
3. Burgin A (2012) The great persuasion. Reinventing free markets since the depression. Harvard University Press, Cambridge
4. Burnay N (2013) Aménagement des fins de carrière : entre reconfiguration des temps sociaux et transformations normatives. *SociologieS* [online]; consulted 08 September 2018. https://journals.openedition.org/sociologies/4440
5. Commission on Global Governance (1995) Our global neighbourhood. Oxford University Press, Oxford
6. Esping-Andersen G (1990) The three worlds of welfare capitalism. Polity Press, Cambridge
7. Esping-Andersen G (2009) The incomplete revolution. Polity Press, Cambridge
8. Hall P (1993) Policy paradigms, social learning, and the state: the case of policymaking in Britain. Comparative Polit 25(3):275–296
9. Kuhn T (1962) The structure of scientific revolutions. The University of Chicago Press, Chicago
10. Laville JL, Nyssens M (2001) Les services sociaux entre associations. L'aide aux personnes âgées, La Découverte/M.A.U.S.S./C.R.I.D.A., Paris, État et marché
11. Nyssens M (2015) L'émergence des quasi-marchés : une mise à l'épreuve des relations pouvoirs publics - associations. Les Politiques Sociales 75(1 & 2):32–51
12. Petersen OH, Hjelmar U (2014) Marketization of welfare services in Scandinavia: a review of Swedish and danish experiences. Scandinavian J Publ Administr 17(4):3–20
13. Pierson P (2000) Increasing returns, path dependence, and the study of politics. Am Polit Sci Rev 94(2):251–267
14. Pollitt C (2007) Convergence or divergence? What has been happening in Europe? In: Van Thiel S, Homburg V (dir.) new public management in Europe. Adaptation and alternatives, Basingstoke Palgrave Macmillan 10–15
15. Vamstad J (2016) Exit, voice and indifference—older people as consumers of Swedish home care services. Ageing Soc 36(10):2163–2181

Long-Term Healthcare System. Development and Future Policies for Romania

Daniela Şoitu ⓘ

Abstract In recent years, countries facing an aging population have been reviewing their long-term care structures and policies. Their goal is to develop a social and health care system that meets the needs, expectations, capabilities, but also challenges. Romania is one of these countries, investing in policies, measures and services. However, the challenges are significant at the level of institutional collaboration for integrated policies and services, at the level of current and future funding. This chapter analyse structural, financial, institutional and legislative challenges from recent documents, researches and analysis, opinions of specialists in the field of long-term health care. The author makes proposals for future policies, for legislative acts dedicated to long-term care, for working mechanisms between institutions and fields and for current and future funding mechanisms. The opportunity of these proposals will be gradually challenged by next years.

Keywords Long-term care systems · Integrated care systems · Care beneficiaries · Costs of care · Healthy ageing · Health care payments · Long-term care legislation

1 Introduction

Long-term care (LTC) is a complex field, defined and approached differently in the legislation in the field of health, insurance and social assistance. Long-term care is defined, in Law 292/2011—Law on social assistance–, in the form of "care with a duration of more than 60 days". According to the same law, LTC is provided at home, in residential centres, in day centres, at the home of the person for whom the service is provided and in the community.

In recent years, theoretical and empirical studies have been developed ([1, 32]; Şoitu 2018; [7, 14, 37]) on the opportunity of long-term care for the elderly, the organization and functioning of the two concomitant systems of evaluation and provision

D. Şoitu (✉)
Department of Sociology and Social Work, Alexandru Ioan Cuza University of Iaşi, Carol I Bvd., 11, 700506 Iaşi, Romania
e-mail: danielag@uaic.ro

© The Author(s), under exclusive license to Springer Nature Switzerland AG 2021   141
D. Soitu et al. (eds.), *Decisions and Trends in Social Systems*,
Lecture Notes in Networks and Systems 189,
https://doi.org/10.1007/978-3-030-69094-6_13

of care—social and health—for the same person. Another challenge comes from the demographic scenarios for the coming years, with implications not only on funding care, but also on preparing the system to address the demands of a larger and sicker elderly population. Romania's population aging is accompanied by another social phenomenon: the emigration of the population, especially young and adult women who emigrate to care for semi-dependent and dependent people, providing long-term personal care for work to countries such as Italy, Spain, the United Kingdom, Germany, Austria (Şoitu 2018).

In Romania the access to services adequate to the needs depends on the offer of the public and private providers, on the map of the availability of the services. A significant imbalance exists, in terms of supply, between the residence environments, the rural one being disadvantaged. An alternative exists through the "professional personal assistant", paid and certified, since 2018, by the relevant institutions in the field. A new methodology for monitoring the quality of care provided by the professional personal assistant [22] comes to balance the observance of the standards of care both in centres and in the families (Şoitu 2018; [34]).

The local public authorities in Romania have set up and finance 114 care centres for the elderly, while the private sector has covered the need for care in centres to an extent three times greater (340 centres) [23]. At the same time, the SHARE data (2019) highlight the fact that over 50% of people over 50 have limitations in the exercise of daily activities: 15.7—very limited, and 35.7%—limited. The share of people over the age of 50 with long-term health problems reaches 45%; of these, 48% are in rural areas and 43% live in urban areas. The same data source (SHARE 2019) shows a sufficient and poor health status among 42% of men and 55% of women over 50 years of age [37].

Europe has shared policy directions and definitions, but the practices and method-ologies are adapted to each state's specificities. There are several sources of critical analysis regarding the responsibility for providing and funding integrated services, adjusted and adapted to the individuals' needs and wishes. The European social models have found or are looking for various versions of an answer to this issue: some states make the individual directly responsible, by requesting the accumula-tion of amounts destined for mitigating the "risk of dependence in old age", while others view the dependence holistically, and others make the individuals indirectly responsible, by collecting and managing globally income tax (Şoitu 2018).

At global level, World Health Organization has warned in recent decades that long-term care needs will continue to grow under the influence of several factors: the evolution of the family structure, the geographical mobility of the active population, the rapid process of demographic aging. Romania had, according to the data of the National Institute of Statistics, on January 1, 2019, a percentage of over 34% elderly population (23.10% in the age group 60–69 years, and 12.93% over 70 years), and these aspects are becoming important in our country, where the care of the elderly was traditionally provided by families [6, 32, 38].

The major problem appears to be the increase in the number of individuals over the age of 70 and above (10.92% of Romania's total population as of 1 January 2019), a stage viewed by the Anglo-Saxon literature as that of "old-old" people (Hooyman and Kiyak 1996), and by the French one as "the fourth age", the "age of dependence" [3, 29, 33].

Thus, social and health policies face a new challenge, that of caring for semi-dependent and dependent elderly people for long-term ([31]; Şoitu 2018). The responsibility of care, according to the current Romanian legislation, goes to the individual, the family, the community. Meanwhile, the poverty rate in Romania and the fluctuations of the job market, the circulatory migration for work, as well as the desire many citizens have for rapidly reaching a high quality of life, are factors that influence negatively the individuals' decision to pay for a LTC insurance.

According to EUROSTAT [5], Romania has the highest percentage of people aged 65 and over who have declared unmet health care needs (22.7% in 2016, in Romania, compared to 5.5% EU average). The unmet health care indicator refers to the person's own assessment of the need for medical examination or treatment, but which has not been met for one or more of the following reasons: (1) financial reasons, (2) long waiting list or (3) too great a distance from the supplier. Healthcare refers to individual health care services (medical examination or treatment, excluding dental care) provided by or under the direct supervision of doctors, in accordance with national health care systems [30, 37].

The SHARE data (2019) show that almost a quarter of Romania's population aged 50 and over negatively assesses their health. Also, the same data indicate a significant differentiation by gender in the (self) assessment of health status. Despite the more positive (self) assessment of the health status of Romanian men (21.3% negatively assess their health status), their life expectancy is, according to the National Institute of Statistics, 72.28 years, while the lifespan of women in Romania is approximately 7 years higher (79.24 years) (27.2% negatively assess their health) [37].

In Romania there are significant differences in the (self) assessment of the general state of health depending on the development regions. Thus, in Macroregion 1 (which includes the North–West and Centre region), one-third of the population evaluates negatively their health status, over 50% declaring that they suffer from various long - term diseases, this being also the region with the lowest life expectancy. Diametrically opposite, in Macroregion 3 (which includes the South and Bucharest region) less than 20% of the population evaluates negatively their health status, 34% declaring that they suffer from various long-term diseases, this being also the region with the highest life expectancy. Rather, this explanation covers a set of hypotheses, from varying degrees of education and awareness of the disease, to easy access to medical and care services, to quality and lifestyle [39].

The SHARE data (2019) show a different assessment of health status depending on the environment of residence in Romania. People in rural areas tend to have a poorer self-assessed health status (26.91%) to a greater extent than those in urban areas (19.88%). The difference is maintained when we refer to the share of people suffering from long-term health problems: 48% in urban areas and 43.4% in urban areas [37].

The quality of LTC is a concern in Romania, but also in other European Union countries. Experts participating in a recent research [20] mentioned the shortage of medical staff, especially geriatricians and nutritionists. In fact, the entire care staff market is deficient in Romania. There are many requests for ongoing staff training, especially to meet the care needs of people with dementia. For the professionalization of human resources in the field of long-term care, the same experts highlight the need to obtain a professional certificate in the field of LTC for the staff of residential centres. Also, guidelines that include basic hygiene actions, and preventive actions to prevent diseases may cover the need for regulated measures regarding the safe provision of medical services.

Prevention of degradation and monitoring of the health of the elderly in centres, exercises and cognitive-motor training are preventive measures in Austria, Germany, along with programs to prevent abuse in LTC provided in centres through staff training and education. Particular attention is paid in Austria to the external social audit of the quality of LTC services in order to obtain the National Quality Certificate (optional). The model is also provided in the Romanian legislation.

Out of the good practices analysed, the case of the United Kingdom stands out; in the period 2014–2018 it carried out programs to organize the operation and financing of the LTC system for the next five years (EHCH). In terms of actions that ensure better monitoring of the elderly who benefit from LTC for the improvement and limitation of deteriorating health, from the good practices analysed, we mention: (1) flexible ways of hiring informal caregivers (vouchers, flexible working hours) in especially for informal immigrant carers (Italy, Austria); (2) adapting the training offer and extending the period of activity for adults and the elderly able to provide care (aging workforce) (Germany); (3) Actions to optimize the working environment for employees (Germany, Austria); (4) major investments in training human resources for the use of assistive technologies (caregivers and the elderly) (UK, Germany, Denmark, the Netherlands), (5) major investments in assessing, preventing and monitoring the health of the elderly in centres using new technologies—Head-mounted technology, Telehealth, Telecare (the UK, Germany, Denmark, the Netherlands, Finland), even in robots that replace part of the activity of the employees of the centres (SARA) (Finland, the Netherlands, Germany) [37].

Although the availability of data has improved and the life expectancy has grown across the EU, an analysis of the recent reforms of national policies indicates that many member stats do not plan adequately for the future the needs of the long-term care system, relying on informal workers and/or in limited cash support for the beneficiaries of care, thus facilitating the use of care provided by illegal migrants. At the same time, many countries expand the offer of services through market mechanisms, contracting states services to not-for-profit providers. This underlines the complexities of developing the LTC sector and proves the need to have the certainty that the obligations of the member states in terms of human rights will remain high on the political agenda (Şoitu 2018).

Romania is one of the countries searching for the best version of response to the threat of population aging, of increased life expectancy, despite it not being a healthy life, on the backdrop of a negative natural increase rate. Throughout the year 2018, the Romanian Government, through its fiscal and medical specialists, has submitted for public analysis the idea of additional contributions to the state budget, corresponding to packages for care in old age or in case of temporary or permanent incapacity. The proposal was not welcomed by the public and was never put into practice. Our analyses support this perspective, within the realisation of the actions of increasing awareness about the determinants of health status and within the national programmes for the yearly assessment of health status.

2 Institutional Long-Term Care Working Mechanisms in Romania

In recent years, in Romanian there have been established inter-ministry and inter-sector committees, commissions and workgroups, with various levels of activity. Also, in area of social and medical services in the long-term care system there exist collaboration structures at inter-ministry level, as well as at the level of the local public authorities. The specialist akin part in a recent Delphi method consultation [20] have stressed the need to have a structure at national level in order to manage the LTC domain (ensuring the regular collection of data, the monitoring and the reporting regarding the LTC service system). The existence of such a structure, which would manage the data regarding LTC would fulfil the need for predictability in the medium and long term development of the LTC system.

Effective and real work mechanisms are needed in LTC, involving the existing inter-ministry commissions and committees such as:

- the National Commission on Social Inclusion (the law governing the commission's operation [9];
- the National Commission for the Supervision of the Health Computer System and for Reporting to the Organization for Economic Cooperation and Development (O.E.C.D.), and inter-ministry structure without legal personality, with a consultative role in the coordination of the National Authority for Quality Management in Health (established through the *Prime Minister's Decision nr. 162 of 25 July 2019*);
- the National Commission for Strategy and Prognosis—a specialised body of the central public administration, with legal personality, subordinated to the General Secretariat of the Government and the coordination of the Prime Minister (the Commission functions and carries out its activity based on the *Government Ordinance no. 22/31.01.2007*);
- the National Commission for Population and Development [12];
- the National Commission for equal opportunities between men and women [11];
- The inter-ministry committee regarding the 112 emergency calls system [8].

The way these inter-institutional structures work can be improved, starting from needs, but also from positive experiences in the United Kingdom and in Norway, where regional system for coordinating social and medical care exists) [35, 36].

The quality of LTC services will increase by establishing functional mechanisms for collaboration between institutions and specialists. The institutions involved in the application of this action are administrative-executive—the Ministry of Health, the Ministry of Labour and Social Protection, the Ministry of Public Works, Development and Administration, local public administrations—, as well as those involved in promoting the rights and responsibilities of citizens, professional associations etc. The main responsibility for organising and running such a national programme could go to an inter-ministry entity (Commission, Committee) for LTC. The Ministry of Labour and Social Protection, together with the Ministry of Health would become involved in the design, development and adjustment of public policies regarding LTC (Şoitu 2018).

The direct beneficiaries of such an action would be institutions (the Ministries of Labour and Social Protection, and of Health, respectively). The indirect beneficiaries are the beneficiaries of long-term care and their families, system staff (caregivers, employees in residential and non-residential services for the elderly), but also local and central authorities. The action could be financed from the budgets of the institutions involved, loans from the EBRD, the World Bank, within the operational programme Administrative Capacity.

3   A Law for Structuring the Long-Term Care System

Since 2000, the social protection system for the elderly in Romania has benefited from a legislative and organizational framework built by the contribution of experienced specialists from the country, but also in accordance with European standards. The Law regarding the social assistance for the elderly [17] was followed by the Gerontological Grid [10], by the Grid for the evaluation of elderly individuals requesting admission to medico-social centres (approved through the Joint Order of the Ministry of Labour and Social Protection and the Ministry of Health and Family, nr. 180/491/2003—for the approval of the *Grid for the medico-social evaluation of elderly individuals requesting admission to medico-social centres*), and by specific laws regarding quality in the area of social services (Law 197/2012 on quality assurance in the area of social services and Order no. 29/2019 for the approval of the Minimum Quality Standards for the accreditation of social services for the elderly, homeless people, young people who have left the child protection system and others categories of adults in difficulty, as well as for services provided in the community, services provided in an integrated system and social canteens, quality standards for care at home and in residential centres). Alternatives to residential care, although not as numerous as those in the child protection system, have allowed steps in this development. The Ministry of Labour and Social Protection has allowed, starting with 2017, the online consultation of updated data regarding the network of services and licensed and accredited providers. In November 2019, beyond those 114 public centres and 340 private centres for the elderly, there were 4 crisis centre, 17 protected living spaces and other licenced socio-medical centres.

The Social Services listing (approved through GD 867/2015) has been updated with new types of services, and the legislation in the field of quality of social services has seen updates of standards, the last entering into force in January 2019 (MLSP Order nr. 29). LTC has been monitored by the European Commission both through the open method of coordination, as well as through the new European platform for preventing and combating poverty and social exclusion.

Romania has taken significant steps in the last 10 years in terms of medium-term priorities that it has established through various documents (*The National Strategic Report 2008–2010*), strengthening the institutional capacity of local communities to provide social and long-term care services. The issue of community and family responsibility, as the main actors in the provision of LTC services, needs clarifications, funding sources, funding prioritization criteria. Sitting among states with a consistent history of informal care systems, Romania now faces, like other European countries, multiple challenges (Şoitu 2018; [38]).

The experience of other states, Slovenia, the United Kingdom, Germany, in structuring the long-term care system can support the process of elaboration and implementation of a piece of legislation in this respect.

The need for LTC will therefore be increasing, and a piece of legislation for structuring the system is absolutely necessary. A piece of legislation dedicated to LTC would establish an integrated approach, responsible institutions, beneficiaries, funding arrangements, as well as various actions for the provision of care services by combining formal and informal services.

Fragmented actions exist, included in documents issued by institutions that have various roles in LTC:

- *The National Strategy for the Promotion of Active Aging and the Protection of the Elderly for 2015–2020 of the Operational Action Plan for 2016–2020 and of the Monitoring and integrated evaluation Mechanism* [15] also provided that "Better coordination is needed in the provision of LTC services between the Ministry of Health, the Ministry of Labour, Family, Social Protection and Elderly people, County and local councils, private providers of medical and social services, hospitals, general practitioners, pharmacists, informal carers, nurses, home caregivers, social work technicians and people in need of care." (p. 26) and also "a unified LTC system" (Sect. 3.3.1).
- *The Institutional Strategic Plan 2018–2021 of The Ministry of Health* [40] includes an action (2.2.3.1.) on: "Development of long-term health care services, rehabilitation/medical recovery, care at home, palliation". The outcome indicators take into account a piece of legislation regarding the medium and long term National Plan regarding medical assistance in rehabilitation, in palliative and long-term care.";

The direct beneficiaries of piece of legislation regarding LTC would be the beneficiaries of long-term care and their families, the LTC service providers, staff in the system (carers, employees in residential and non-residential services for the elderly), as well as local and central authorities. The indirect beneficiaries would be the subsystems in the fields of health, labour and social protection, administration and interior, respectively the departments, services from the Ministry of Labour, and Social Protection, the Ministry of Health, the Ministry of Public Works and Development. Also: professional training providers, universities and post-secondary schools, which will design their training offer in accordance with the needs and the new institutional framework offered by this piece of legislation; providers of medical products and services; associations attached to religious organisations.

The institutions involved in the implementation of the action are those with a legislative role, the Romanian Parliament, the administrative-executive ones in the central apparatus, the Ministry of Health, the Ministry of Labour and Social Protection, the Ministry of Public Works, Development and Administration, the Insurance Administration Authority, the administrative-executive ones in the local apparatus, local public administrations, as well as those involved in promoting the rights and responsibilities of citizens: trade unions, business owners associations, professional associations, associations of families of caregivers, associations of the elderly, associations of private providers of care services.

4   Funding the LTC Services

Romania is one of the European countries that allocates the lowest percentages from GDP for health (4%), the European average being 7.1% of GDP. Moreover, Romania allocates insignificant percentages of GDP for research and development in health, while the European average is 0.1%. Austria is a clear leader in public health spending, allocating 8% of GDP [5]. Romania's public spending on social protection (11.6%) is well below the European Union average of 19.1% of GDP. As a consequence, Romania allocates one of the lowest percentages of GDP for the elderly (8.4%), the European average being 10.2%. In fact, the most substantial budgetary allocations in terms of public spending across the entire European Union on social protection go to aging (approximately 53% of public spending on social protection, mostrly insured pensions actualy). Due to the small percentage of GDP allocated to social protection and to the pressure generated by the issue of aging, in Romania expenditures on the elderly represent over 70% of total social protection expenditures [5].

The inventory of current LTC expenditures in Romania involves the selection of several categories of medical, recovery, social, palliative services. Amounts from official documents [15] present Romania's total public expenditures for LTC as being 0.69% of GDP (in 2011).

The data obtained from specialist by researchers in area (Matei and Ghenţa 2019) highlights the following aspects that support actions for financing the current LTC subsystems, the health component: creation of a special funding mechanism for social and medical services for the elderly, significant co-financing of services from the state budget, investing in supportive technologies specially designed for the elderly, establishing a long-term insurance that recognizes the incidence of neuro-degenerative diseases and of other categories of diseases with an increased incidence among the elderly.

According to EUROSTAT [5], Romania faces one of the highest rates of poverty risk among people aged 65 and over (19.1%), compared to the EU average of 14.6%. The gender gap in the risk of poverty among people aged 65 and over is also among the highest in Europe, at 11.5% against women, compared to the EU average of 4.7%. The average equivalent net income per household in Romania is extremely low (2746 eur, EUROSTAT [5]), compared to the European Union average (18,837 eur, EUROSTAT). The equivalent average net income per household is almost seven times lower in Romania than in the European Union. This situation is also underlined by the volume of GDP per capita index expressed in standard purchasing power. The rate of severe material deprivation in the case of people aged 65 and over, i.e. the share of the total population of people who, due to lack of financial resources, can not afford to purchase or pay for basic goods or services such as utilities, staple foods, appliances etc., is very high in Romania, 22.5%, compared to the EU average of 5.8% [5]. Also, 73.8% of Romanians aged 65 and over are unable to cope with an unexpected expense, compared to the EU average of 41%.

The analysis of data collected by the *European Quality of Life Survey* (Eurofound, EQLS 2016) supports a correlation of data on satisfaction with current life and appreciation of the LTC system, but also a group of rather hypercritical countries. Although people aged 65 and over in Romania (self) rate their satisfaction with life as very low (5.7 points), the evaluation of the quality of care services in Romania tends to be more in an area of social benevolence (6.0 points).

EUROSTAT presented, in 2016 (according to Spasova et al. 2018), the situation of payment possibilities for LTC services: over 80% of households in ten Member States reported difficulties in covering the costs of professional home care services, especially in Slovakia (95%), Lithuania (94%) and Greece (93%). In contrast, the largest share of households claiming to be able to make these payments easily was in Finland—75%, Sweden—73% and Denmark—69%. According to EUROSTAT data from 2016, Romania is one of the countries with the lowest rate of use of professional home care services (6%), along with Bulgaria and Estonia (both 5%). The highest share of households using professional home care services was found in Luxembourg (88%), the Czech Republic (58%) and Denmark (54%). At European level, the average household with people in need of help due to long-term health problems and who have used professional home care services is 20%.

A recent survey [20] conducted in residential centres with permanent and semi-permanent care for the elderly in the country has indicated that two thirds of the elderly can not take care of themselves, and one in four individuals is dependent on care. Meanwhile, the supply of residential services is not evenly distributed across the country. Economically disadvantaged areas, such as the North-East and South-East Region, also have the fewest residential services for the elderly. This creates inequities in accessing services, as the waiting lists in these areas are longer.

In order to ensure the quality of health care for the elderly, to respect human dignity and increase the responsibility of family and institutional actors, a new legal framework is neccessary. The flexible funding programs from other European countries can be positive experiences for sustainable frameworks for securing the financing of LTC services.

The Ministry of Labour and Social Protection would be an indirect beneficiary, as these services reduce and mitigate some of the social vulnerabilities and risks. The Ministry of Health would be an indirect beneficiary, as the funds from the state public budget and from the insurance fund could also be allotted for national programmes for the prevention of disease and for promoting a healthy society. The central and local public authorities will be able to design active policies and budget chapters, investing, in the short, medium and long term, in the development of the employment market and of the quality of life of the inhabitants.

The financing of long-term care services, in the short term, could follow the model of the national programs of the Ministries of Health and that of Labour and Social Protection. Another source of funding would be the National Health Insurance Fund (NHIF), other social insurance types, contributions from the local public authorities, contributions from people in need of care and their families. The best impact will be reached through a programme with a rapid start, but with long-term objectives and plans, run for as long a necessary. The costs of managing the insurance should be included in the budgets of the Ministry of Health and of the Ministry of Public Finance, as well as of the Ministry of Labour and Social Protection. The most rapid impact will be reached by recognizing or equating the private insurance already contracted by citizens and private insurers for LTC or for diseases with incapacitation potential.

Institutions involved in the application of this action can be: The Romanian Parliament, the Ministry of Health, the Ministry of Labour and Social Protection, the Ministry of Public Works, Development and Administration, the Ministry of Public Finance, the Insurance Administration Authority, trade unions, business owners associations, professional associations, associations of families of caregivers, associations of the elderly, social services providers, providers of complex services and care products, religious organisations.

5 The Long-Term Care Insurance

The fact that in Romania, according to the SHARE (2019) data, almost one-third of the population aged 50–60 assess their own health status as "sufficient" (20.1%) and "poor" (12.7%) indicates a failure of the health and social protection systems, as well as the need for integrated prevention actions. The higher the age, the more negative the health status evaluation is. The health status of individuals over the age of 70 is self-assessed as "very poor" by 43.52% of them. The life interval 50–70 is a critical, difficult one, dominated by an increase in dependence. Health status is also influenced, according to the SHARE data, by the individual's marital status.

Thus, married individuals state that they have a better health status (almost 55%) than single individuals (about one third) (unmarried or widowed individuals) [27]. As a good practice example, Germany has mandated yearly mandatory health assessment programmes as primary and secondary prevention activities. The citizens have thus increased responsibility for monitoring their own health status, they are invited as partners in ensuring their own wellbeing. Not taking part in such programmes results in sanctions such as covering out of their own pocket the cost of medical services for diseases whose signs could have been identified early through a yearly check-up.

In Romania, the answers of the experts who took part in a Delphi methodology organised within a recent project [20] support the design of mechanisms that would support the predictability of the development of the long-term care system, insisting, for the health component, on the establishment of a long-term care insurance that would recognize the incidence of neurodegenerative diseases and of other categories of diseases with high incidence among older individuals.

The analysis of good practices shows Germany to stand out due to the fact that, through its 1994 *Long-term Care Act* has established a legal scheme for insuring LTC, specifying the benefits and the coverage for all those who require LTC. The long-term care insurance (LTCI) was also introduced in the United Kingdom in 1991. The insurance may be granted in those cases in which care will be required at some point in the future (pre-financed) or just once, when needed (point of need or immediate care). The long-term care insurance (LTCI) is designed in such a manner that a number of amounts are paid regularly, usually every month, whenever an individual (as a rule an older individual) requires long-term care. Long-term care is required whenever an individual cannot look after themselves, for instance as a result of a stroke or after becoming very fragile. These amounts are used for purchasing services either in the home, or in a residential care institution. Most policies are designed in order to have the payments made while the individual who requires care is alive.

The funding mechanisms are needed now, for limiting the deterioration and improving the health status of the elderly receiving LTC and for the future, as un insurance.

The legislation in the field of quality assurance in social services, respectively the conditions for the licensing and accreditation of social service providers, the protocols of the Ministry of Health in the field. The direct beneficiaries of LTC insured services would be the current beneficiaries and their families, as well as the local public authorities. The Ministry of Labour and Social Protection would be a direct beneficiary, supporting the quality standards in LTC services. The Ministry of Health is also a direct beneficiary, with the framework for monitoring the quality of care services and preventing and limiting the deterioration of the health of long-term patients. The local public authorities are the direct beneficiaries, having the responsibility to ensure and monitor the quality of care.

Potential funding sources: public or private funds, of the bodies that manage the care services; NHIF, programs for the development of human capital, for digitization and development.

Institutions involved in the application of this action: The Romanian Parliament, the Ministry of Health, the Ministry of Labour and Social Protection, the Ministry of Public Works, Development and Administration, the Ministry of Public Finance, the Insurance Administration Authority, trade unions, business owners associations, professional associations, associations of families of caregivers, associations of the elderly, social services providers, providers of business services and of LTC products.

Potential funding sources for LTC insurance: taxes managed by the state or by the private system; private health insurance, a budget for managing insurance. The best impact will be reached through a coherent programme run for a minimum of ten years, reprised and adapted for as long a necessary. Based on the current institutional structure, the costs of managing the insurance should be included in the budgets of the Ministry of Health and of the Ministry of Public Finance, as well as of the Ministry of Labour and Social Protection. The resources collected through this form of insurance will require separate management, being destined for the insured risk. The most rapid impact will be reached by recognizing or equating the private insurance already contracted by citizens and private insurers for LTC or for diseases with incapacitation potential. The main responsibility for organising and running such a national programme would go to the Ministry of Health. It would have as partners the local public authorities, as well as bodies that manage health insurance in both the public and the private system, trade unions, business owners' associations, other associations from the civil society, social services providers. The Ministry of Labour and Social Protection would be an indirect beneficiary, supporting the access to this programme of vulnerable individuals, groups and communities.

6 Conclusions

Establishing ways of working and collaborating between the institutions involved in the long-term care system will directly contribute to increasing the quality of services, but also to the design, development and adjustment of public policies regarding the improvement of the quality of LTC services. The experience of ongoing collaboration of the Ministries of Labour and Social Protection, respectively of Health in programs financed for the development of the social assistance sector provides frameworks for the development of an efficient and effective structure also in the field of LTC.

The direct beneficiaries of the LTC insurance would be adult individuals, today's children and youngsters. The Ministry of Labour and Social Protection would be an indirect beneficiary, the LTC insurance supporting, in the medium and long term, the design, development and adjustment of public policies regarding the improvement of the quality of services for vulnerable individuals and groups. The Ministry of Health would be an indirect beneficiary, as the funds from the state public budget and from the insurance fund can also be allocated for national programmes din for preventing diseases and for promoting a healthy society. The central and local public authorities will be able to design sustainable economic policies, with lower healthcare costs for their own budgets and with higher potential of investment in local and regional development.

Many of the troubles with a significant weight among the causes of mortality and morbidity, once assessed and identified early, can be managed, eliminating the causes or limiting their aggravation. Also, the correlation of the results of the yearly health checks with socio-economic risk factors can support an early intervention, resulting in effective activities of primary and secondary prevention.

References

1. Bălaşa A (2005) Îmbătrânirea populaţiei. Provocări şi răspunsuri ale Europei. (Population ageing. Challenges and answers of Europe) Calitatea vieţii, XVI(3–4):273–288
2. Börsch-Supan A (2019) Survey of health, ageing and retirement in Europe (SHARE) Wave 7. Release version: 7.1.0. SHARE-ERIC. Data set. https://doi.org/10.6103/share.w7.710
3. Caradec V (2001) Sociologie de la vieillesse et du vieillissement. Ed. Nathan, Paris
4. Eurofound (2018) EQLS 2018. Quality of life and public services
5. EUROSTAT (2018) Data Available at https://ec.europa.eu/eurostat/data/database
6. Gîrleanu-Şoitu D (2006) Vîrsta a treia [Third Age]. Institutul European, Iaşi. ISBN 973-611-391-4
7. Glendinning C (2018) ESPN thematic report on challenges in long-term care. ESPN, European Commission, United Kingdom
8. Government Decision nr. 166 of 27 July 2019 regarding the establishment, organisation and functioning of the 112 emergency calls system
9. Government Decree nr. 1217 of 6 September 2006 regarding the establishment of a national mechanism for promoting social inclusion in Romania
10. Government Decree nr. 886/2000 for the approval of the National grid for the evaluation of the needs of the elderly
11. Government Decree Nr. 933 of 27 November 2013 for approving the Regulations for the organisation and functioning of the National Commission for equal opportunities between men and women—CONES
12. Government Decree Nr. 997 of 2 September 2009 regarding the establishment, organisation and functioning of the National Commission for Population and Development
13. Government Ordinance no. 22/31.01.2007, republished, with subsequent amendments and completions
14. Greve B (ed) (2017) Long-term care for the elderly in Europe: development and prospects. Routledge/Taylor and Francis
15. Guvernment Decision (2015) Decission no 566 from 15 July 2015 refering the approuval of The National Strategy for the Promotion of Active Aging and the Protection of the Elderly for 2015–2020, of the Operational Action Plan for 2016–2020 and of the Monitoring and integrated evaluation Mechanism for these
16. Joint Order of the Ministry of Labour and Social Protection and the Ministry of Health and Family (2003) Order nr. 180/491/2003 for the approval of the Grid for the medico-social evaluation of elderly individuals requesting admission to medico-social centres
17. Law 17/2000 on Social Assistance for the Elderly
18. Law 197/2012 on quality assurance in the area of social services

19. Lupu A, Şoitu D (2017) Resource allocation, personal and practical experiences of providers and beneficiaries of social and health care services for elderly and disabled persons. A qualitative analysis. Analele Ştiinţifice ale Universităţii "Alexandru Ioan Cuza" din Iaşi (serie nouă). Sociologie şi asistenţă socială, tom X no.1. Editura Universităţii "Alexandru Ioan Cuza". Iaşi, pp 67–78, ISSN print 2065-3131. Available at: https://anale.fssp.uaic.ro/index.php/asas/article/view/469
20. Matei A (ed), Ghenţa M, Mladen-Macovei L (2019) Monitorizarea şi evaluarea calităţii îngrijirii în instituţiile care furnizează servicii de îngrijire de lungă durata destinate persoanelor vârstnice [Monitoring nd evaluation of the quality of care in long-term care facilities for elderly]. Editura Universitara, Bucureşti
21. Ministry of Labour and Social Protection (2008) The National Strategic Report 2008–2010 on social protection and social inclusion
22. Ministry of Labour and Social Protection (2018) Order no. 1690 of 21.05.2018 for the approval of the procedure for monitoring and inspecting the activity of the professional personal assistant
23. Ministry of Labour and Social Protection (2019) Licenced and accredited providers for long-term care services in centres. Available at www.mmuncii.ro. Accessed November 2019
24. Ministry of Labour and Social Protection Order no. 29/2019 for the approval of the Minimum Quality Standards for the accreditation of social services for the elderly, homeless people, young people who have left the child protection system and others categories of adults in difficulty, as well as for services provided in the community, services provided in an integrated system and social canteens, quality standards for care at home and in residential centres
25. National Institute of Statistics (2019) Demographic data 2019
26. Prime Minister's Decision nr. 162 of 25 July 2019 regarding the establishment of the National Commission for the Supervision of the Health Computer System and for Reporting to the Organization for Economic Cooperation and Development. Official Gazette, issue 619 of 25 July 2019
27. Râşcanu M, Şoitu D (2016) Etape de viaţă, calitatea relaţiei maritale şi starea de bine a vârstnicilor. în Şoitu D, Rebeleanu A (eds) (2016) Noi perspective asupra cursului vietii. Cercetări, politici şi practici. [New perspectives on the life-course. Researches, policies and practices]. Editura Universităţii Alexandru Ioan Cuza din Iaşi, pp 82-114. ISBN 978-606-714-272-3
28. Social Protection Committee of European union and European Commission (2015) Social Europe—aiming for inclusive growth—annual report of the social protection committee on the social situation in the European Union (2014). Available at https://ec.europa.eu/social/main. jsp?catId=738&langId=ro&pubId=7744&furtherPubs=yes
29. Şoitu D, Bârsan C (2012) Giving and doing: identity perceptions and images of elderly women in the rural environment. Postmodern Open. 3(3):69–88
30. Şoitu D, Rebeleanu A (2011) Vulnerabilities of the socio-medical legal framework for the elderly. Revista de Asistenţă Socială (3/2011):137–152. ISSN 1583-0608
31. Şoitu D, Rebeleanu A, Oprea L (2013) Social and medical public policies, between "Vulnerabilities" and "Risks". Revista Română de Bioetică XI(1):123–133
32. Şoitu D (2014) Being healthy means being educated and acting accordingly. Procedia Soc Behav Sci 142(2014):557–563
33. Şoitu D (2015) Social vulnerabilities and ageing. Analele Ştiinţifice ale Universităţii "Alexandru Ioan Cuza" din Iaşi (serie nouă). Sociologie şi asistenţă socială, tom VIII(2), pp. 73–82. Editura Universităţii "Alexandru Ioan Cuza", Iaşi. ISSN print 2065-3131. Available at: http://www.phil.uaic.ro/index.php/asas/article/view/394
34. Şoitu D (2020a) Social work system and social innovations in Romania. In: Sarasola Sanchez Serano JL et al (eds) Challenges and opportunities in qualitative and quantitative models in socio-economic systems and social work. Springer Nature, Switzerland AG, pp 17–24. https://doi.org/10.1007/978-3-030-18593-0_2, ISBN 978-3-030-18593-0
35. Şoitu D (2020b) In: Sarasola Sanchez Serano JL et al (eds) Researching ageing by the life course perspective in qualitative and quantitative models in socio-economic systems and social work. Springer Nature, Switzerland A, pp 83–88. https://doi.org/10.1007/978-3-030-18593-0_7, ISBN 978-3-030-18593-0

36. Şoitu D, Johansen K-J (2017) The space of innovation and practice on welfare, health and social care education and practice in Romania and Norway. Analele Ştiinţifice ale Universităţii "Alexandru Ioan Cuza" din Iaşi (serie nouă). Sociologie şi asistenţă socială, tom X(1). Editura Universităţii "Alexandru Ioan Cuza", Iaşi, pp 5–17. ISSN print 2065-3131. Available at: https://anale.fssp.uaic.ro/index.php/asas/article/view/456

37. Şoitu D, Matei A (eds) (2020) Îngrijirea de lungă durată. Practici, măsuri, politici (Long-term care. Practices, measures, policies). Editura Universităţii Alexandru Ioan Cuza din Iaşi, Iaşi

38. Şoitu D, Şoitu CT (2020) In: Romania in Ní Léime Á et al (eds) Extended working life policies. International gender and health perspectives, pp 385–394. https://doi.org/10.1007/978-3-030-40985-2, ISBN 978-3-030-40984-5

39. Svarcova I, Hoskova-Maycrova S, Navratil J (2016) Crisis management and education in health. In: The European proceedings of social and behavioural sciences EpSBS, vol. XVI, pp 255–261. http://dx.doi.org/10.15405/epsbs.2016.11.26

40. World Bank (2018) The institutional strategic plan 2018–2021 of The Ministry of Health. WB

Welfare, Rights, Decisions

Antonella Sciarra ⓘ

Abstract The chapter illustrates welfare state pathways related to the recognition of social rights in various historical cycles whose main stages are reconstructed showing a constant increase in social protection offered by the state. But recent changes in political values in the society of the globalized economy pose obstacles and deviations to the advancement of social rights protected by the state, to which the rights of the private market are opposed. The study presents a variety of ethical-legal cases of these conflicting rights between state and market on which a crisis and a deconstruction of the welfare system depends. This chapter explores a proposal to solve the dilemma between the two antagonistic decision models, on the one hand the reasons for solidarity welfare, on the other the calculation of profit in the market, advancing a solution of their conflictual relationship through a rational model of the ecological decision, which regulates sustainable limits for their complementarity and circular integration.

Keywords Welfare state · Social rights · Conflict of rights · Rational welfare decisions · Rational market decisions · Ecological decisions

1 Introduction

Welfare is based on the rationality of distributive equity against utilitarian rationality. The utilitarian rationality claims the right of each individual to decide the maximization of his profit in competition with others in the market, on equal terms so that he wins the merit. In truth there are no equal conditions and selection of merit in market competition, because for example being born rich or poor or intelligent or handicapped are conditions that depend on the fate that guides the results of the market. Each person may find themselves at risk of advantageous or disadvantageous conditions in the lottery of life.

A. Sciarra (✉)
University of Teramo, Via Renato Balzarini n.1, 64100 Teramo, Italy

© The Author(s), under exclusive license to Springer Nature Switzerland AG 2021   157
D. Soitu et al. (eds.), *Decisions and Trends in Social Systems*,
Lecture Notes in Networks and Systems 189,
https://doi.org/10.1007/978-3-030-69094-6_14

In the face of risk, a more reasonable and advantageous solidarity decision is emerging than the individual utilitarian choice. The solidarity decision concerns a social contract which foresees to redistribute the natural and social resources available to the community, in order to face the risk of the disadvantaged conditions in which everyone can find themselves. A distributional equity is asserted whereby resources are redistributed by the state to protect those who are most disadvantaged.

The rational decision of the social contract against the risk of disadvantage that can affect anyone in the lottery of life, is an insurance to the essential rights of life of every possible disadvantaged person under the solidarity of the welfar state. The welfare state regulates the uncertainty of the economic market with the redistribution of wealth to protect the lives of the most disadvantaged, in order to restore fairness of conditions of freedom, equality, solidarity, among all members of the community [34].

Welfare is therefore the product of a rational insurance decision against the risks of social disadvantage that fate can reserve for anyone. The social contract that expresses the rational political motivation to ensure the essential right to life for the most disadvantaged is codified in the Constitutional Charter of the rights of the welfare state.

The welfare of the Italian Constitutional Charter of 1948 protects the inviolable and universalistic right to the integrity of every person's life, proclaiming the fundamental right to health, which in article 32 involves the commitment of the Republic to free care: "the Republic protects health as a fundamental right of the individual and an interest of the community, and guarantees free care to the poor". The welfare of the Constitution governs the social pact on the values of freedom, equality, solidarity, three principles of modern public ethics of social justice, which the Constitution translates into substantial economic, political and social rights and obligations [22]. Constitutional welfare prescribes mandatory duties of substantial solidarity and in fact the Republic in article 2 "requires the fulfillment of mandatory duties of political, economic and social solidarity". In article 3, the Republic declares the rights of freedom and substantial equality, assuming the task of "removing the obstacles of an economic and social order, which, by effectively limiting the freedom and equality of citizens, prevent the full development of the human person. and the effective participation of all workers in the country's political, economic and social organization". In this way, the Constitution finalizes welfare to the rights of freedom from need, to the reduction of inequalities, to the duties of solidarity, operating for the priority objective of the full development of the human person and for his effective economic, social and political participation, in freedom, equality, solidarity.

The resources of the welfare policies are drawn from the contributory capacities of all citizens with a principle of social equity, being the tax system set with progressive criteria according to article 53, so that in proportion the richest taxpayers pay more for the welfare services for the indigent, according to a model of democratic solidarity [18].

2 Development Cycles of Welfare State Rights

The welfare state development cycles have been marked by innovative rights recognized by socio-political changes, that promote values for the expansion of life rights of marginal classes [41]. The growth of the welfare state has been a gradual expansion of laws and constitutions for the protections of the weakest sectors of society, the evangelical care of the poors, in the various forms of rights for necessities of life, from income support to the social securities of work, from health to education, from housing to indigence.

Social assistance in Europe was set up with the *Act for the Relief of the Poor* by Elizabeth I of England in 1601, introducing locally based taxes for the poor, aimed at allowing community and parishes to support their poor through a tribute instead of religious charity. Elizabeth introduced this innovative law because she was aware that the crown was now responsible for both the State and the Church of England, after the political revolution of the Anglican secession from the papacy.

The welfare interventions for the poorest sections of the population were maintained by protecting small groups and in discretionary forms, up to the organic formulation of the *New poor law* of 1834, which subtracted the assistance to the control of the individual parishes and provided for a centralization of equal aid for the needy.

With reference to the first industrial revolution, social insurance is also promoted to guarantee workers first on a voluntary basis, afterwards compulsory, to protect accidents at work, illness, old age. "*Workhouses*" for poor unemployed people are opened, also for the purpose of having low-cost workers. The motivation for legislative reforms was to ensure social peace in the conflict between workers and industrialists.

The English legislation of 1834 remained in force until 1948 with the *National Assistance Act*, a landing for the welfare state to repair the rubble of the Second World War, nourished by the experiences of the nineteenth-century mutual aid societies and unions [23].

In 1883, in Germany, a land that had just been politically unified under William I, the chancellor Otto von Bismarck established the first forms of social security and workplace accident insurance, for the protection of the working people with the aim of their integration into the unified state. The industrialists themselves pushed to introduce social protection with compulsory payments of their workers, in order not to bear the full cost of social security, as well as to extinguish the conflict raised by the workers' claims through the benefits of the welfare state.

Similarly, after the Italian unity and independence under the constitutional monarchy of Vittorio Emanuele II, mandatory and free education of primary school was introduced thanks to *Coppino law* in 1877, to create the politically based cultural unity of the nation. After an investigation, on behalf of a parliamentary commission, on thousands of existing *Opere Pie*, institutions with substantial private resources due to contributions and bequests mostly run by the Church, (from charitable subsidies to boarding institutions for fallen women, from hospitals to schools for the blind and deaf-mutes, from nursery schools to re-education homes for wayward minors, from hospices of motherhood to mental hospitals), the *Crispi law* of 1890 requires their transformation from private entities to *Public Institutions of Assistance and Beneficence* (IPAB), designed to supply free goods and services to people in difficult conditions.

After the Great War the fascist regime develops a strong social policy for large families on which depends the population growth of the national power, creating highly centralized institutions as, for an example, in 1925 the National Opera for Motherhood and Childhood (ONMI), or societies for the assistance of special categories such as in 1923 the Italian Blind Union (UIC), or in 1937 the National Opera for Orphans of War (ONOG), to end up in 1937 in abolishing charitable congregations and replacing them with the Municipal authorities of Assistance (ECA), financed by the state to provide the necessary goods for subsistence not only in conditions of poverty, but also in conditions of need.

In 1933 the National Institute of Social Security (INPS) was created as a social security institution of the Italian public pension system, which transformed the National Pension Fund for the invalidity and old age of workers, created in 1898 as a voluntary association supported by a state contribution and a contribution from entrepreneurs. Also in 1933, the Institute for Industrial Reconstruction (IRI) was born, a public body in favor of banks overwhelmed by the great crisis of 1929 and to protect Italian companies for support to employment [35].

The foundations of the Welfare State in America are thrown by the democratic President Roosevelt after the election in 1932, opening a New Deal of state intervention in the market economy in accordance with the perspectives of the English economist Keynes.

The US government asserts protective welfare state policies against the devastating outcomes of the Great Depression of 1929, after the financial failure of the Wall Street that caused the collapse of banks, businesses, mass unemployment.

Roosevelt promote large investments in public works to revive the internal market and pursuing a Keynesian policy of full employment, instituted in 1935 the *Work Progress Administration* for the allocation of a salary to the unemployed, giving occupations even provisional instead of humiliating charitable checks.

An important step in the development of welfare in the United States is the *Social Security Act,* a law promoted by Franklin Delano Roosevelt in 1935. The law introduced unemployment allowance and an old-age pension. The planned sickness benefit was removed due to opposition from the doctors lobby. The law was financed by contributions from workers, employers and the federal budget [28].

After the great suffering of the war won by the Americans, President Truman favored the approval of the *Employment Act* of 1946 with the aim of achieving full employment by orienting the market.

The rush of the welfare state continued in the United States in the 1960s and 1970s when public spending on supporting disadvantaged groups of the population increased with laws and rights that developed earlier democratic conquests such as 1935's Roosevelt *Social Security.*

President Lyndon Johnson as part of his Great Society program promoted reforms hampered for years by the powerful lobby of doctors, by promoting the approval of *Medicare*, a medical care system for the elderly, and *Medicaid*, which helped states with federal funding to partial coverage of spending on health care for the most deprived.

The process of welfare development under democratic administrations came into crisis in the 1980s under the neoconservative administration of President Ronald Reagan, who made heavy cuts in social spending, particularly to the detriment of education, public housing, subsidies to poor families, in the neoliberal wave that spread also in Europe with the British premier Margaret Thatcher, who made cuts in the expenses of the solidarity of the Welfare State, in favor of the profit of the private market of social services considered more efficient and less expensive [20].

The manifesto of contemporary Welfare State is the Beveridge Report published in England in 1942, where the economist Lord Beveridge, in line with the principles of the liberal tradition of the market economy, outlines the state social policies such as redistribution of income, not only for solidarity ends, but also to correct the capitalist competition that produces growing inequality and poverty, blocking expansive development, interrupting full employment, producing misery that generates hatred and conflict, expensive and undesirable.

In the first Beveridge Report of 1942, "Social Insurance and Allied Services", citizens' health protections stood out by establishing a national health service, child benefits, job guarantees, support for the personal needs of the disadvantaged. The Second Beveridge Report of 1944 "Full Employement in a free Society" designed a welfare state that implemented policies of intervention in the market economy in a free society of private capitalism, having as its aim full employment on the Keynes line, so as to guarantee every citizen a secure income as the right to a life of personal and social well-being.

Beveridge's program is compatible with Keynes' public policies to stimulate demand to reactivate and expand the market, in continuation with the interventionist plans of the state inaugurated in the New Deal of Roosevelt in the previous decade devastated by the great depression due to the failure of the financial system.

Citizens, independently from paid insurance contributions, must be able to take advantage of universalist services, at least for the minimum of health welfare, pension, housing, education, etc., welfare for minimum levels of life security, from the public funding financing through a progressive tax system. Public expenditure on aid services to the discomfort of the weaker redistributes income to people, who in their turn support the market demand for goods that feeds the production, correcting the economic system towards a steady expansion, with the maintenance of riches and full employment, just like a Keynesian multiplier in a democratic state intervening in regulating the market economy [27].

Beveridge rector Oxford University College was a liberal who had also collaborated with Sidney and Beatrice Web, sociologists of work and founders of the Fabian Society which inspired the reform of the Labor party. The welfare state was born in fact to face the devastation of the war when the Labor party was in coalition with the Conservative party to plan a better future for all citizens after the miseries of the war.

Beveridge's Liberal-Labor welfare service plan was developed under the government of the conservative Churchill, then approved by the government of Labor Clement Attlee premier since 1945, through several acts of Parliament.

National health service act of 1946 activates a national health system. The National Insurance Act of 1946 guarantees many social security protections, orphan's allowances, unemployment benefits, death grants, old-age pensions, in exchange for a contribution from all workers. National assistance act of 1948 abolishes the ancient law on poverty and establishes a safety net for all those who have not paid national insurance contributions, such as homeless, handicapped, single mothers, elderly without means, providing housing for the homeless or sick without means, aid to voluntary organizations.

The Western countries of the capitalist freemarket won the Second World War but in the '50 were involved in the block policy, opposed to statist communism of the Soviet Union, allied in the war against Nazism. The Beveridge's model of the liberal state that regulates the market economy to protect the disadvantaged, in the Cold War between the 2 blocks represents a third Liberal-Labor welfare way between market capitalism and communism statism. The welfare third way found fair implementation in the social democratic states of Northern Europe, especially in Sweden, the first country to introduce in 1948 popular retirement based on birthright, in a universal welfare from cradle to grave [25].

After the Second World War, the new republican and democratic Italian state promulgated the Constitution of 1948 in which the protection of the welfare state becomes social security as the right of all citizens with the mandatory nature of protective measures. The Republic recognizes and guarantees the inviolable rights of man both as an individual and as a social person and requires mandatory duties of political, economic and social solidarity in article two of the Constitution.

Article three states equal social dignity and equality of all citizens, declaring that "it is the duty of the Republic to remove all economic and social obstacles which, by limiting the freedom and equality of citizens, prevent the full development of the human person".

In particular the article 4 protects the fundamental right to work, the article 32 protects the right to health, the article 34 the right to mandatory and free education, the article 38 outlines the rights of the social security and welfare system for the disabled and workers, who must be assured of adequate means for their life needs in the event of an accident, illness, disability, old age, involuntary unemployment. Article 36 recognizes the right of every worker to a minimum wage which is in any case sufficient to ensure himself and his family a free and dignified existence. Article 21 recognizes the right of everyone to freely express their thoughts by word and writing, while the press is free and cannot be censored, likewise for article 33 art and science are free and free it is the teaching.

Article 41 provides that private economic initiative is free but cannot harm the security, freedom, human dignity of citizens, as it cannot be carried out in contrast with social utility because private economic activity must be coordinated for social purposes. For article 42, public and private property is guaranteed by law although with limits to ensure its social function and make it accessible to all. For article 43, companies of essential public services, energy sources, monopolies, which have a pre-eminent general interest, may be originally reserved or transferred by way of compensation to the State, public bodies, workers' communities, and users.

In Italy, thanks to the political mood of the students and the workers libertarian movement of 1968 as well as the feminist movement of the 70 s, new rights and social gains are developed.

Since the post-war reconstruction, the economic boom that places Italy among the first industrial countries in the world arises in Italy. The constantly growing domestic product (PIL) allows social spending for welfare, without too much burdening the public debt, developing in Italy an efficient welfare state.

Public expenditure for assistance services to the disadvantage of the weaker groups redistributes income to people, who in turn support the demand for goods that feeds production in the market, correcting the economic system with a constant expansion of the disadvantaged classes towards the middle class, distributing wealth and expanding employment, on the Keynesian model in an interventionist democratic state regulating the market economy. The welfare state expands citizens' rights of life with continuous reforms supported by the financing of public intervention through a progressive tax system.

In 1970 a statute of workers's rights is passed. At the same time the regions, institutions established in 1970, are given skills thanks to the decentralization of social care. From 1968 to 1975 laws on support for working mothers, the public nurseries, on family planning consultation are passed. In 1978 the law that regulates the voluntary termination of pregnancy, providing medical, social, psychological services is passed. Moreover, in 1978 the law establishing the National Health Service that introduced, together with the phase of cure, the prevention and rehabilitation phases, is passed. A conquest that places Italy among the top of the world panorama for the adoption of universalistic welfare in the protection of the health of all citizens dependent on the welfare state [3].

As a result of the founding of the National Health Service, the National Institute of Social Security (INPS) absorbs other national institutions for compulsory health insurance as the INAM, and manages mandatory contributions paid by workers and employers to the hospital and medical care of the employees. In 1983 the law governing the adoption and custody of children is passed. In 1988 and 1998 laws for the elderly aid project are passed. In 1992 the framework law for the assistance, the social integration and the rights of disabled persons is passed.

In 2000 the framework law on the integrated system of interventions and social services, activated between families, government, private and voluntary social organizations is passed. This law outlines complementary roles between state and non-profit sector, in the current welfare mix situation that goes beyond the traditional model of public social security welfare state. The complementary roles may indicate an integration, but also a crisis of the public welfare system, which must turn to private social organizations to survive. In fact in the globalization of the last decades, the welfare state sufferes more than one cut, loses socio-political protections, becomes more precarious, faces a crisis in the same capacity to protect constitutional rights [24].

3   Welfare Crisis and Conflicting Rights Between Welfare State and Market Economy in the Society of Globalization

In the 90 s the socio-political context rapidly changes. With the fall of the Berlin Wall in 1989 the communist bloc of Eastern Europe Soviet Union and of its satellite states flakes. What remains to the government of the world is only the Western capitalist bloc led by America and its allies which requires the unique model of free market globalization, involving China, Russia, Japan, Germany, but also new emerging countries from India to Brazil, in a world that is more and more multipolar, multicultural.

The free market has no more space limitations, does not want rules by states, and imposes, through financial capitalism and corporates, transnational economic agreements governed by supranational bureaucracies not elected but appointed such as the WTO in Geneva, the World Trade Organization. The rule of these supranational organizations in the economic globalization is the competition in a free market aimed at the highest profit and the lowest cost, with no constraints due to social welfare rights respecting the constitutions of national states. The economic interventions of the states aimed at protecting welfare are indeed considered an alteration of the freedom of the market, and the welfare state is deemed an unbearable cost because of the widespread public debt of sovereign states, indebted to private finance and banking.

In globalization we see a deconstruction of the welfare state, just as unemployment and poverty increase, the middle class becomes precarious, there are increasing inequalities between the few rich and many destitute people without social protections. The continuation of the progressive expansion of welfare rights, documented in the previous socio-political cycles, seems to have lost its driving force.

The liquid world of globalization is hell for people with limited rights, more precariousness, insufficient public protections, dragged into the utopia of easy gains in the risky financial games where very few make money and many get into debt [5].

The deconstruction of the welfare state is testified by the many constitutional rights of social security which come into conflict with the rights of the competitive market. The complex, liquid and ambiguous society of globalization accentuates the co-presence of equally legitimate but antagonistic values.

The conflict of alternative rights is still open on the decision of their priority, which may determine the choice of the primacy of welfare rights for the protection of the weakest people, or the choice of the primacy of the rights of the strongest competitor, which to obtain the utmost utility makes precarious the constitutional rights of welfare.

The problem of welfare is posed by the comparison between solidarity versus utilitarianism as selection criteria for the still open decision, a comparison that we want to highlight with reference to exemplary cases of conflicting rights. The bioethical cases are particularly significant. Let's think of the case of the rules for the social protection of maternity and voluntary interruption of pregnancy of the law 194/1978.

A pregnant woman, victim of a rape, wants an abortion. The interpretation of the law is not easy, since it involves the objection of conscience of the personnel of the clinic who, as doctors, have the right to safeguard the life of the future child. Women can thus exercise their right to the interruption of pregnancy within three month, in the presence of health personnel who are not objectors of conscience. These are particularly rare because up to 90% of gynecologists are objectors. So the decision criterion between conflicting rights, abortion rights of raped women and the right to life of the unborn, is pragmatically solved in the choice of a gynecologist who is not objector.

The principle governing the Welfare to ensure the life of the weakest is in crisis, because in fact the abortion law guarantees the well-being of the mother, and does not guarantee the life of the son. The ruling of the Constitutional Court n. 27 of 1975 declared the partial unconstitutionality of the rules against abortion of the Criminal Code of 1930, not denying the right to life of the unborn, but also considering the rights of the mother in a state of necessity to the interruption of pregnancy. Even the ruling of the Constitutional Court n. 35 of 1997 that interprets the law 194/1978, reaffirms the right to life of the unborn as an inviolable human right with reference to article 2 of the Italian constitution, but also reaffirms the right of the parent who wants the abortion, maintaining open a fluid ambiguity between conflicting rights. The Law 194/1978 serves as an example of an ambiguous and complex competition between equally legitimate and yet antagonist values, so alternative as not to have solutions, including well-being of mother and son's death by abortion. The voluntary interruption of pregnancy as a social maternity protection of the law 194/1978 appears as the application of the decision criterion mors tua vita mea. The basic criteria of the welfare state dealing with the protection of the life of the weaker party, here sacrificed for the welfare of the stronger party, are factually deconstructed [39].

Always in the field of bioethics, let's consider the case of medically assisted procreation rules of the law 40/2004. The right of the infertile couple to form a family with children using the gametes, eggs and sperm from a donor, (the case was forbidden by law), is authorized by the Constitutional Court ruling of 10 June 2014 even in Italy. What remains is the ban (Art. 5 of the law 40/2004) of access to medically assisted procreation for singles and same-sex couples, but they use uterus for rent permitted in other countries. According to the judgment of the Court, the right of the sterile or infertile couples to form a family with children using heterologous fertilization is considered dominant with respect to the rights of the child to his/her biological parent represented by the donor.

Thus the need to protect the unborn is considered by the Court of a minor rank, even if it is the weaker party that, according to the principles of welfare, should prevail. The criterion of greater or lesser rank who presides over the choice between alternatives of rights leads the Law 40/2004 to the decision of sacrificing the weaker party, contradicting the principle of the welfare state. Of course, the manipulation of cells and human embryos in artificial insemination is driven by strong commercial and industrial interests in the complex society of globalization, pushing to applications of increasingly uncontrolled reproductive technologies against the principles of the welfare state, in favour of market principles, with 60,000 women in Italy that call on medical facilities of procreation [37].

In particular in the field of economic market of globalization, the principles of the welfare state are opposed, even by placing in a conflict the rights of the market against the health rights. Italian people for many years ate cattle fattened with hormones, imported from countries where they are not forbidden as the United States. When it was made sufficiently clear that this type of meat produced health damages Italy stopped importing it. As a consequence we were punished to pay hundreds of millions of euros by the WTO in Geneva, the World Trade Organization that oversees the numerous trade agreements between 164 Member States. This supranational bureaucracy of globalization that includes over 95% of world trade in goods and services, has issued a sanction to a sovereign nation for breaking the law to free market recognized by the Italian government in international trade treaties, which take precedence over domestic law on health protected by our Constitution.

The welfare of the Italian constitutional right to health contrasts with the right to free trans-national market in an obvious conflict between opposing rights, on the one hand the less expensive supply in the competitive market governed by international treaties, on the other health rights in line with the Constitution and the current governmental regulations. We need to choose and decide between the health value and the cost value. In the conflict of rights Italy decided in the case of foods containing hormones the prevalence of the right of people to health, on the right to profit from the resulting low cost in free trade. But in many other cases, they allow for economic benefit imports of goods from countries with serious environmental pollution, or cheap foods with chemical preservatives and toxic additives and massive pesticide, with high intake of health risk. Often European and domestic laws do not support full traceability of these goods, so in the global competitive market, protecting the health and welfare is overshadowed and eroded by ambiguities that do not even allow the transparency of information to decide.

In the free market globalization the greatest damage to welfare occurs in the field of labour protection. Globalization had promised welfare, but has only led to economic insecurity for workers and to the loss of welfare rights. In globalization there exists the competition for which the lowest cost wins in the market even at the expense of the rights. The example is China's competitiveness in the global market that lowers costs with the removal of rights (underpaid workers, exploiting working hours, damage to health because of the polluted environment, etc.). This is unfair competition because it is not on equal terms with the production costs due to our constitutionally protected workers.

The Italian welfare for what concerns work provides, in accordance with Art. 36, an "adequate remuneration ensuring them and their families a free and dignified life", a maximum of daily working hours, a weekly rest, a remunerated annual leave. The Article 37 guarantees equal rights for working women to whom conditions to fulfill their role in the family are given. The Article 37 also guarantees the protection of child labor. The Art. 38 guarantees assistance to the disabled, insurance for workers in case of injury, sickness, disability, retirement, involuntary unemployment. The Art. 39 guarantees collective agreements by means of trade unions representatives.

The Art. 40 guarantees the right to go on strike. All these rights are gradually reduced or eliminated by the liberal principle that for the maximum profit, the lowest cost of production pursues, to win in the downward competition of the price of goods in the global market.

The liberal principle places the rights of workers in conflict with the consumers' rights. If the workers are entitled to welfare and decent wage conditions, consumers are entitled to the lowest prices. A rational decision between these conflicting rights should give precedence to the rights of workers, because the less consumers (being workers too) are paid in order to reduce production costs, the less they consume, depressing thus the economy of the entire system in a vicious downward cycle. On the contrary, the economic system of the advanced Swedish welfare works at its best with well-paid workers that stimulate the consumption demand, which in turn increases the offer of production of the goods, in a virtuous upward cycle. However, in the globalization of free market, profit interests prevail with downward competition of the production prices, using the ideological defense of the right of consumers to lower prices. In this way the work becomes increasingly precarious, and the welfare state is dismantled [1].

Let's consider some exemplary cases in Italy where it was decided to consider the prevailing rights of the private competitive market with respect to the public rights gained by workers. One such example is Uber, a Californian Corporate, present in most Italian cities that allows, via app, to hire driver and car from any consentient private car owner. Hence, costs for consumers are lower. Uber is causing the taxi workers, owners of a license for public services, to get poor and poor, by practicing an unfair competition. Uber did not pay to have a license, does not pay taxes, and pays very little sums to hits private drivers for each race. Uber does not recognize labor rights such as sickness, retirement, strike, a maximum daily working hours, etc. Uber is actually abusive, but since it offers the lowest cost service, in line with the principle of the free market, asks for a form of recognition from the Italian government, which is willing to change laws to allow a convenient innovation for consumers, instead of asking Uber an adjustment to the Italian regulations. The taxi workers are violently protesting but are considered a corporation closed to competitive innovations [13].

The German Corporate Foodora pays a few Euros private bikers who bring home food ordered at restaurants for each delivery commissioned via mobile phone. A provider of casual work, the biker does not enjoy any rights. Nevertheless, in Turin, some bikers asked for a contract with better conditions and pay on the basis that such work had lasted for a long time. Foodora never called them again. They were sacked without any rights, let alone without a trade union representation. The bikers who replaced them, being in need, have agreed to all.

The situation is similar for workers of the Amazon Corporate dealing with purchases online. Amazon like all high-tech Corporates, from Apple to Google, operating on the network with fabulous profits, does not pay taxes in Italy and does not recognize labour protections of italian welfare.

In the case of workers paid a few hundred euro per month by Almaviva who had contracts by ministries, it was decided to lay off thousands of call center operators. In the new downward contract it was preferred to displace the service in Albania, where call center operators cost less, though they may pay a lower quality service. Even the state and the public administration prefer the market low cost then the defense of employment and quality of service.

Precarious work due to its lower cost overrides the welfare of workers without rights who paid with vouchers. A voucher for payment born for odd and accessories jobs for families, which is now applied to millions of employees for companies, from building to fast-food Corporates. Vouchers were designed for students and retired persons activities to supplement a small gain in leisure time. The former students are now in their forties, jobless and desperate, and they accept anything for a few Euros. Since they are required a number of hours work without limits and without any insurance such as sickness and retirement, they are temporary workers without protection [40].

In southern Italian countryside Italian agriculture labourers and immigrants work for a few Euros without any Welfare protection, under the exploitation of a fee extorted from a "corporal" who procures work and that has modernized his enterprise in temporary agency to escape the law. To handle the export of wine and table grapes competition, new slaves are born for seasonal work in Puglia countryside. Paola Clemente was 49 and had three children. She earned €25–27 a day, a sum of money necessary for her family. Actually she earned €15 a day, because the corporal extorted from her money for having procured her job as well as her bus ticket. Paola died in the camps killed by fatigue while picking grapes for 12 consecutive hours. This is the story of a labourer, this is the story of slavery under the corporals. Paola had to pay for the bus ticket to go every day 300 km away from her home in San Giorgio Ionico to the Andria country side, leaving at two in the morning to go back in the late afternoon [16].

The market without rules that practices downward competition in costs and rights tends to deconstruct the entire welfare system. Considering the extreme consequences of the free market globalization that makes the rights of people precarious, it is no surprise that some national governments are back to choose and decide for social protection and economic protectionist policies. Globalization today is faced by the sovereignty of the nation state which recovers the identity of its people because it promises to protect the citizen abandoned to solitude in the global liquid worldì [6].

4 The Paradigm of the Ecological Decision and the Circular Complementarity Between the Rational Decision of the Welfare State and the Rational Decision of the Economic Market

The welfare state entered a crisis when its social rights of solidarity in support of the disadvantaged classes were opposed by the utilitarian rights of the global economic market, in a competition between social rights and private rights equally legitimate but conflicting. The decision-making rationality of maximizing individual competitive profit in the free market of globalization has overwhelmed the decision-making rationality of social insurance of the constitutional rights that had marked constant growth in the history of the modern state.

However, we have also historically seen a collaborative possibility operate between social security rights and market economy rights. The market with the optimization of private profit has created wealth and the welfare state with the tax levy on that wealth has created social insurance measures not only for the poor, but also to support the free market in its cyclical crises, according to a model mutual enhancement circular between welfare state and market economy [36].

New collaborative forms between welfare and the market indicate ways out of the welfare crisis in the sustainability context of the new ecological rationality which favors a circular balance of complementarity between opposites. We intend to deepen an ecology of circular complementarity between the decision-making rationality of welfare and the decision-making rationality of the market, on which the solution of a conflict depends which still prevents the balance between economic freedom and social security.

The economic decision-maker from the ecological point of view of the relationship with the environment aims at maximizing profits through the systematic and unlimited exploitation of natural and human resources, and therefore oppresses the reactions that the environment opposes by applying the competitive logic of making more and more profit than to costs. The economic decision maker thus establishes an asymmetrical relationship whereby he takes more resources from the environment than he returns and in this difference constitutes the unequal individual profit that produces social wealth.

In the logical scheme of market rationality, the decision maker chooses among the possible options the action whose consequences imply the greatest profit at the lowest cost. It is believed that this type of rational choice expresses the nature of man, who is therefore represented as a substantially selfish and individualistic being who always chooses the most useful action for his own advantage. The rational person decides on the basis of the calculation of costs and profits assessable in the consequences of his choices, following the criterion of maximizing profits and minimizing losses [29]

But Weber [43] had already pointed out that people decide rationally also in a coherent relationship with social values and ethical principles, with choices independent from the calculation of the consequences. Therefore it is possible to distinguish two forms of rational decision, the economist one that follows the logic of advantageous consequences and the value one that follows the logic of adequacy to ethical principles and rules of social equity [32]. The utilitarian logic of the consequences belongs to a businessman in the financial securities market in response to the question: "what action should I decide to choose from the various possible ones, in relation to the given situation, the consequences of which can maximize my profit?".

Instead, the value logic of appropriateness is that of the captain of the ship in distress in response to the question: "which decision is appropriate to the identity of my values and to the ethical rules of my social role for a choice of the best action towards my crew, among the many possible choices in the given situation?".

Since the commander obeying the rules of internalized values and his social role, chooses in the danger of saving the crew even at the expense of his own safety, this type of ethical decision shows that man it is a social being also capable of altruistic and supportive actions towards its social environment, without calculating the advantageous consequences to maximize one's personal utility.

From the point of view of the relationship with the environment, the ethical decision maker also establishes an asymmetrical relationship because in his altruism he gives more resources to others than he takes, a behavior opposite to that of the economic decision maker who in his selfishness takes more resources from its environment than it returns [26]. If the asymmetries of the two opposing decision rationalities are related to each other, they tend to compensate each other because the ethical decision maker tends to donate resources to the environment as much as the economic decision maker tends to take. It is only a matter of defining the proportions of this compensation to find the balance between altruism and selfishness, between social state and market utility, between solidarity and profit, a balance that ecology resolves within limits of mutual sustainability between the man who it gives to the environment and environment that it gives to man.

Ecological rationality therefore develops a continuous circular adjustment between man and the environment to find the exchange that tends to equalize withdrawals and transfers of reciprocal resources, saving the balance of life of both man and the environment in their complementary relationship [4].

If the rational economic decision shows man as individualistic and selfish, the ethical rational decision shows man as a social and altruistic being. Obviously, different decision models discover different and somewhat opposite dimensions of human nature, a complex system that does not allow itself to be framed in one type, but which in its complexity benefits from its versatility of antagonistic functions. In fact, both decision-making models perform a synergy function for the ecological balance of the social system. The rational economic decision of the individual with the aim of maximizing profit by taking resources from the natural and social environment produces wealth in the competition of the free market. The preliminary accumulation of this wealth allows the ethical decisions of the welfare state to redistribute the wealth to the marginal groups of society, to be recovered to balance the loss of human capital expelled from the market in the liabilities of the social system, allowing inclusion and the social cohesion of new recovered productive forces, which in turn stimulate the production of other private wealth and other public solidarity, in a virtuous circle between private economic freedom and public social security. The profit market and the state of welfare are not antagonists but in synergy in the balance of complementary functions of the rational choice of life of the ecological system.

The rational ecological decision makes the oppositions between economic egoism and social altruism complementary by developing a complex dynamic system first of accumulation and then redistribution of environmental resources, so as to ensure the integrity and well-being of individual and social life in the overall circular balance of interactions between profit and solidarity, between market and welfare. The balance of the complex man-nature system develops through self-regulatory feedback between subsystems of differentiations and integrations of opposite functions of market accumulation and redistribution of welfare in the synergistic organization between man and the environment [33].

The ecological paradigm welcomes the market and the welfare as internal antagonistic but complementary subsystems, setting a sustainability limit among them, with the aim of guaranteeing the safety conditions of individual and social human life against environmental risks.

While economic decision-making rationality pursues the objective of maximizing individual profit with the intensive and unlimited exploitation of environmental resources, ecological rationality sets a sustainability limit to the economic withdrawal of resources from the natural and social environment, because the planet it is a finite living system with exhaustible resources, which responds with perverse effects against the man who forces the ecological balance to maximize his profit. The perverse effects of the environment produced by man range from pollution to climate warming, from desertification to damage to health.

In order to avoid the risk of perverse effects harmful to human life due to over-intensive use of the environment, an ecological sustainability limit prescribes to withdraw environmental resources at the minimum necessary human life, in order not to exceed the threshold of irreversibility of the environmental change produced by man, given that only reversibility allows to correct errors at the first indication of perverse effects. Ecological decision-making rationality prescribes a minimal reversible and responsible intervention to allow the environment to replenish the resources of its altered balances, in order to still be able to return to the future to take advantage of the environment with a new moderate withdrawal, instead of burning irrationally in a single solution resources ended with intensive exploitation [30].

In the ecological paradigm there is also the rationality of welfare as a precautionary principle responsible for the maximum protection of life from environmental risks.

Ecological rationality on the one hand corrects the economic rationality of the maximum individual profit that plunders nature and triggers harmful consequences for life, preferring the most advantageous choice of the economics of life that cultivates renewable and sustainable resources to avoid the environmental risk of perverse effects. Ecological rationality on the other hand supports the rationality of welfare solidarity which protects every variety of biological forms of the environment by respecting the natural cycle of reconstitution of natural resources on which the safety of human life itself depends.

Market economics and welfare state are complementary by limiting human intervention on the environment to ensure sustainability limits for the production and reproduction of both human and environmental life. The precautionary principle adopted by the ecological movement in the 70 s and now introduced in European treaties since 2000, configures a right to protect the renewable balance between environment and society as an extension of health security and social protection for life [11].

The precautionary principle prescribes maximum caution in conditions of uncertainty in order to avoid environmental hazards that are not yet fully known but likely through evident indications. The precautionary principle expands the field of welfare prudence by setting sustainability limits to intensive human economic intervention on the environment, in order not to trigger the risk of potential damage to human life, also protecting with life the economic resources of the environment necessary for life. Ecological rationality through the precautionary principle limits the excesses of environmental exploitation and makes compatible the circular balance between economic market freedom, environmental protection, social security of human life, in complementary collaborative reciprocity.

Recent trends confirm a growing success of the ecological model in the revitalization of new forms of welfare to rebalance economic inequalities that have grown to alter the global ecosystem since the nation state has no longer been able to protect citizens from the globalized market that corrodes the rights of the welfare. The inequalities that alter the balance of the ecosystem in the world of the market economy without rules are unprecedented: the private income of the 10 richest families in the world equals the entire public GDP of Germany, the United Kingdom, France; 2000 super rich people possess as much as 4 and a half billion people who are half the world's population; 1% richest in the world owns more than the remaining 99% [2].

Inequalities make marginal groups grow and therefore the need for welfare increases. Since the state indebted to the globalized market does not have sufficient public investments, a new "shared welfare" is growing which involves the whole environment of civil society in an ecological solidarity network. A horizontal ecological network of private actors assumes responsibility for subsidiary interventions in the areas of welfare left uncovered by the vertical public intervention of the state. A participatory proximity welfare system is growing, developed with new alliances between public and private, between companies and community services, between third sector associations and insurance companies, between banking foundations and confessional welfare of churches, between mutual aid associations and large philanthropic enterprises [31].

The inequalities are so large and involve growing masses of citizens, that a "citizenship income" is needed outside the market, in order to guarantee the most urgent basic needs to the marginalized people and then recover them to productive activities [42].

In cases of serious crisis in the social system, neo-Keynesian public investment policies have been implemented to correct the market, to guarantee welfare protection for all citizens and recover those expelled into the production system in order not to lose the human capital of the ecological system [12].

In cases of emergency, monetary bonus policies were also made to families and businesses drawing on extraordinary issues from central banks, as even liberal economists had foreseen and advised in the form of helicopter money to support the collapses of the system [21].

An ecological reconversion expands not only in the field of new forms of welfare but also in the field of new forms of the economy. Circular economies are affirming for the sustainable growth of a system of common goods as collaborative relational goods with the human and natural environment, instead of an economy of maximizing individual profit in market competition that produces winners and losers and that tends to growth unlimited to the detriment of a finished environment. According to the circular economy, the best rational choice regards the solidarity decisions of reciprocal collaborations rather than the decisions of the maximum individual profit. Thus we have an economy of cooperation [15], an economy of the common good [17], a civil economy of public equity [9], an economy of circular solidarity of intangible relational goods [14]. An ecological model that applies not only to the world of non-profits such as self-help and mutual aid groups, but also to economic companies that are on the market with corporate social responsibility towards the human and natural environment.

In fact, new models of sustainable and responsible investments in economics and finance are being developed, with rational decisions oriented to the social use of resources respecting the needs and capacity to reproduce the environmental system, in order to ensure the future capacity of resources to continue to produce value for future generations. The sustainable and responsible ecological decision develops a rationality of greater advantage than the rationality of decision of the maximum immediate profit, because it does not immediately burn the environmental resources to the detriment of its future and the future of its heirs. The rational ecological decision feeds new forms of responsible contamination between ethics and economics [38], between ethics and finance [10], between social values and competitive enterprise [8], between corporate social responsibility and fair trade in solidarity [7]. Economy and welfare collaborate in an ecological dimension in an attempt to reduce the economic inequalities that are decompensating the system. The economy produces wealth and welfare activates redistributive policies for the sustainable development of the system.

5 Conclusion

This conclusion outlines my personal version of the re-proposed welfare, after its crisis due to the conflict of the market economy oriented towards the rationality of profit, against the welfare state oriented towards the rationality of solidarity rights. I advance the proposal to re-evaluate and develop welfare according to the reasons of the ecological paradigm because it is capable of integrating social state and market economy in sustainable circularity, complementary with mutual benefit in the management of environmental risk.

The risk consists in the production by the nature of perverse effects to the detriment of the economy and the state that have forced the dynamic balance between man and the environment with irreversible intensive interventions. Risk management involves the solution of integrated interventions of the economy and the state that decide not to force the environment with irreversible changes that generate perverse effects, but instead decide sustainable, responsible, non-intensive changes that support the reversibility of the balance of the altered environment by human intervention. The balance is sustainable when it guarantees the safety of the reproducibility of the mutual life of both man and the environment.

An ecological system represents a field of antagonistic but synergistic elements such as human society and its natural environment which mutually change in dynamic equilibrium and mutually adapt with circular feedback. Society and the environment are circularly related so that a random fluctuation in one point of the ecological field will have a corresponding random fluctuation in another point of the field. Random fluctuations can balance the ecological system, but they can also unbalance it with unexpected perverse effects, presenting a risk of collapse for the state and the market. Risk management of random fluctuation involves a rational ecological decision to ensure the certainty of maintaining the balance of the sustainable and reproducible system for any unexpected random fluctuation. A rational ecological decision is made to maintain the balance which consists in setting aside a reserve of resources produced by the private market to be transferred to the public welfare, for its redistributive function of resources to immediately support the life of subsystems for a precautionary principle as soon as they present risks of collapse, whatever the social, environmental, state, market subsystems, in order to maintain production and reproduction of the global life of the ecological system. The ecological decision-making rationality to support the life of the global system from the risk of perverse random fluctuations makes the decision-making rationality of the economic market that produces resources, complementary with the decision-making rationality of the social state that redistributes resources. Complementary integration resolves the ancient conflict between them that has put welfare in crisis. The welfare state can be reborn as a regulator that ensures the global sustainable ecological balance between human life and the life of the environment.

Ecological decision-making rationality shifts the objective of the most advantageous choice among the possible ones to maximize the safety of mutual life between man and the environment, in their sustainable circular collaborative competition, different from the conflictual economic competition of the market in search of the maximum dominant profit of man on environmental resources with the risk of generating perverse effects. In the challenge against the risks of the environment in which man can win or lose, ecological decision-making rationality seeks the most beneficial result for the life of the system, which consists in ensuring that both always win, instead of risking winning or losing in dependence on chance. The Harvard negotiation project presents an optimal collaborative rational decision model in transactions in which all competitors win according to the Win and never Win Lose model, even if each wins to a greater or lesser extent in proportion to the capabilities of the parties involved [19].

The natural environment is a living organism that metabolizes materials and energy, produces resources, emits waste, but with a permanent circular redistribution mechanism it recycles waste into resources, because none of the production potential of the system is lost for its reproduction [30].

The decision of social protection for every disadvantaged person does not arise from charitable reasons only, but from the rational ecological decision to recover every expelled from the productive system, in order not to lose any human labor capital for the life of the system. Welfare security for all is combined with the productive and reproductive security of the global social and environmental system [12].

No player knows in advance whether he will be expelled as a loser from gambling at risk with environmental challenges. So it is rational that each player takes the ecological decision to control uncertainty by choosing to maximize his safety with the protection of a systemic regulator, so as never to lose in the lottery of life in the face of environmental risks. For this reason every competitor at risk agrees to invest a part of his wealth as an insurance policy in a public system regulator of redistribution of wealth, in order to protect and recover the group of disadvantaged losers in which he could find himself.

This ecological decision-making rationality for the management of the uncertainty of personal life within the sustainable balance of the environmental system, involves in a complementary way both the function of public welfare and the function of the private economy. The economy deals with the decision-making rationality of the production of the maximum individual profit that distributes unequal wealth, while welfare is responsible for the decision-making rationality of the redistribution of that wealth to guarantee equal opportunities for the security of each individual life. A synergistic decision between economy and welfare not only as a recognition of social justice rights, but as a rational decision of ecological economy useful to ensure the sustainable equilibrium of production and reproduction of mutual life between man and the ecological system, in the face of unpredictable risks of random fluctuations in the natural and social environment [34].

Welfare can be reborn and affirm its irreplaceable social function as an ecological regulator of the relationship between the economic system and the environmental system, promoting sustainable development and responsible citizenship with its decision-making rationality.

References

1. Acocella N (1999) Globalizzazione e stato sociale. Il Mulino, Bologna
2. Alvaredo F, Chancel L, Piketty T, Saez E, Zucman G (2019) Rapporto mondiale sulle diseguaglianze nel mondo 2018. La nave di Teseo, Milano
3. Ardigò A (1997) Società e salute. Franco Angeli, Milano
4. Bateson G (1993) Mente e natura. Un'unità necessaria. Adelphi, Milano
5. Bauman Z (2007) Dentro la globalizzazione. Le conseguenze sulle persone, Laterza, Bari
6. Bauman Z (2009) La solitudine del cittadino globale. Feltrinelli, Milano
7. Becchetti L, Paganetto L (2003) Finanza etica. Commercio equo e solidale. La rivoluzione silenziosa della responsabilità sociale. Donzelli editore, Roma
8. Bicciato F (ed) (2000) Finanza etica e impresa sociale: i valori come fattori competitivi. Il Mulino, Bologna
9. Bruni L, Zamagni S (2004) Economia civile. Efficienza, equità, felicità pubblica. Il Mulino, Bologna
10. Capriglione F (1997) Etica della finanza e finanza etica. Laterza, Bari
11. Comandé G (ed) (2006) Gli strumenti della precauzione: nuovi rischi, assicurazione e responsabilità. Giuffrè, Milano

12. Caffè F (2014) In difesa del welfare state. Rosemberg & Sellier, Torino
13. Comito V (2016) La sharing economy. Dai rischi incombenti alle opportunità possibili. Ediesse, Roma
14. Donati P, Solci R (2011) I beni relazionali. Bollati Boringhieri, Torino
15. Dorigatti M, Zamagni S (eds) (2017) Economia è cooperazione. Città Nuova, Roma
16. Fana M (2018) Non è lavoro è sfruttamento. Laterza, Bari
17. Felber C (2012) Economia del bene comune. Tecniche Nuove Edizioni, Milano
18. Ferrera M (1993) M_odelli di solidarietà. Politica e riforme sociali nelle democrazie. Il Mulino, Bologna
19. Fisher R, Ury W (1995) L'arte del negoziato. Mondadori, Milano
20. Flora P, Heidenheimer AJ (eds) (1983) Lo sviluppo del Welfare State in Europa e in America. Il Mulino, Bologna
21. Friedman M (1981) Per il libero mercato. Sugarco, Milano
22. Giovanola B (ed) (2016) Etica pubblica, giustizia sociale, diseguaglianze. Carocci, Roma
23. Girotti F (2001) Welfare state. Storia, modelli e critica. Carocci, Roma
24. Gori C et al (2014) Il welfare sociale in Italia. Carocci, Roma
25. Hills J et al (eds) (1994) Beveridge and social security. Clarendon Press, Oxford
26. Hoskova-Mayerova S, Maturo A (2016) Fuzzy sets and algebraic hyperoperations to model interpersonal relations. Recent trends in social systems: quantitative theories and quantitative models. In: Studies in systems, decision and control, vol 66. Springer, pp 211–223. https://doi.org/10.1007/978-3-319-40585-8_19
27. Keynes JM (1991) La fine del laissez- faire e altri scritti. Bollati Boringhieri, Torino
28. Kudlak A, Urban R, Hoskova-Mayerova S (2020) Determination of the financial minimum in a municipal budget to deal with crisis situations. Soft Comput 24(12):8607–8616. https://doi.org/10.1007/s00500-019-04527-w
29. Lindley D (1990) La logica della decisione. Il Saggiatore, Milano
30. Lovelock J (2011) Gaia. Nuove idee sull'ecologia. Bollati Boringhieri, Torino
31. Maino F, Ferrera M eds (2019) Nuove alleanze per un welfare che cambia. Quarto rapporto sul secondo welfare in Italia 2019. Giappichelli, Torino
32. March JG (2002) Prendere decisioni. Il Mulino, Bologna
33. Morin E (2017) La sfida della complessità. Le Lettere, Firenze
34. Rawls J (2008) Una teoria della giustizia. Feltrinelli, Milano
35. Ritter G (2007) Storia dello stato sociale. Laterza, Bari
36. Sabattini G (2009) Welfare state. Nascita, evoluzione e crisi. Le prospettive di riforma. Angeli, Milano
37. Scarpelli U (1998) Bioetica laica. Baldini e Castoldi, Milano
38. Sen AK (1986) Etica ed economia. Il Mulino, Bologna
39. Sgreccia E (1998) Manuale di bioetica. Vita e Pensiero, Milan
40. Staglianò R (2018) Lavoretti. Così la sharing economy ci rende tutti più poveri. Einaudi, Torino
41. Urban R, Hoskova-Mayerova S (2017) Threat life cycle and its dynamics. Deturope 9(2):93–109
42. Van Parijs P, Vanderborght Y (2017) Il reddito di base. Una proposta radicale. Il Mulino, Bologna
43. Weber M (2004) Il lavoro intellettuale come professione. Einaudi, Torino

Residential Centers. Accreditation and Licensing

Simona Ioana Bodogai

Abstract Quality and efficiency in Social Work? Do we assure the preconditions for quality through licensing, regulation, accreditation and standards? Should we focus more on the efficiency of social workers' activity? Entry into "legality" of social service providers by obtaining accreditation as providers and by licensing for social services offered is a process that is not easy to go but mandatory in present. Our research analyses the accreditation of social service providers and the licensing process of residential centers for elderly in Romania. Taking into consideration the legal minimal standards we examined the way these are put in practice and we identified different ideas for improvement.

Keywords Long-term care of the elderly · Accreditation of social service providers · Licensing of social services

1 Introduction. Elderly Population and Residential Care Services

According to the forecasts, in the European Union the share of elderly (65+) in the total population will gradually increase up to 40% in 2060 (doubling the current percentage), while the proportion of the adult population (active) will decrease till 52% [3]. The middle-aged children of the elderly appear to face with difficulty the two generations in their lives: parents and children. The phenomenon was called the "sandwich generation", the elderly's kids being their main caregivers [6]. Due to the growing restriction of extended family involvement in elderly care, it will be necessary to further develop home care and residential care services (Fig. 1).

Romania seems to follow the same trend of increasing the share of the elderly population in the total population, but growth will not be so high (28.9% in 2060) [7] (Fig. 2).

S. I. Bodogai (✉)
Faculty of Social and Human Sciences, Department of Sociology and Social Work, Oradea University, No. 1 University Street, Oradea City 410087, Bihor County, Romania

© The Author(s), under exclusive license to Springer Nature Switzerland AG 2021   177
D. Soitu et al. (eds.), *Decisions and Trends in Social Systems*,
Lecture Notes in Networks and Systems 189,
https://doi.org/10.1007/978-3-030-69094-6_15

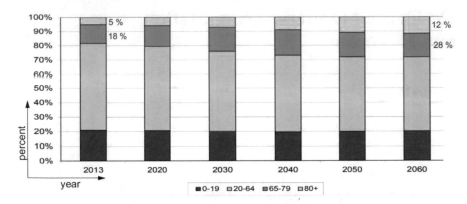

Fig. 1 Projection of changes in the structure of the EU population by main age groups. *Source* Eurostat, EUROPOP2013—apud European Commission [3], p. 25

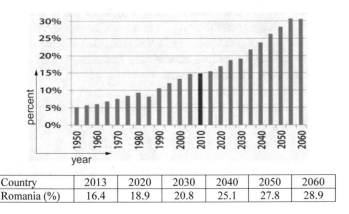

Country	2013	2020	2030	2040	2050	2060
Romania (%)	16.4	18.9	20.8	25.1	27.8	28.9

Fig. 2 Romania—elderly population (65+) as percent of total population. *Source* World Bank [7], p. 41, European Commission [3], p. 219

On 1st of January 2018, Romania's resident population was 19,523,621 people, of which 18% were elderly (65+). As it can be seen in the chart below, we are witnessing a feminisation of aging.

Demographic aging continues to worsen, the elderly population (65+) exceeding with 513,000 the young people (0–14 years old). The demographic aging index was 116.9 elderly per 100 young people. The average age of the population was 41.9 years and the median age was 42.2 years. Demographic dependency ratio was 50.9 young and elderly people (under 15 and over 64 years old) to 100 adults in working age (15–64 years old) [4] (Figs. 3 and 4).

Analysing the distribution of forms of care and support for the elderly in the European Union, we find that the Czech Republic and the Netherlands have the most extensive residential care systems. In Romania, there is a balanced distribution between cash benefits and services; the residential care sector is slightly less

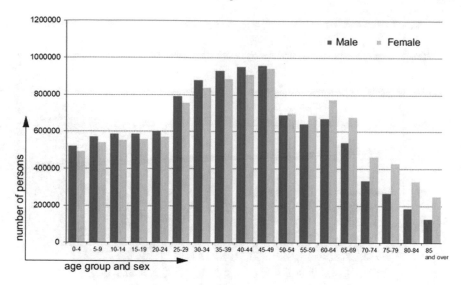

Fig. 3 Population by residence by age group and sex, January 1, 2018 (Romania). *Source* National Institute of Statistics [4]

developed than home care services [3]. The situation of residential centers for the elderly is as follows in Romania (2015): public centers 118 (total capacity 7693 places; average monthly number of beneficiaries 6615; waiting list 2797), private centers 194 (total capacity 7778 places; average monthly number of beneficiaries 6530) [5].

2 Providers' Accreditation and Licensing of Social Services in Romania

According to Government Ordinance No. 68 of 2003 on Social Services [8] and Law No. 197 of 2012 on Quality Assurance in Social Services [10], *providers can only grant social services if they are accredited for this purpose.* On the one hand we refer to the *accreditation of the supplier* (attested by the Accreditation Certificate), and on the other hand we refer to the *accreditation of each social service* (attested by an Operating License) that the supplier wishes to grant. The Accreditation Certificate authorizes the supplier for an indefinite period to provide licensed social services.

Providers of social services as well as social services granted by them *are accredited under the Law No. 197/2012 on Quality Assurance in Social Services* [10], as subsequently amended, and are registered in the Unique Social Services Electronic Register according the Law No. 292/2011, article 43 [9]. In order to highlight the degree of excellence of social services, it may be required (and is optional) to evaluate and rank them in quality classes.

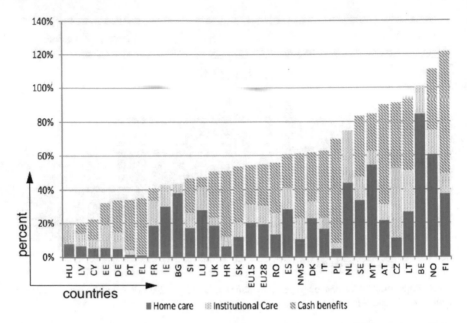

Fig. 4 Country-specific coverage rates of long-term care recipients, as % of dependent population (Median coverage rates between 2009-2013 in the EU and Norway; coverage estimated as ratio between recipients and potentially dependent population; recipient data, as provided by Member States; coverage may be above 100%, as some recipients may receive cash benefits and in-kind benefits at the same time, which is not corrected for in this graph. Population of potentially dependent based on EU-SILC data on "self-perceived longstanding limitation in activities because of health problems [for at least the last 6 months]" is used). *Source* European Commission, EPC—apud European Commission [3], p. 146

Within the accreditation process, the *Ministry of Labour and Social Justice* and the public institutions under its authority (the National Authority for the Protection of Child's Rights and Adoption, the National Authority for Disabled Persons, the National Agency for Equal Opportunities for Women and Men, the National Agency for Payments and Social Inspection) have the responsibility to develop the standards, criteria and indicators stipulated by the law as well as to organize, coordinate and implement the activities of assessment, certification, monitoring and quality control in the field of social services [11].

Quality in social services is the set of requirements and conditions that are met by suppliers and the social services they provide to meet the needs and expectations of the beneficiaries. The main *conditions for the operation of social services* are: providers can only grant social services if they have an Accreditation Certificate, and social services can operate on the territory of Romania only if they have an Operating License. Quality *assessment* in the field of social services consists in analysing how providers and social services meet the requirements of the law, as well as the specific quality requirements highlighted by a set of *standards, criteria and indicators.* The quality *monitoring* is aimed at maintaining the level of quality found and the level of beneficiaries' satisfaction and taking actions to further improve social services. Quality Assurance *Control* in the field of social services represents the activities carried out in order to verify compliance with the standards, criteria and indicators underlying the accreditation of social service providers and services, as well as the recognition of quality levels; this control is carried out by *social inspectors* (for accreditation/re-accreditation), respectively by assessment teams consisting in social inspectors and experienced personnel in the field within the Ministry of Labour (for qualitative classification).

Supplier accreditation is based on specific criteria that highlight its *ability to set up, administer and provide social services*. The Accreditation Certificate is granted *indefinitely*; if within 3 years the provider does not establish social services or fails to receive an Operating License for any of his social services, then the Accreditation Certificate is withdrawn and the deletion is made from the Unique Social Services Electronic Register with the possibility of requesting a new accreditation after at least 2 years. The supplier's accreditation involves the following *steps*: supplier's assessment based on criteria established by law; approval or rejection of the supplier's accreditation request; the issuance of the Accreditation Certificate or, as the case may be, the Notification of Rejection of the accreditation application; registration of the accredited supplier in the Unique Social Services Electronic Register, set up and administered by the Ministry of Labour and Social Justice [10].

The file required to apply for accreditation as a social service provider includes:

1. application for accreditation;
2. supporting documents on the establishment and status of the organization that differ according to: the status of the provider (legal person under public or private law) and his managerial experience in the field of social services;
3. public and private providers who did not set up a social service at the time of filing the application must have the following documents: CV of a person with experience/qualification in social services management or social work; copy of the employment contract or the service contract concluded with that person/the person just mentioned; commitment to meet the obligation to set up social services within maximum 3 years after obtaining the Accreditation Certificate.

Supplier's credentials file is filed directly at the registry office of the Ministry of Labour and Social Justice or can be sent by mail with acknowledgment of receipt or electronically in a format that cannot be edited [11].

Social Service Licensing: beyond the accreditation of the provider that certifies his ability to set up, administer and provide social services, the organization needs to accredit/license *separately each social service* it wishes to offer (meeting the minimum quality standards in providing that social service). The basis for the licensing of any social service is: possession of the Accreditation Certificate as a social service provider (earlier obtaining thereof); submitting the request 60 days prior to the start of the social service or the expiry of the previous Operating License. Licensing a social service involves the following *steps*:

a.   provider's request for social service licensing (for each social service);
b.   administrative evaluation of the social service licensing application file (of supporting documents and the self-assessment sheet completed by the supplier) according to the minimum standards provided by the law;
c.   the issuance of the *Provisional Operating License* for *one year* or, as the case may be, the Notification of Rejection of the license application;
d.   enrolment of the licensed social service in the Unique Social Services Electronic Register;
e.   planning by the County Agency for Payments and Social Inspection of the field evaluation of the social service, within maximum 30 days from receipt of the documents from the accreditation department of the Ministry of Labour and Social Justice (Provisional Operating License and the Social Service Self-Assessment Sheet);
f.   notification by the County Agency for Payments and Social Inspection of the social service provider of the date on which the on-site evaluation will take place (at least 15 days prior to it);
g.   the on-site evaluation of the social service by a team of 2 social inspectors and the drawing up of the assessment report, where it is recorded as the case may be: if the minimum standards are met, the Operating License is granted; if the minimum standards are respected only at 75%, it is decided to maintain the Provisional Operating License until the expiry of the 1st year period, with the condition that the minimum standards will be fully met; in case the minimum standards are not respected, the contravention sanction (1000–5000 lei) is applied, the provisional license is withdrawn and the deletion from the Unique Social Services Electronic Register takes place;
h.   the issuance of the *Operating License* for *5 years* or, as the case may be, the Notification of Rejection of the license application [10].

The file required to apply for accreditation of a social service (obtaining the Operating License) includes:

1.   Social Service Licensing Application (Annex 7 of Government Decision No. 118/2014);
2.   Supporting documents (in a copy certified by the person making the request):

   • Accreditation Certificate of the social service provider;
   • A document attesting the right of administration, concession or use over the space in which the social service operates, such as: Land Book Extract for Information (valid on the date of filing the application), contract for renting, concession, administration, exchange etc.;
   • Legal document regarding the decision of establishment of the social service;
   • Organization and Operation Rules of the social service (respecting the framework model approved by the Government Decision No. 867 of 2015);
   • Sanitary Authorization for Operation or, as the case may be, the document provided in the current health regulation procedure for the operation of activities at risk for the health of the population;

- Veterinary Sanitary Authorization for services providing food preparation and distribution activities;
- Fire Prevention and Extinction Opinion or the document that shows this was requested, registered at the depositing institution;
- Tax Registration Certificate, only in case of social services with legal personality;
- Collaboration Convention between the private provider and the County Council/Local Council on whose administrative-territorial range the social service operates;

3. Provider's commitment to notify any changes to the accredited social service that occurred after the granting of the Operating License;
4. Self-Evaluation Sheet completed by the provider (information in line with field reality).

The file for obtaining a social service license is submitted directly or sent to the Ministry of Labour and Social Justice in a sealed envelope, in A4 format, containing the name and address of the consignee and of the sender, with the mention "Social service accreditation request". The Social Service Licensing Application and the Self-Assessment Sheet are also sent to the e-mail address (acreditare@mmuncii.ro) at the same date [11].

In order to obtain the Operating License, residential centers for the elderly should obtain a score of between 104 and 114 points [1].

The following acts constitute offences and financial penalties are imposed: the lack of the Social Service License; lack of the Supplier's Accreditation Certificate; provision of non-compliant information; failure to notify the discontinuation/closure of the social service; failure to report changes; not requesting the accreditation of the social service 60 days prior to commissioning [10].

3 Minimum Quality Standards for Elderly Care Services in Residential Centers (Romania)

At the time when our research took place, "Order No. 2126 on 5th November 2014 of the Minister of Labour, Family, Social Protection and the Elderly on the Approval of Minimum Quality Standards for the Accreditation of Social Services for the Elderly, the Homeless, the Young People Who Left the Child Protection System and Other Adult Categories in Difficulty, as well as for Services Rendered in the Community, Services Provided in an Integrated System and Social Canteens" regulated in Annex 1 the minimum quality standards for the care in residential centers of elderly people. These standards were structured in 6 modules:

a. *Access to services*—this module includes standards for: information on the services offered; organizing informative materials (leaflets, website); access to the center, prior to the beneficiary's admission, based on a visit schedule; procedures for admitting the beneficiaries to the center, the necessary documents for admission and the conditions for termination of services. Required documents: informational materials, Visits Register, Record Keeping Register on Informing the Beneficiares, Beneficiary's Guide, Admission Procedure, Service Contract, personal files of the beneficiaries kept in closed cabinets, Record keeping register of archived files, The Cessation/Temporary Cessation Procedure of Services and Registry of the Exits of the Beneficiaries from the Center.

b.   *Assessment and planning* includes requirements for the assessment of the beneficiary prior to as well as after his/her placement in the center (social inquiry, socio-medical assessment). Annual reassessment of beneficiaries is carried out through the Re-evaluation Sheet, which must have a heading for the signature of the beneficiary or legal representative, in order to prove his involvement in the re-evaluation.

All services granted to the beneficiary are provided in accordance with an Individualized Assistance and Care Plan/Intervention Plan, in which the beneficiary/legal representative is also involved. This plan is adjusted once a year with the reassessment of the beneficiary.

The services granted are monitored on paper basis for each service provided: health status and treatments, functional recovery services and social integration/reintegration services [12].

Every elderly person should have an individualized care plan. Before providing any care, specialists should read this plan and other relevant records available to make the most accurate picture of the elderly, his specific needs, and the problems that have arisen in providing care. Before undertaking anything for the elderly, it's important to consider how you can do this to promote the elderly's independence.

A written plan is crucial for all staff engaged in elderly care. It grants a framework and structure for the help provided, offers a record and a reminder of the care provided, the points reached and the issues that need to be addressed. It will highlight changes in the situation of the elderly; it can also be used as a communication tool with the elderly and the other specialists involved [2].

c.   *Activities/Services*—this module includes standards for care provided to the beneficiaries: personal hygiene, medical assistance, recovery/rehabilitation (physiotherapy, ergo-therapy), socialization activities.

In order to maintain contact with the family and not only, there is a special room for visitors. Depending on the category of beneficiaries and its purpose/functions, the center has to carry out information, counselling, extracurricular education, occupational therapies, that address diverse needs: knowledge of social rights and legislation, community values, rights and obligations as citizens, training for independent living, facilitating access to housing and adapting it, access to employment, access to vocational and professional training etc., which are embodied in a social integration/reintegration program.

Centers have the obligation to set up their own procedure for terminal assistance or death, and the services provided in case of death are established together with the beneficiary/legal representative. The centers must place a partition curtain or, as the case may be, transfer the beneficiary to a room equipped with the appropriate equipment and medical equipment in case of death. Also, centers must provide spiritual assistance and inform the beneficiary's relatives within 2 h in case of death.

d. *The living environment*—this module establishes minimum requirements related to: location of the center and the necessary spaces, telephone and internet communication, video surveillance of the premises, accessibility of the premises for all beneficiaries, accommodation conditions (maximum 4 beds/room; minimum 6 m²/beneficiary), storage areas of the beneficiaries, keeping cleanliness.

e. *Rights and ethics*—the rights of the beneficiary are presented as well as the obligation of the center to inform him of the Charter of Rights through training sessions recorded in the Information Register. The Center has the obligation to measure the degree of satisfaction of the beneficiaries through a questionnaire or other own procedure.

The center activity is based on a Code of Ethics and there is an obligation to train the staff about it (Registry on Continuing Personnel Training).

In order to avoid or detect abuses, the Center has the obligation to establish a procedure for the identification, reporting and resolution of cases of abuse and neglect, to inform beneficiaries about types of abuse, to educate staff to prevent abuse and to apply questionnaires to beneficiaries and persons with whom they come into contact outside the center in order to prevent abuse. The Center provides a mailbox where they can submit written complaints about the negative aspects found, as well as proposals for improving the center's work.

The Center has the obligation to inform the competent institutions about all the special incidents in the center. The Center maintains a special register of special incidents that affect the beneficiary's physical and mental integrity (illnesses, accidents, aggressions etc) or other incidents in which he was involved (unauthorized departure, theft, immoral behaviour etc).

f. *Management and human resources*—the head of the center is called director when the social services have legal personality and is called chief center when the social services does not have legal personality (a graduate with a bachelor's degree or equivalent); staff are subject to regular medical check-ups established by law; contracts of employment/service provision/volunteering are available at the headquarters and employees are enrolled in the General Employee Records Registry [12].

In present (2020), Order No. 2126 on 5th November 2014 [12] was replaced by Order No. 29 on 3th January 2019 [13], who tries to improve the reglemetantion of minimum quality standards for the care in residential centers of elderly people. Specialists who applies these standards (social service providers and assessors/social inspectors) say that they are higher than the old ones, unfortunately things that were signaled as wrong or needing adjustments were not improved. We believe that these issues can be overcome when the legislator will take into account the opinion of specialists directly involved (at the moment there is no real consultation).

4 Conclusions

In Romania, the coverage of the needs of elderly is done through home care, residential care, day care centers, clubs specifically dedicated to this age group, social housing etc. Although home care is a viable alternative for some elderly, we find that we still have uncovered demand for residential care.

The situation of placement of people with dementia in centers is delicate. Elderly hospitalization in a residential center is subject to their consent or, in the absence of discernment, with the consent of the guardian/legal representative. In some European countries, the so-called "Biological Testament" is drawn up whereby the elderly person decides beforehand what protection measures will be taken in the future when he/she will no longer have a discernment due to medical problems. In Romania, families of elderly people with dementia either do not want to carry out the necessary procedures for the parent to be deprived of his rights, or they do not allow themselves from financial reasons or pressure of time. The service contract concluded with the beneficiary is considered null if a person with dementia has no legal representative. Specialists are thus confronted with families who urgently need support for their parents, but this support cannot be granted until legal proceedings are completed. We therefore see the need for preventive counselling and information to citizens about the procedures that apply in such situations.

The compulsory licensing of social services is considered useful, thus ensuring a minimum level of quality in all residential centers. In Romania, although there is good collaboration between suppliers and evaluators, the problems encountered are not so closely related to the licensing process and the required standards, although they have some gaps (signaled by suppliers as well as evaluators), as well as the financial support required to achieve a quality level considered decent.

A lot of the specialists directly involved in the protection of the elderly in Romania draw attention to the many documents that keep track of the services provided and emphasize that they reduce the time spent on direct work with the elderly in the centers.

The financial problem of institutionalizing older people is a major one. Although the aim is to ensure the quality of life of elderly people in residential centers in Romania, we note the impossibility of paying the costs of living in a center only from the beneficiary's pension. The very poor financial involvement of the state in supporting these social services (through subsidies) leads to an increase in the contribution of the beneficiary and the relatives.

In Romania, we note the need for a stronger involvement of the social work departments within the mayoralties in the analysis of the needs of the elderly and in the development of adequate social services.

By comparing the situation of Romania with other European states, we propose some solutions for improving the social protection system of the elderly:

- easier access by all the professionals directly involved (both practitioners and theorists) to relevant target group databases (example: population distribution by age group, demand for social services, social services offer, social services beneficiaries etc.);
- developing a unified database with all social service beneficiaries (clear evidence, avoiding overlapping, target group analysis and improvement opportunities);
- the current analysis of the needs of older people and the adaptation of services;
- reducing bureaucracy, simplifying procedures and managing documents (computer system, electronic signatures, limitation to maximum on using of physical documents);
- adapting social legislation to the feed-back of specialists directly involved in its application;
- increasing public funds allocated to social services, but also better management of these funds (correct and objective setting of priorities, strategic organization);
- development of prevention services (greater emphasis on prevention and not just on recovery or crisis intervention).

The provision of quality social services for any target group, including the elderly, is based on permanent and open communication between all stakeholders, as well as ongoing analysis of the dynamics of this system: creation of unitary databases, facilitation of researchers' access and use of research results in improving services.

References

1. National Agency for Payments and Social Inspection (2018). Accessed at 20-04-2018. http://www.mmanpis.ro/inspectie-sociala/conditii-de-acreditare-furnizori/
2. Bradshaw A, Merriman C (2007) Caring for the older person: practical care in hospital, care home or at home. Wiley, West Sussex, England
3. European Commission (2015). The 2015 ageing report: economic and budgetary projections for the 28 EU member states (2013–2060), p 25. Accessed at 15-03-2018. http://ec.europa.eu/economy_finance/publications/european_economy/2015/pdf/ee3_en.pdf. ISSN 1725-3217 (online), ISSN 0379-0991 (print)
4. National Institute of Statistics (2018) Comunicat de presă ["Press Release"]. Accessed at 01-10-2018. http://www.insse.ro/cms/sites/default/files/com_presa/com_pdf/comunicat_persoane_varstnice_2.pdf
5. National Institute of Statistics (2015) Accessed at 13-08-2016. http://statistici.insse.ro/shop/
6. Stern SH, Wolford B (2001) Putting the «Elder Care» in Elder Law. J Gerontol Soc Work. 34(3):55–59. https://doi.org/10.1300/J083v34n03_08

7. World Bank (2014) "Viață lungă, activă și în forță. Promovarea îmbătrânirii active în România" ["Long Life, Active and Forceful. Promoting Active Aging in Romania"], p 41. Accessed at 25-04-2018. http://www.mmuncii.ro/j33/images/Documente/Familie/DGAS/IA-RO/Raport_Act ive_Aging_RO.pdf

8. GO No 68 (2003) "Ordonanța Guvernului nr. 68 din 28 August 2003 privind serviciile sociale" ["Government Ordinance No. 68 on 28th August 2003 on Social Services"]. Published in Official Journal of Romania No. 619 on 30th August 2003

9. Law No. 292 (2011) "Legea nr. 292 din 2011 a asistenței sociale" ["Law No. 292 of 2011 on Social Work"]. Published in Official Journal of Romania Ist Part No. 905 on 20th December 2011

10. Law No. 197 (2012) "Legea nr. 197 din 1 noiembrie 2012 privind asigurarea calității în domeniul serviciilor sociale" ["Law No. 197 on 1st November 2012 on Quality Assurance in Social Services"]. Published in Official Journal of Romania No. 754 on 9th November 2012

11. GD No. 118 (2014) "Hotărârea de Guvern nr. 118 din 19 februarie 2014 pentru aprobarea Normelor metodologice de aplicare a prevederilor Legii nr. 197/2012 privind asigurarea calității în domeniul serviciilor sociale" ["Government Decision No. 118 on 19th February 2014 for the Approval of the Methodological Norms for the Application of the Provisions of Law No. 197/2012 on Quality Assurance in Social Services"]. Published in Official Journal of Romania No. 172 on 11th March 2014

12. Order of MLFSPE No. 2126 (2014) "Ordin nr. 2126 din 5 noiembrie 2014 al ministrului muncii, familiei, protecției sociale și persoanelor vârstnice privind aprobarea Standardelor minime de calitate pentru acreditarea serviciilor sociale destinate persoanelor vârstnice, persoanelor fără adăpost, tinerilor care au părăsit sistemul de protecție a copilului și altor categorii de persoane adulte aflate în dificultate, precum și pentru serviciile acordate în comunitate, serviciilor acordate în sistem integrat și cantinelor sociale" ["Order No. 2126 on 5th November 2014 of the Minister of Labour, Family, Social Protection and the Elderly on the Approval of Minimum Quality Standards for the Accreditation of Social Services for the Elderly, the Homeless, the Young People Who Left the Child Protection System and Other Adult Categories in Difficulty, as well as for Services Rendered in the Community, Services Provided in an Integrated System and Social Canteens"]. Published in Official Journal of Romania No. 874 on 2th December 2014

13. Order of MLSJ No. 29 (2019) "Ordin nr. 29 din 3 ianuarie 2019 al ministrului muncii și justiției sociale pentru aprobarea Standardelor minime de calitate pentru acreditarea serviciilor sociale destinate persoanelor vârstnice, persoanelor fără adăpost, tinerilor care au părăsit sistemul de protecție a copilului și altor categorii de persoane adulte aflate în dificultate, precum și a serviciilor acordate în comunitate, serviciilor acordate în sistem integrat și cantinele sociale" ["Order No. 29 on 3th January 2019 of the Minister of Labour and Social Justice on the Approval of Minimum Quality Standards for the Accreditation of Social Services for the Elderly, the Homeless, the Young People Who Left the Child Protection System and Other Adult Categories in Difficulty, as well as for Services Rendered in the Community, Services Provided in an Integrated System and Social Canteens"]. Published in Official Journal of Romania No. 121 on 15th February 2019

Disenfranchised Grief, Burnout and Emotional Distress Among Nursing Home Workers in Spain: A Validation Study

Ana Vallejo Andrada⬝, José Luis Sarasola Sánchez-Serrano⬝,
Evaristo Barerra Algarín⬝, and Francisco Caravaca Sánchez⬝

Abstract In Spain, as well as in the majority of Europe, elderly people are a social group which has been increasing during the last few decades, consequently our society has created new specific resources orientated towards them, an example of this is nursing homes. Nursey Homes constitute not just a new area for seniors, also a new working environment, with new risk and challenges. With the aim to study this new environment we have created a questionnaire composed with the most presented variables in previous researchers among nursing home workers: Depression, Anxiety, and Stress Scales; The Maslach Burnout Inventory; Grief Support in Health Care Scale; Disenfranchised Grief. With a sample of 90 nursey home workers this study objective is probe the association between these variables and the possibility to used them harmonically in a single questionnaire.

Keywords Nursing homes · Grief · Burnout · Emotional distress · Disenfranchised grief

A. Vallejo Andrada (✉)
Pablo de Olavide University, Utrera Road 1, 41013 Seville, Spain
e-mail: avaland@upo.es

J. L. Sarasola Sánchez-Serrano
Jaen University, Campus Las Lagunillas, S/N, 23071 Jaén, Spain
e-mail: jlsarsan@upo.es

E. Barerra Algarín
Department of Social Work and Social Services, Pablo de Olavide University, Seville, Spain
e-mail: ebaralg@upo.es

F. Caravaca Sánchez
Department of Psychology, Social Work and Social Services Area, Jaén University, Jaén, Spain
e-mail: caravaca@ujaen.es

© The Author(s), under exclusive license to Springer Nature Switzerland AG 2021
D. Soitu et al. (eds.), *Decisions and Trends in Social Systems*,
Lecture Notes in Networks and Systems 189,
https://doi.org/10.1007/978-3-030-69094-6_16

1 Introduction

In Spain, as well as in the majority of Europe, elderly people are a social group which has been increasing during the last few decades, consequently our society has created new specific resources orientated towards them, an example of this is nursing homes.

Nursey Homes constitute not just a new area for seniors, also a new working environment, with new risk and challenges. Which is normally compose for a wide range a professional: Social Workers, nurses, nursing assistants, psychologists, occupations therapists, physiotherapists, who must work as a single group in order to give the elderly and their families the best attention.

This new working environment have certain characteristics that make it different to others, some authors have made some comparison between nursey homes environments and hospitals, as in both cases they professionals must deal with the residents and their families and death.

With the aim to analyse nursey home working sphere, we have decided to create a questionnaire which englobe ones of the most common problematic that could interfere in the professional's labour.

1.1 Burnout

Burnout, first described by Freudenberger [13], is a psychological condition involving a prolonged response to enduring interpersonal stressors [17]. Fredenberger [12] summarized the signs of burnout including the hopeless, fatigued, bored, resentful, disenchanted, discouraged, confused quickness to anger, instantaneous irritation, frustration responses, totally negative attitude, etc. Burnout was further developed independently by Maslach [21, 22] to be characterised by three domains: emotional exhaustion (EE), depersonalisation (DP), and a diminished sense of personal accomplishment (PA).

Emotional exhaustion occurs when the level of emotional resources becomes depleted such that individuals feel unable to give anymore of themselves emotionally. Depersonalisation describes feelings of cynicism towards clients, whilst reduced personal accomplishment refers to forming a negative evaluation of one's work ability [21]

1.2 Depression Anxiety and Stress

The World Health Organization (WHO) in 2011—"Mental health is defined as a state of well-being in which every individual realizes his or her own potential, can cope with the normal stresses of life, can work productively and fruitfully, and is able to make a contribution to her or his community." More than 450 million people suffer from a type of mental disorder [25]. Among the mental health disorders; depression, anxiety, and stress form a large proportion [28].

1.3 Disenfranchised Grief

Much has been written on general patterns of grief. Though Elisabeth Kübler-Ross's five stages of grief denial, anger, bargaining, depression, and acceptance still dominate most teaching and training, many aspects of her theory have been challenged, particularly the notion of progressive stages that end in acceptance [5, 30, 31].

While these general patterns serve as an important backdrop, the concept of disenfranchised grief, first advanced by Doka in 1989 and refined since [11], carries relevance to home care workers. Health care workers' grief is often described as "disenfranchised" [3, 24] or grief "that is not openly acknowledged, socially validated, or publicly observed" [11].

Within the context of mourning the death of another person, the primary distinction that sets disenfranchised grief apart from a normal range of grief experience is the lack of social recognition and validation of the significance of the bereaved person's relationship with the deceased [29, 34].

1.4 Burnout and Grief

Several researchers have already demonstrated the relation between Burnout and Grief, one of these examples is Adwan research: Paediatric Nurses' Grief Experience, Burnout and Job Satisfaction 1 in which he surveyed 120 paediatric nurses recruited from a large midwestern academic medical centre's paediatric patient care units and paediatric float pool RN's. For this sample, he created a questioner composed by: The Revised Grief Experience Inventory (RGEI); The Maslach Burnout Inventory (MBI) and Index of Work Satisfaction for Nurses (IWS).

His analysis revealed a statistically significant positive correlation between the RGEI total score and MBI's EE subscale ($r = 0.38$, $p = 0.001$), a significant positive correlation with the MBI's DP subscale ($r = 0.19$, $p = 0.04$), and a negative correlation with the MBI's PA subscale ($r = 0.244$, $p = 0.009$). Most RGEI subscales are individually correlated to the subscales of the MBI subscales in the same direction of the total RGEI scores [1].

Another example of these variable associations could be Anderson and Ewen study, which is specifically focuses in the nursing home area, titled Death in the Nursing Home, where the researchers surveyed 380 NAs working in 11 nursing homes in Midwestern city.

Their questioner was composed by the 21-item Texas Revised Inventory of Grief (TRIG), the 12-item Personal Growth subscale of the Hogan Grief Reaction Checklist (HGRC); 25-item Maslach Burnout Inventory (MBI) and 8-item Physical Well-Being and the 8-item Psychological Well-Being subscales from the Perceived Well-Being Scale-Revised.

Whit it, the obtained several conclusions, the most relevant ones for our study are: that grief maybe an important contributor to emotional, physical, and occupational well-being in this group of health care workers; NAs who are having problems processing grief would have higher levels of emotional exhaustion and a more difficult time interacting with and humanizing residents; which NAs who experienced greater difficulties with their grief reported higher levels of burnout (Aderson and Even 2011)

1.5   Burnout and Depression Anxiety and Stress

The association between Burnout and stress have been demonstrated by many authors, for instance, recently Hunter et al. [16] in their research: Midwives in the United Kingdom: Levels of burnout, depression, anxiety and stress and associated predictors, surveyed a total of 1997 midwives with a questionnaire compose by Depression, Anxiety and Stress Scale (DASS-21) and Copenhagen Burnout Inventory (CBI). From which they obtained as results significant levels of emotional distress recorded by most participants. Eighty three percent (n = 1464) of participants scored moderate and above for personal burnout and 67% (n = 1167) recorded moderate and above for work-related burnout. Over one third of participants scored in the moderate/severe/extreme range for stress (36.7%), anxiety (38%) and depression (33%). Which probe an association between burnout and depression anxiety and stress [16].

Another example of this association is Martínez-Monteagudoa, Inglés, Granadosc, Aparisia and García-Fernándeza´s research title Trait emotional intelligence profiles, burnout, anxiety, depression, and stress in secondary education teachers. I which we can appreciate a questionnaire created from Trait Meta-Mood Scale-24 (TMMS-24); Maslach Burnout Inventory (MBI) and Scales of Depression, Anxiety, and Stress-21 (DASS-21).

With this instrument, they conclude with a sample of different teachers of different school levels, that the surveyed who shows significantly higher scores in Emotional Exhaustion, Depersonalization, Depression, Anxiety, also seems to have a higher level of burnout but lower Personal Accomplishment scores [20].

1.6   Depression Anxiety and Stress and Grief

With the aim to show the relation with depression, anxiety, stress and grief, we are going to reference some previous research that have been done in this file.

Boelena, Reijntjesc and Smid in their article Disorders Concurrent and prospective associations of intolerance of uncertain twitch symptoms of prolonged grief, posttraumatic stress and depression after bereavement, show the relation between these variable, using a sample of 265 participants, with a questionnaire compose by Prolonged grief disorder scale (PGD scale); PTSD symptom scale-self-report version (PSS-SR); Hospital anxiety and depression scale-depression scale (HADS-D); Intolerance of uncertainty scale-12 items version (IUS-12); Penn state worry questionnaire-abbreviated (PSWQ-A); Ruminative response scales (RRS) brooding scale and ten item personality inventory (TIPI) neuroticism scale. Providing as a conclusion that there is a strongly and significantly concurrently associated with grief anxiety and depression [4].

As well as Carmassi and colleagues [7] their study title Adult separation anxiety differentiates patients with complicated grief and/or major depression and is related to lifetime mood spectrum symptoms, where they used Inventory of Complicated Grief (ICG); Adult Separation Anxiety Symptom Questionnaire (ASA-27); Work and Social Adjustment Scale (WSAS) and the Mood Spectrum Self-Report (MOODS-SR) to create a questionnaire for 70 people. As one of this study conclusions they prove the relation between depression, anxiety, stress and grief [7].

1.7 Burnout: Depression Anxiety and Stress and Grief

Experiences like Lyckholm's research in clinic professionals who deal with cancer patient death, associate professional grief for patient death and stress with burnout symptomatology. In her study, she collected several patient death experiences and relate them with stress and burnout symptomatology proving their association [19].

Considering, the previous researches in this area and the theory already explained we propose the following hypothesis:

H1 Depression, Anxiety, and Stress Scales (DASS-21); The Maslach Burnout Inventory (MBI); Grief Support in Health Care Scale and Disenfranchised Grief Scales could be used in a single questionnaire to describe these phenomena among nursey home workers.

To this hypothesis, we propose the objective to verify that the Maslach Burnout Inventory (MBI); Grief Support in Health Care Scale and Disenfranchised Grief Scales, could be used in a single questionnaire to describe theses phenomenon among nursey home workers.

2   Method

2.1   Design and Participants

A cross-sectional methodology was used between December 2019 to March 2020, applying several demographic, occupational and psychological measurement instruments to a sample of nursing home workers in the province of Seville (southwestern Spain). Current research was approved by the Ethics Committee of the Pablo Olavide University (Spain), and by the board of executives of all facilities. The sample for current validation study was obtained from 90 working in two nursing homes located in the province of Seville, composed by 76 female (with a mean age of 34.6 years, SD = 8.7) and 14 male (with a mean age of 33.9 years, SD = 7.4) nursing staff who ranged in age from 22 to 60. Regarding work characteristics, over half were staff workers (61.1%) and mostly working in private companies (82.2%)

2.2   Procedure

Recruitment of the participants and data collection was conducted in several steps. Firstly, two facilities were randomly selected from the province of Seville (near to the Pablo Olavide University, where the current research was conducted). Secondly, the Director of each facility were contacted by phone or email by the Principal Investigator (PI) who were informed about: (a) about the nature and purpose of the research; (b) that data would be exclusively used for academic proposes; and (c) that participation was totally anonymous and voluntary; and (d) that their decision to participate or not would have no consequences for the facilities or nursing home workers. All nursing homes that were approached agreed to participate. Thirdly, and once approval to proceed was obtained for each facility, facility administrators identified an employee in each facility who would serve as the point of contact (mainly social workers). These points of contact in each facility, contacted directly with the workers and provide detailed information about the study, characteristics, consent forms, thus, workers who were interested in participate in the study were provided with the PI email and also a web address with the link address to the questionnaire.

In the fourth stage, data collection, was conducted using questionnaires online via Google Form, this tool has been proven to be methodologically viable and particularly useful for ensuring satisfactory statistical power by previous studies as a collection tool [14, 15]. At the beginning of the online questionnaire, the purpose of the investigation was found, and information to the participant about the possibility to stop participating in the study at any time. The questionnaire was voluntary and anonymous (without collecting any personal information) in order to guarantee the confidentiality of the participants. In addition, the research complied with the ethical principles for the investigations of the Helsinki statements [26].

2.3 Measurement Instruments

For the validation among nursing workers in Spain, demographic and work variables designed "ad hoc" specifically for the current research and four validated psychological assessment instruments were analysed.

Demographic and work information. Participants completed background information on their age (coded as continuous variable), gender (female or male), position in the nursing home (staff or supervisor), work sector (public administrator or companies), working hours (working full time or working part time) years of nursing home experience (coded as continuous variable), and duration of employment at the present nursing home (coded as continuous variable).

Depression, Anxiety, and Stress Scales (DASS-21). Adapted by Lovibond and Lovibond [18], DASS-21 is a short form of 42-item self-report measure of Depression, Anxiety, and Stress (DASS) developed by Brown and colleagues [6]. DASS-21 was designed to measure three subscales of emotional distress: symptoms of Depression (e.g., loss of self-esteem/incentives and depressed mood), Anxiety (e.g., fear and anticipation of negative events), and Stress (e.g., persistent state of over arousal and low frustration tolerance) over the past week. Each subscale is composed of 7-item subscales (thus is composed by a total of 21 items) measured using 4-point Likert scales ranging from 0 ("did not apply to me at all") to 3 ("applied to me very much or most of the time"), and each subscale was score individually (ranging from 0 to 21). In the current research, the Spanish version of DASS-21 validated by Daza and colleagues [9] was used.

Burnout. The Maslach Burnout Inventory (MBI) developed by Maslach and Jackson [21] was used. The MBI is a self-reported 22-item scale composed by three subscales: nine items measuring emotional exhaustion (extent to which one feels overextended and exhausted by one's work), five items focus depersonalization (negative attitudes and behaviors toward care recipients) and eight items measuring personal accomplishment (feelings of competence and achievement at work). Each item is measured on a seven-point Likert scale ranging from 1 ("completely disagree") to 7 ("completely agree"), thus the maximum score of the total burnout scale is 154, reporting a higher score a higher overall level of burnout. This Spanish version of the MBI has been found to have strong psychometric properties [10].

Grief Support in Health Care Scale. Developed by Anderson et al. [2], the Grief Support in Health Care Scale (GSHCS) is a self-reported 15-item scale composed by three subscales (five items per subscale): (1) recognition of the relationship (defined as "support for grief is predicated on the recognition and authentication of the relationships that exist between healthcare workers and those who are under their care"), (2) acknowledgement of the loss (it is theorized that support for grief is marshaled only when losses are acknowledged), and (3) inclusion of the griever (Facilitating and including healthcare workers in such grief rituals, e.g. remembrance ceremonies, funerals, can be critical elements of support in their grief processes). Each item is measured on a seven-point Likert scale ranging from

1 ("strongly disagree") to 5 ("strongly agree"). A higher score on the GSHCS indicate higher levels of grief specific support. The Spanish version of GSHCS adapted and validated by Vega et al. [32] was used, reporting strong psychometric properties.

Disenfranchised Grief. In the current research, disenfranchised grief were assessed used the validated version by Anderson and Gaugler [3] among nursing homes in the United States. The Disenfranchised Grief scale is a 15-item self-report questionnaire was designed to determine the presence and strength of factors that may disenfranchise the grief, measured using 5-point Likert scales ranging from 1 ("strongly disagree") to 5 ("strongly agree"). Reverse scores on items 11 through 14. The Grief Symptoms Scale has been used with a diverse variety of sample and countries, including Spain [27].

2.4 Analysis Plan

The data were statistically analyzed using the SPSS software (version 22.0 for Windows) in three steps: firstly, the mean score (M) and Standar Desviation (SD) were calculated for each scale used (overall and sub scales). Additionally, internal consistency of each scale used (overall and sub scales) was estimated using the Cronbach's alpha coefficient, as well as the 95% bootstrap percentile confidence interval (ICBP95%) is estimated, for which 1000 re-samples were performed in each estimate [8]. Secondly, Pearson's correlation was calculated (in order to avoid to analyze multicollinearity between overall scales and subscales) to analyze the correlation between overall scores of the DASS21, the MBI-GS, the GSHCS, and the Disenfranchised Grief scales. Finally, a second Pearson's correlation was calculated between DASS21 subscales (depression, anxiety and stress), MBI subscales (emotional exhaustion, depersonalization and personal accomplishment), GSHCS subscales (Recognition of the relationship, Acknowledgement of the loss and Inclusion of the griever). A significant level of p < .05 were adapted for all the contrasts.

3 Results

Table 1 shows a summary of the descriptive statistics and the values of the total Cronbach's alpha coefficient and for each of the 4 scales and subscales. Regarding DASS21, the mean score for the full scale was 13.6 (SD = 10.6), and 3.5 (SD = 4.0), 4.0 (SD = 4.1) and 67 (SD = 4.5) for the depression, anxiety and stress subscales, respectively. The Cronbach's alpha for the overall score was $\alpha = 0.92$, with $\alpha = 0.84$, $\alpha = 0.84$ and $\alpha = 0.86$ for the depression, anxiety and stress subscales, respectively. Exploring descriptive data for burnout scale (MBI), the mean score for the full scale (22 items) was 65.3 (SD = 21.5), ranging from 10.2 (SD = 5.6) in depersonalization subscale to 31.4 (SD = 9.2) to personal accomplishment. The Cronbach's alpha

Table 1 Descriptives for the scales

Protective factors	Items	M	SD	α	ICα 95%
DASS21-sum	21	13.6	10.6	0.92	0.92–0.94
DASS-depression	7	3.5	4.0	0.85	0.80–0.89
DASS-anxiety	7	4.0	4.1	0.84	0.79–0.88
DASS-stress	7	6.7	4.5	0.86	0.81–0.90
MBI-sum	22	65.3	21.5	0.87	0.83–0.90
MBI-emotional exhaustion	9	19.0	14.3	0.93	0.91–0.95
MBI-depersonalization	5	10.2	5.6	0.72	0.65–0.76
MBI-personal accomplishment	8	31.4	9.2	0.90	0.87–0.93
GSHCS-sum	15	50.1	10.4	0.88	0.84–0.91
GSHCS-Recognition of the relationship	5	20.0	4.2	0.88	0.83–0.91
GSHCS-Acknowledgement of the loss	5	17.0	4.5	0.91	0.87–0.93
GSHCS-Inclusion of the griever	5	13.0	4.7	0.79	0.71–0.85
Disenfranchised grief	15	48.7	8.5	0.71	0.62–0.79

Note SD: Standard Desviation; α the Crombach: Crombach Alfa Coeficcient; CI: Confidence Interval

for the MBI scale was 0.87 (full scale), with $\alpha = 0.93$, $\alpha = 0.72$ and $\alpha = 0.90$ for emotional exhaustion, depersonalization and personal accomplishment. Regarding grief symptoms (measured by the GSHCS scale), the full scale shown a mean score of 50.1 (SD = 10.4), and 20.0 (SD = 4.2), 17.0 (SD = 4.5) and 13.0 (SD = 4.7) for recognition of the relationship, acknowledgement of the loss and inclusion of the griever, respectively. The Cronbach's alpha for the GSHCS (full scale) was 0.87 and for the subscales ranged from 0.79 (inclusion of the griever) to 0.91 (acknowledgement of the loss). Finally, the Disenfranchised Grief was characterized by a mean of 48.7 (SD = 8.5), showing strong psychometric properties (Cronbach's alpha $\alpha = 0.71$). Overall, all the scales and subscales used to show strong psychometric properties with high internal consistency, ranging from $\alpha = 0.71$ (Disenfranchised Grief) to $\alpha = 0.93$ (MBI-emotional exhaustion).

Table 2 shows the Pearson correlations obtained in the working nursing homes dimensions between the overall scores of the DASS, the MBI-GS, the GSHCS, and

Table 2 Correlations between scales explored

Variables	1	2	3	4
1. DASS21-sum	1			
2. MBI-sum	0.55*	1		
3. GSHCS-sum	−0.29	−0.36	1	
4. Disenfranchised grief	−0.16	−0.30	0.67*	1

*p < .05

the Disenfranchised Grief scales. As indicated, DASS21 were significantly intercorrelated positively MBI (r = 0.55; p < 0.01). Regarding grief symptoms, GSHCS presented greater significance with Disenfranchised Grief (r = 0.55; p < 0.01). It is interesting that there were not any negative or positive statically significant correlation between DASS21 with grief (measured with GSHCS or Disenfranchised Grief) and between MBI with grief (GSHCS or Disenfranchised Grief).

Table 3 presents bivariate Pearson correlations for the study variables, including DASS21 subscales (DASS-depression, DASS-anxiety and DASS-stress), MBI subscales (MBI-emotional exhaustion, MBI-depersonalization and MBI-personal accomplishment), GSHCS subscales (GSHCS-Recognition of the relationship, GSHCS-Acknowledgement of the loss and GSHCS-Inclusion of the griever) and Disenfranchised Grief.

DASS-depression was positively and moderately correlated with DASS-anxiety and DASS-stress (r = 0.59; p < 0.01 and r = 0.69; p < 0.01, respectively), with each of the three MBI subscales, as MBI-emotional exhaustion (r = 0.77; p < 0.01), MBI-depersonalization (r = 0.34; p < 0.01) negative associated with MBI-personal accomplishment (r = -0.30; p < 0.01) and with Disenfranchised Grief (r = 0.25; p < 0.05). DASS-anxiety showed positive correlations with DASS-stress (r = 0.74, p < 0.01) and with MBI-emotional exhaustion (r = 0.40; p < 0.01) and negative correlations with MBI-personal accomplishment (r = -0.29, p < 0.001). DASS-stress showed positive correlations with two measures of burnout, specifically MBI-emotional exhaustion (r = 0.54; p < 0.01) and MBI-personal accomplishment (r = 0.33, p < 0.001) and with GSHCS-Acknowledgement of the loss (r = 0.24; p < 0.05).

Regarding MBI subscales, MBI-emotional exhaustion was positive associated with MBI-depersonalization (r = 0.54; p < 0.01), GSHCS-Recognition of the relationship (r = 0.24; p < 0.05) and negative with associated with GSHCS-Inclusion of the griever (r = -0.37, p < 0.01) and Disenfranchised Grief (r = -0.33, p < 0.05). MBI-depersonalization was exclusively positive associated with MBI-personal accomplishment (r = 0.39, p < 0.01). MBI-personal accomplishment was positively associated with GSHCS-Recognition of the relationship (r = 0.34; p < 0.01) and Disenfranchised Grief (r = 0.40; p < 0.01). Finally, the subscales of GSHCS were positively intercorrelated, showing the strongest correlations between GSHCS-recognition of the relationship with GSHCS-Acknowledgement of the loss (r = 0.55, p < 0.01) and between GSHCS-Inclusion of the griever with Disenfranchised Grief (r = 0.58; p < 0.01).

4 Discussion and Conclusion

To our knowledge, this is the questionnaire to be designed for nursing home workers which associate these variables in Spain and one of the few exceptions worldwide [19]. As all the scales used in this research (DASS-21; MBI; GSHCS and DG) have a Cronbach's alpha coefficient superior than 0.70 we can affirm they have a high reliability.

Table 3 Correlations between subscales explored

Variables	1	2	3	4	5	6	7	8	9	10
1. DASS-depression	1									
2. DASS-anxiety	0.59**	1								
3. DASS-stress	0.69**	0.74**	1							
4. MBI-emotional exhaustion	0.77**	0.40**	0.54**	1						
5. MBI-depersonalization	0.34**	0.55	0.12	0.54**	1					
6. MBI-personal accomplishment	−0.30**	−0.29**	0.33**	0.05	0.35**	1				
7. GSHCS-Recognition of the relationship	−0.11	0.16	−0.05	0.24*	0.08	0.34**	1			
8. GSHCS-Acknowledgement of the loss	0.08	0.16	0.24*	0.09	0.03	0.16	0.53**	1		
9. GSHCS-Inclusion of the griever	−0.26*	0.30	0.02	−0.37**	0.20	0.17	0.42**	0.34**	1	
10. Disenfranchised grief	0.25*	−0.03	−0.10	−0.33*	−0.11	0.40**	0.55**	0.45**	0.58****	1

*p < .05; **p < .01; ***p < .001

In which concern to our hypothesis and our objective: verify that the Maslach Burnout Inventory (MBI); Grief Support in Health Care Scale and Disenfranchised Grief Scales, could be used in a single questionnaire to describe theses phenomenon in nursey home workers. We can affirm that our hypothesis has been confirmed and our objective fulfil.

In concordance with previous studies ([16, 20]. We have found an association between MBI and DASS 21 as well as between GSHCS and DGS, however we have not found previous studies to associate these two scales are very innovative instruments.

Moreover, we have also found association between the subscale of DASS 21, DASS-depression with the DGS scale and the subscale of GSHCS, Inclusion of the griever. And the subscale DASS 21, DASS-stress with the GSHC subscales GSHC-Recognition of the relationship and GSHCS-Acknowledgement of the loss. Which in concordance with other researches probe the association between stress and Grief [4, 7].

An association between MBI subscales and GSHCS or DG has also been found, in concordance with other authors [2, 1]. MBI-emotional exhaustion is associated with GSHCS-Recognition of the relationship, GSHCS-Inclusion of the griever and Disenfranchised Grief. And MBI-personal accomplishment is associated with GSHCS-Recognition of the relationship and Disenfranchised Grief.

For this reason, as a final conclusion we can verify that the Maslach Burnout Inventory (MBI); Grief Support in Health Care Scale and Disenfranchised Grief Scales, could be used in a single questionnaire to describe theses phenomenon in nursey home workers.

References

1. Adwan J (2014) Pediatric nurses' grief experience, burnout and job satisfaction. J Pediatr Nurs 29(1):329–336
2. Anderson KA, Ewen HH, Miles EA (2010) The grief support in healthcare scale: development and testing. Nurs Res 59(6):372–379
3. Anderson KA, Gaugler JE (2007) The grief experiences of certified nursing assistants: personal growth and complicated grief. OMEGA J Death Dying 54(4):301–318
4. Boelen P, Reijntjes A, Smid G (2016) Disorders Concurrent and prospective associations of intolerance of uncertainty with symptoms of prolonged grief, posttraumatic stress, and depression after bereavement. J Anxiety Disord 41(1):65–72
5. Bonanno GA, Boerner K (2007) "The stage theory of grief": comment. J Am Med Assoc 297(1):2693–2693
6. Brown TA, Chorpita BF, Korotitsch W, Barlow DH (1997) Psychometric properties of the depression anxiety stress scales (DASS) in clinical samples. Behav Res Ther 35(1):79–89
7. Carmassi C, Gesi C, Corsi M, Pergentini I, Cremone IM, Conversano C, Perugi G, Shear MK, Dell'Osso L (2015) Adult separation anxiety differentiates patients with complicated grief and/or major depression and is related to lifetime mood spectrum symptoms. Compr Psiquiatryc 58(1):45–49
8. Davison AC, Hinkley DV (1997) Bootstrap methods and their application. In: Cambridge series in statistical and probabilistic mathematics

9. Daza P, Novy DM, Stanley MA, Averill P (2002) The depression anxiety stress scale-21: Spanish translation and validation with a Hispanic sample. J Psychopathol Behav Assess 24(1):195–205
10. De las Cuevas C (1994) El desgaste profesional en atención primaria: presencia y distribución del síndrome de burnout. Laboratorios Servier, Madrid
11. Doka KJ (2002) Disenfranchised grief: new directions, challenges, and strategies for practice. Research Press, Champaign, IL. ISBN 978-0-87822-427-2
12. Fredenberger HJ (1975) The staff burn-out syndrome in alternative institutions. Psychother Theor Res Prac 12(1):73–82
13. Freudenberger HJ (1974) Staff burn-out. J Sociol 30(1):159–165
14. Gosling SD, Vazire S, Srivastava S, John OP (2004) Should we trust web-based studies? A comparative analysis of six preconceptions about internet questionnaires. Am Psychol 59(2):93–104
15. Horton JJ, Rand DG, Zeckhauser RJ (2011) The online laboratory: conducting experiments in a real labour market. Exp Econ 14(3):399–425
16. Hunter B, Fenwich J, Sidebothan M, Henley J (2019) Midwives in the United Kingdom: levels of burnout, depression, anxiety and stress and associated predictors. Midwifery 79(1):1–12
17. Leiter MP, Maslach C (2009) Nurse turnover: the mediating role of burnout. J Nursi Manag 17(3):331–339
18. Lovibond P, Lovibond S (1995) Manual for the depression, anxiety, stress scales. Psychology Foundation of Australia, Sydney, NSW
19. Lyckholm L (2001) Dealing with stress, burnout, and grief in the practice of oncology. Lancet Oncol 2(12):750–755
20. Martínez-Monteagudo MC, Inglés CJ, Granados L, Aparisi D, y García-Fernández JM (2019) Trait emotional intelligence profiles, burnout, anxiety, depression, and stress in secondary education teachers. Personal Indiv Differ 142:53–61
21. Maslach C, Jackson SE (1981) The maslach burnout inventory. Consulting Psychologist Press, Palo Alto, CA
22. Maslach C, Jackson SE (1981) The measurement of experienced burnout. Maslach C. Burnout: A multidimensional perspective. In: Schaufeli WB, Maslach C, Marek T (eds) Professional burnout: recent developments in theory and research 1993. Taylor and Francis, Washington DC, pp 19–32
23. Maslach C (1993) Burnout: a multidimensional perspective. In Schaufeli WB, Maslach C, Marek T (eds) Series in applied psychology: social issues and questions. Professional burnout: recent developments in theory and research, pp. 19–32. Taylor & Francis
24. Moss SZ, Moss MS (2002) Nursing home staff reactions to resident deaths. In: Doka KJ (ed) Disenfranchised grief: New directions, challenges, and strategies for practice. Research Press, Champaign, IL, pp 197–216
25. Murthy RS, Haden A, Campanini B (eds) (2001) Mental health: new understanding, new hope. World health report. ISBN 92 4 156201 3
26. Organización Medica Mundial (2008) Declaración de Helsinki de la Asociación Médica Mundial. Principios éticos para las investigaciones médicas en seres humanos. Colegios Oficiales de Médicos, Madrid
27. Pascual ÁM, Santamaría JL (2009) Proceso de duelo en familiares y cuidadores. Revista Española de Geriatría y Gerontología 44(1):48–54
28. Rao S, Ramesh N (2015) Depression, anxiety and stress levels in industrial workers: A pilot study in bangalore, india. Ind Psychiatry J 24(1):23–28
29. Silva DW (2000) A grief recovery for gay men. Infinity Publishing, West Conshohocken, PA
30. Silver RC, Wortman CB (2007) The stage theory of grief. J Am Med Assoc 297(24):2692–2694
31. Stroebe M, Stroebe W, Schut H, Boerner K (2017) Grief is not a disease but bereavement merits medical awareness. The Lancet 389(1):347–349
32. Vega P, Melo J, González R (2015) Validación de escala de apoyo en duelo en atención de salud para población hispanoparlante. Psicooncologia 12(2–3):355–366

33. WHO Statistics, India, World Health Organisation, 2010, Geneva. [online 10-04-2020]. Available online at http://www.who.int/countries/ind/en/
34. Whipple V (2005) Lesbian widows: invisible grief. New York, NY. J LGBT Issues Couns 7(1):87–116

Informal Care of Dependent Older People—Support Services and Funding Schemes

Luise Mladen-Macovei and Andra-Bertha Sănduleasa

Abstract This chapter is dealing with aspects regarding informal care of dependent older people. Based on the analysis of the information obtained from the good practice and the informal care regulatory status in Romania, the authors proposed a set of support services and funding schemes that could be implemented in our country as follows: care leave and care allowance; palliative care leave; prolonged care leave; the employment of a personal assistant under the terms of the Framework Law no. 153/2017; allowance for long-term care; pension for severe dependence; flexible working arrangements for the informal caregiver; the possibility of early retirement up to 2 years before reaching the standard retirement age, without penalizing the amount of pension rights; respite care. The methodology was based on the good practices research, using qualitative and quantitative information from bouth primary and secondary sources of information.

Keywords Informal care · Long-term care · Dependent older people

1 Introduction

Informal care of people, especially of the older people, represents from the historical perspective the most traditional form of organization of care and social welfare in a society.

Informal carers are generally family members, close relatives, friends, neighbours, unprofessional/untrained people in the field of long-term care and who often provide care without a well-established timetable for contact with the care-receiver, the care activity being often outside the limits imposed by a paid activity.

L. Mladen-Macovei (✉) · A.-B. Sănduleasa
National Scientific Research Institute for Labour and Social Protection, Povernei 6-8, 010643 Bucharest, Romania
e-mail: lmladen@incsmps.ro

A.-B. Sănduleasa
e-mail: andra.sanduleasa@olt.insse.ro

© The Author(s), under exclusive license to Springer Nature Switzerland AG 2021
D. Soitu et al. (eds.), *Decisions and Trends in Social Systems*,
Lecture Notes in Networks and Systems 189,
https://doi.org/10.1007/978-3-030-69094-6_17

203

Support services for carers such as care leave accompanied by the financial benefits received by the informal carer can help to better reconcile work and family life. Care arrangements also have a positive impact on the division of care responsibilities within the family and can contribute to a more equal sharing of care tasks among family members. Such measures shaped to enable carers to better reconcile work and family life are key elements for family organization in the future in most European Union Member States [1, 16].

Although the contribution of informal care to reduce government spending with long-term care is recognized, the adverse effects on carers (stress, fatigue, etc.) as well as on cared persons who may not always benefit from the best care should not be ignored. In cases where family members can provide care for the elderly, they must be supported by measures such as the conclusion of flexible employment contracts to prevent early withdrawal from the labour market, provision of care leave for elderly people paid at least partially, identification of informal carers and provision of counselling services [5].

According to the researchers in the field, there are serious reasons for encouraging the growth of the informal care phenomenon. However, an increased level of care (for more than 20 h per week) is associated with a decrease in labour market participation, the probability of informal carers to find a job being lower than that of non-carers. Moreover, carers who have a job work on average 2 h less than non-carers and have in a higher extent part-time jobs [22]. However, lower labour market participation and fewer hours worked are associated with a higher risk of poverty and increased prevalence of mental health problems among carers compared with non-carers, as care for a family member is stressful, particularly in the case of carers who live with cared persons. In addition, caring for a family member can increase expenditures with heating, medication, transport, etc., and for these reasons governments need to find solutions to support informal carers.

Numerous studies highlight the difficulty of combining paid work with the role of informal carer for an older person. Some of the studies show that the number of hours of care puts pressure on the balance between work and family life [14], while others argue that the number of hours spent at work has overloaded informal carers [13]. To help reconcile work and family life, governments have made it possible for carers to opt for care leave or for a flexible work program [33]. However, the degree of use of such support measures depends on several factors. One of these factors is the sense of awareness: the positioning of one's own person as an informal carer [24] and the congruence of the different roles the carer has—caregiver, son/daughter, husband/wife [21]. In the case of chiefs from the workplace, the sense of aware-ness refers to the knowledge of the persons in their subordination who care for dependent elderly people and their needs related to the arrangements at the work-place. Workplace culture, workplace values and attitudes of colleagues are important factors regarding the option for an arrangement or other at work so that the duties of informal carer for a family member to be successfully covered.

Some studies show that the likelihood of using public care is higher for the elderly living alone. Other determinants for the choice of the type of long-term care are age, health status, degree of dependency and the existence of a descendant who can care for the dependent older person [2, 20].

In social research as well as in political decision-making process, the situation of informal carers who have a job has attracted attention relatively late [6, 18]. The care of the older persons has for many years been neglected in the discourse on long-term care but also on the balance between working and family life. In the light of the increase in women's participation in the labour market, of the prolonged labour market participation as a result of reforming pension systems and of the demographic change, the situation of employed people who care for a family member has become a central topic of public policy agenda. Thus, it can be noticed that the last years' speech includes in the theme of reconciling professional life with family life not only the topic of childcare as dependent persons, but also of older people [6].

The challenge for governments in the coming years will be to cope with a growing number of older people in need of long-term care, amid a drop in the working population, which will lead to a lower availability of carers for the older persons.

2 Good Practices Regarding Informal Care

Assuming that we cannot talk about a universal character of best practices in a field, and that the applicability of good practice is time-dependent, but also socially and politically dependent, we appreciate that examples of good practice have proved to be extremely useful in formulating national policies, being considered smart practices [3], innovative initiatives [4] or inspirational landmarks for decision-making [17].

The ultimate goal of the analysis of best practices is to identify and propose support services and funding schemes to support informal carers of dependent elderly people in Romania, starting with the analysis of support services developed at the level of five European countries identified as models of good practice.

The analysis of good practices took into account both primary sources of information and secondary sources of information (desk research), appreciating that it is not enough to focus only on the analysis of the legislative provisions in the investigated countries, given that the identification of some long-term care policies that to cover also the needs of informal carers of older persons are a challenge even for states recognized for policies in the field of long-term care, in the context of the demographic aging phenomenon.

In addition, the analysis of good practices focused mainly on the identification of the measures/support services oriented to informal carers, respectively access to information, training, exchange of experience, care leave, flexible working conditions, allowances for care, health insurance and social insurance (pensions) etc. However, the analysis did not rule out presentation of indirect measures, which are not directly targeted at caregivers, but which were implemented in order to support informal care at the level of the analyzed countries.

Out of the multitude of approaches regarding informal care for the older persons, the experiences of five European countries were selected, as follows:

- **Czech Republic**—a former communist country with increased use of informal care and increased support for informal care, policy-oriented support for informal care such as retirement schemes and paid leave for informal carers;
- **The Netherlands and Sweden**—countries with lower use of informal care but with increased support for informal care, with national policies aimed at the informal sector, oriented towards generous, accessible, formalized support schemes; with a system for locating informal carers at national health and social assistance levels;
- **Spain and Austria**—countries with increased use of informal care and increased support for informal care, known for the combination of support services provided to informal carers, including counselling, information, training and relaxation schemes for informal carers.

The ultimate challenge in building support services for the informal care of the elderly is to take into account both the needs of informal carers and of the beneficiaries of care—elderly people, given that the needs of informal carers do not coincide always with the needs of dependent elderly people, and policy-makers need to clarify their needs by supporting policies and avoid potential conflicts that may arise. Practically, as far as informal carers are concerned, we are talking about their dual role, both care providers as informal carers and beneficiaries of long-term care policies in relation to their own needs of support.

Based on these considerations, it is important to review the specifics of each of the analyzed countries, enforcing the following issues:

- What is the role of the family in the care of the older persons?
- At what level is offered the care of the older persons by informal carers as a share in the total of the long-term care services?
- What is the link between the care of the family members and the participation of informal carers on the labour market and how difficult is to combine the paid work with the care responsibilities, including gender aspects?
- Public spending on long-term care, including expenditures by type of long-term care (residential/home/other).

Long-term care is seen as a family responsibility in the *Czech Republic*. The long-term care system in the Czech Republic is going towards moving from an institutional care system to a system that ensures the aging of people in a family environment, at home. Public policies therefore are focused on the development of social services accessible at the local level, particularly non-residential services, and provide support for families that care for the older persons. The strategy is consistent with the fact that the majority of the informal carers have a job, with 80% of them having a full-time job [28]. In the Czech Republic, three quarters of the informal carers are women, informal carers generally taking part from the age group of 55–64.97% of the occasional care needs and 78.5% of the daily care needs are covered by family members or friends [8].

More than 80% of long-term care is covered by family members (informal carers). Indeed, it has also been found that women's employment rates drop significantly in the 55–59 and 60+ age groups (compared to other age groups), the age at which most women become informal carers for their family members, while employment rates for men remain high. Thus, women in the Czech Republic become more often economically inactive than males as a result of taking over care tasks [8, 9, 30].

Recent studies show that most informal carers in the Czech Republic provide care for moral and emotional reasons. Financial problems, the risk of job loss, health deterioration, mental and physical fatigue, social isolation are key elements in the decision of informal carers to continue to take care of family members. Informal caretakers consider that the financial support from the authorities is low and does not cover all the needs of the dependent family member [19]. At the time of 2014, more than 80% of the dependent persons benefiting from the care allowance were over 75 years of age [15].

The Netherlands has a long tradition of institutionalized long-term care services. Although this country is known as a good provider of public care services, in 2015 a novelty element was introduced into the Dutch long-term care system, according to which people should be able to live longer in their own homes. As a result, informal care now plays an important role alongside public care services, which is why Spain, like Spain, is one of the highest rates of informal care—18% of the total population (Fig. 1).

In order to cover both elderly care needs and financial sustainability control, the Dutch government promotes both labour market participation and the provision of informal care for relatives and friends. In the Netherlands about one fifth of informal carers have a paid job. The number of informal paid jobseekers has increased in recent years and is expected to grow even more. However, some of the informal carers who also have a job say they feel overloaded [29]. At OECD level, the Netherlands ranks second after Sweden in the highest spending countries in the long-term care sector—3.7% of GDP (Fig. 1).

Sweden has a comprehensive long-term public care system, and family member care is seen as an additional service to public services, without being part of the Swedish family policy. Besides, there is no legal obligation to take care of family members. However, with the reform of the social services system in 2009, local public authorities are obliged to provide support to informal carers. Thus, in conjunction with a gradual decline in the formal care system, elderly care is increasingly focused on the family. For this reason, issues related to the reconciliation of professional life (care of family members) with professional life are a relatively new phenomenon in this country.

Life expectancy in Sweden is above the EU average for both women and men. On the other hand, it is worth noting that in the Swedish vocabulary the word "addiction" is not used, as neither the well-known international expression "long-term care" is used. Instead, policies are defined as elderly care policies, in the context in which they are different from policies for people with disabilities in Sweden. The main care services for the elderly are home and residential services. On average, a home care recipient benefits on average 7 h a week [31], but the intensity of care varies a great deal, from a visit per month to a few visits in one day. In recent years, there has been a strong decline in the coverage of elderly care services in Sweden. Thus, if in 1980

	CZECH REPUBLIC	SPAIN	AUSTRIA	SWEDEN	NETHERLAND
Public expenditure on long-term care (social and medical component) (% of GDP) in 2015, OECD (2017)	1.3%	0.8%	1.2%	3.2	3.7%
Expenditure on long-term care by type of services, OECD (2017)	82% - Residential care 14% - Home care 4% - Others (day centres, outpatient)	66% - Residential care 23% - Home care 11% - Others (day centres, outpatient)	42% - Residential care 57% - Home care 1% - Others (day centres, outpatient)	64% - Residential care 31% - Home care 6% - Others (day centres, outpatient)	86% - Residential care 11% - Home care 3% - Others (day centres, outpatient)
Employment rate among female population by age groups (2017), EUROSTAT (lfsa_ergan)	24.3% - 15-24 years 77.5% - 25-49 years 84.9% - 50-59 years 29.8% - 60-64 years 4.7% - 65+	19.7% - 15-24 years 68.3% - 25-49 years 57.2% - 50-59 years 32.9% - 60-64 years 1.5% - 65+	49.0% - 15-24 years 81.2% - 25-49 years 72.6% - 50-59 years 17.2% - 60-64 years 3.3% - 65+	46.0% - 15-24 years 83.7% - 25-49 years 84.5% - 50-59 years 66.0% - 60-64 years 7.0% - 65+	63.6% - 15-24 years 79.2% - 25-49 years 71.7% - 50-59 years 44.7% - 60-64 years 3.7% - 65+
Informal carers as% of total population (2016), European Commission (2018)	9% - T 9% - M 9% - F 6% - 18-34 11% - 35-64 9% - 65+	16% - T 13% - M 19% - F 14% - 18-34 19% - 35-64 11% - 65+	10% - T 8% - M 12% - F 4% - 18-34 11% - 35-64 15% - 65+	12% - T 10% - M 15% - F 12% - 18-34 14% - 35-64 10% - 65+	18% - T 13% - M 23% - F 10% - 18-34 23% - 35-64 17% - 65+
The difficulty of combining paid work with care responsibilities (difficult and very difficult), Eurofound (2016)	56.5% - T 58.6% - M 54.7% - F	51.3% - T 52.3% - M 50.4% - F	26.1% - T 28.5% - M 23.9% - F	29.8% - T 24.6% - M 35.2% - F	22.5% - T 20.9% - M 24.0% - F
Distribution of working age population (18-64 years) by occupational status (2016), Eurofound (2016)	91.5% - Non-Carers 3.6% - Carers with a paid job 4.9% - Carers without a job 8.5% - All carers	87.6% - Non-Carers 6.6% - Carers with a paid job 5.8% - Carers without a job 12.4% - All carers	94.9% - Non-Carers 2.6% - Carers with a paid job 2.5% - Carers without a job 5.1% - All carers	95.7% - Non-Carers 3.6% - Carers with a paid job 0.7% - Carers without a job 4.3% - All carers	90.7% - Non-Carers 6.3% - Carers with a paid job 3.0% - Carers without a job 9.3% - All carers
Percentage of people taking care of family members, neighbours or friends with disabilities or infirmities aged 75 or over at least a few days a week, Eurofound (2016)	4% - T 3% - M 5% - F 2.7% - Employed 5.5% - Not employed	7.8% - T 5.4% - M 10% - F 6.7% - Employed 8.7% - Not employed	4.1% - T 3.1% - M 5.2% - F 2% - Employed 6.6% - Not employed	3.2% - T 4% - M 2.4% - F 2.3% - Employed 4.6% - Not employed	5.6% - T 4.8% - M 6.3% - F 4.5% - Employed 6.9% - Not employed

Fig. 1 Trends in public spending for long-term care and reconciliation between work and family life

Percentage of people not in employment who find it difficult or very difficult to combine paid work (10 hours per week) with care responsibilities, Eurofound (2016)	47.4% - T 54% - M 43.9% - F	34.3% - T 26% - M 39.3% - F	35.0% - T 29.3% - M 38.3% - F	25.0% - T 22.6% - M 26.8% - F	34.4% - T 26.2% - M 40.3% - F
Population 65+ living in multigenerational households (2017) - Female / Male, EUROSTAT (ilc_lvps30)	8.6 % - T 12.5% - M 5.7% - F	17.5% - T 23.4% - M 12.9% - F	9.0% - T 12.4% - M 6.3% - F	2.3% - T 3.6% - M 1.2% - F	2.8% - T 4.2% - M 1.5% - F
Population 65+ living alone (2017) – Female / Male, EUROSTAT (ilc_lvps30)	31.9% - T 18.1% - M 42.2% - F	24.6% - T 17.7% - M 29.9% - F	33.0% - T 20.9% - M 42.4% - F	39.0% - T 28.5% - M 48.0% - F	31.2% - T 19.7% - M 41.2% - F
Population 65+ living without children (2017) - Female / Male, EUROSTAT (ilc_lvps30)	49.8% - T 65.1% - M 38.3% - F	39.9% - T 49.3% - M 32.6% - F	49.4% - T 62% - M 39.6% - F	57.2% - T 66.9% - M 48.9% - F	64.7% - T 75.4% - M 55.4% - F
Family role, Eurpean Commission (2007)	54% would prefer to be cared by a family member in their own home 36% considers that the best care solution for older parents is to live with their children	48% would prefer to be cared by a family member in their own home 39% considers that the best care solution for older parents is to live with their children	39% would prefer to be cared by a family member in their own home 17% considers that the best care solution for older parents is to live with their children	34 would prefer to be cared by a family member in their own home 4% considers that the best care solution for older parents is to live with their children	33% would prefer to be cared by a family member in their own home 4% considers that the best care solution for older parents is to live with their children

Fig. 1 (continued)

16% of people over 65 years of age had formal home care, their share dropped to 9% in 2012 [25]. Over the same period, the share of persons over 80 years of formal home care has decreased from 34% to 23%. Moreover, since 2000, every fourth bed of the residential care system has disappeared, without this decline being compensated for by home care services [32]. People in the 45–65 age group are generally informal carers for their parents, while people over the age of 65 are informal carers for their spouses/life partners.

Even though in Sweden more daughters than sons provide support to dependent parents, and working-class daughters are most affected by the reduction in elderly care services, it is important to emphasize that informal care provided by the family is less intensive and more evenly distributed between family members compared to other European care regimens [26].

The model of long-term care in **Spain** is family-based. Most informal carers in Spain are women, men rarely opting for care leave for dependent relatives, for reasons that are part of the country's cultural model. Spain has one of the highest rates of informal care (16% of the total population).

In addition to family members' carers, in Spain the older people are often cared by non-family carers, immigrant women. If the person in need of care is a man, he is generally cared for by his wife/partner. If a woman needs care, her primary carer is the daughter (36%).

The informal carer's profile in Spain is that of a 55-year-old woman, married, with primary education and no paid occupation. The number of informal care hours in Spain is among the highest in the OECD: more than 20 h a week [7].

Long-term care in *Austria* is characterized by an increased share of informal care. Financing "24-hour care for 24 h" ("24-Stunden-Betreuung") as a benefit in kind is considered to be a feature of the long-term care system in Austria. Care leave is intended to enable informal carers to identify new care solutions for a dependent relative and does not target informal carers for more than 3 months. The reconciliation of work and family life is encouraged through part-time work arrangements and benefits that substitute for salary.

In Austria, about 80% of elderly people and people with disabilities are cared for by informal carers, family members, most of whom are women. Although the average age of these informal caregivers is high, informal caregivers of active age are often met. Approximately 77% of informal carers suffer from mental disorders, while 24% feel tired, 19% feel pressured and 11% feel financially burdened. For hired workers, conflicts between the requirements of the two spheres are added to these strains [27].

3 Aspects Regarding Informal Care in Romania

In Romania, the responsibility for the care of dependent older people is mainly attributable to family members in the context in which there is no structure of formal support for them. The role of informal carers and their related support mechanisms have been scarcely addressed by legislation. In addition to the possibility of 15 days' unpaid care leave, which serves primarily for short-term emergencies, the only existing support mechanisms for carers of family members are the employment of the local council as a "personal assistant" of relatives for people with severe disabilities and reducing the working hours of informal carers who have a paid job and supporting the salary rights for that part from the local budget, corresponding to the monthly gross wages of the home caregiver.

The factors that hinder the development and implementation of measures to support informal carers in Romania, factors that resulted from the pilot survey, are the following:

- The lack of financial resources for the development of home care services, but also for the payment of informal carers;

- The lack of qualification/training of potential informal carers, on the one hand, and of specialists with long-term care responsibilities (geriatric doctors, counsellors, psychologists, social assistants), on the other hand;
- Reluctance to the authorities: both the reluctance of the elderly to be monitored by SPAS specialists, and the reluctance of family members to be monitored for their care of dependent elderly people in their care.

Factors that could stimulate the development and implementation of support measures for informal carers who provide home care to elderly people in Romania identified by investigation are:

- Funding the local authorities for the development and licensing of social services to enable support measures for informal carers;
- Hiring additional staff at the SPAS level, as well as training of the existing staff;
- Improving the legislative framework and simplifying procedures;
- Organizing local information campaigns through which both dependent elderly people and their families understand the importance of social assistance and the role of monitoring informal/formal care at home.

4   Conclusions and Recommendations

Based on the analysis of the information obtained from the good practices and the informal care regulatory status in Romania, a set of support services and funding schemes that could be implemented in our country were proposed as follows:

- **Leave and allowance for care**: the right to care leave for dependent elderly family members for a period of 1–3 months. During care leave, caregivers will benefit from a care allowance amounting to 85% of the amount of the country's gross minimum wage guaranteed. Care periods are considered as periods contributing to the social security system, so they are taken into account in the calculation of the pension rights. Paid care leave is a useful measure at the beginning of the care period, when the elderly is accommodated in the dependence situation and is more in need of the presence of a family member around him.
- **Prolonged care leave**: Employees are entitled to up to two-year care leave for the care of a family member, an elderly person, who is in a dependency of grade IA, IB or IC (corresponding to the dependent persons the provisions of GD No 886/2000). Employees are guaranteed the right to return to work on the same job in the first year or in a similar position in the following year, and the period of care leave is considered as working time when calculating the pension in the first year of leave. Leave may be unpaid if the employee has not earned income taxed in the last 24 months for 12 consecutive months. If they have earned professional earnings during the last 24 months for 12 consecutive months, then during leave, carers are entitled to an allowance amounting to around 85% of the average net income earned over the last 12 month, however, provided with a higher threshold. Thus, the government takes over the retirement insurance and health insurance

payments, and periods of care are considered as periods contributing to the social security system, so they are taken into account in the calculation of the pension rights. Care leave can also be taken in a part-time system.

- **Respite care**: the right of informal carers to take a holiday every year in order to recover from the stress accumulated during the care of the elderly and during the holidays the care of the elderly to be ensured through the public service of social assistance. For this purpose, a period of accommodating of the dependent elderly person with the professional caregiver should be ensured and the latter should be monitored by the sending unit so that the care needs are adequately covered.

- **Long-term care allowance**: dependents who need care and are enrolled in grades IA, IB or CI on the basis of the National Evaluation Grid for the elderly according to GD no. 886/2000, are entitled to a monthly care benefit irrespective of income if the care services in the home town are not available or are not suitable for their individual needs. The amount of the allowance may vary depending on the degree of dependence, the number of necessary hours of care, the type of care and will be determined by reference to the Social Reference Indicator (ISR). By choosing to receive the long-term care allowance directly by the dependent elderly person, she/he assumes responsibility for providing herself with the basic and instrumental activities of her daily life or paying her/him formal/informal care from the allowance received. Beneficiaries of the allowance should be monitored by the staff of public social service to ensure that the allowance is used for the purpose for which it was granted and that the elderly receive the care they need. The measure would be very useful, especially for elderly people whose carers have a job and cannot take care of people with a high degree of dependence.

- **The right of workers to adapt their working hours, to opt for a reduction in work schedules and/or to choose the place where it is easier to carry out their paid work (telework)** when they prove that they have dependent elderly people in a situation of addiction to grade IA, IB or IC (corresponding to the dependent persons according to the provisions of GD no 886/2000). In the case of reduced working hours, the working time cannot be less than 10/20 h a week. For employers to be open to such practices, they should also have some benefits, such as tax incentives.

- **Personnel assistant under the terms of the Framework Law no. 153/2017 (similar to the personal assistants of persons with disabilities)**: public services of social assistance concludes a contract for the employment of a family member or other non-professional caregiver for the care of a dependent elderly person (assigned to one of the grades IA, IB or IC according to the provisions of Government Decision no 886/2000). The caregiver (similar to the personal assistant of a disabled person) has the status of insured person in the social security system. Caregivers acquire pension entitlements and contribute to the health insurance scheme during their care period, and less paid contributions may be taken into account than other insured persons (currently there is the possibility of employing a carer under the conditions of Article 13 paragraph 1 of Law 17/2000).

- **The possibility of early retirement up to 2 years before reaching the standard retirement age, without penalizing the amount of pension rights, for**

the purpose of caring for an elderly dependent person classified in IA, IB or IC grades. The standard retirement age from which the population can obtain a retirement pension is currently 65 years for men and 63 years for women. Persons who care for dependent elderly people in grades IA, IB or IC may benefit from a reduction in the standard retirement age by up to two years if they have reached the minimum contribution period (currently 15 years) without to incur penalties for the amount of pension rights. The minimum contribution period does not include assimilated periods.

Other proposals:

1. Regulating the state of the informal caregiver so that it can benefit from support services and funding schemes. Obtaining the status of informal caregiver will lead to the acquisition of specific rights but will also lead to the engagement of certain obligations;
2. Provision of training courses in the field of health care for informal carers (first aid and not only) in an organized setting;
3. Provide for informal carers the right to be regularly evaluated by specialized staff and to receive the care they need, be it psychological counselling or health care;
4. Conduct wider information campaigns on the rights of informal carers to benefit from support services and funding schemes.

Acknowledgements The work presented in this chapter has been supported by The Ministry of Labour and Social Justice, through The Research and Development Programme for the period 2018–2020 (according to *Ordin nr. 2042/2017 al ministrului muncii și justiției sociale privind aprobarea Planului sectorial de cercetare dezvoltare al Ministerului Muncii și Justiției Sociale pentru perioada 2018-2020)*—Contract No. 3209/20.08.2018.

References

1. Bakx P, De Meijer C (2013) The influence of spouse ability to provide informal care on longterm care use. SSRN Electron J Available from: https://www.researchgate.net/public ation/272244827_The_Influence_of_Spouse_Ability_to_Provide_Informal_Care_on_Long-Term_Care_Use
2. Bakx P, Schut E, Van Doorslaer E (2013) Can risk adjustment prevent risk selection in a competitive long-term care insurance market? Tinbergen institute discussion paper, TI 2013-017/V. Available from: https://www.econstor.eu/handle/10419/87476
3. Bardach E (2000) A practical guide for policy analysis: The eightfold path to more effective problem solving. Chatham house Publishers, New York
4. Bendixsen S, Guchteniere P (2003) Best practices in immigration services planning. J Policy Anal Manag 22(4)
5. Bonsang E (2009) Does informal care from children to their elderly parents substitute for formal care in Europe? J. Health Econ 28(1):143–154

6. Bouget D, Spasova S, Vanhercke B, European Social Policy Network—ESPN (2016) Work-life balance measures for persons of working age with dependent relatives in Europe. A study of national policies. European Commission, Brussels
7. Colombo F, Llena A, Mercier J, Tjadens F (2011) Help wanted? Providing and paying for long-term care. OECD Health Policy Studies, OECD Publishing, Paris
8. Dudová R (2018) Care allowance as 'special money': the meyearsngs and uses of the care allowance in close relationships. Gend Res 19(1):58–80
9. Dudová R (2016) Position of caregivers in the Czech Republic: analysis of the concept and implementation of social policy in the Czech Republic with focus on care for seniors. Alternativa 50+, Praga. Available from: http://alternativaplus.cz/wp-content/uploads/2013/02/Position-of-caregivers-in-the-Czech-Republic_e-version_final.pdf
10. Eurofound (2016) European quality of life survey 2016. Available from: https://www.eurofound.europa.eu/data/european-quality-of-life-survey
11. European Commission (2018) Informal care in Europe. Exploring formalisation, availability and quality. Available from: https://ec.europa.eu/social/BlobServlet?docId=19681&langId=en
12. Eurpean Commission (2007) Eurobarometer "health and long-term care in the European Union" (Eurobarometru Special 283), pp 67, 97. Available from: http://ec.europa.eu/commfrontoffice/publicopinion/archives/ebs/ebs_283_en.pdf
13. Fredriksen-Goldsen KI, Scharlach AE (2006) An interactive model of informal adult care and employment. Commun Work Fam 9(4):441–455
14. Henz U (2006) Informal caregiving at working age: effects of job characteristics and family configuration. J Marriage Fam 68(2):411–429
15. Hirose K, Czepulis-Rutkowska Z (2016) Challenges in long-term care of the elderly in central and Eastern Europe. International Labour Orgyearszation. Available from: https://www.ilo.org/wcmsp5/groups/public/—europe/—ro-geneva/—sro-budapest/documents/publication/wcms_532427.pdf
16. Hoyer S, Reich N (2016) Leave and financial support for family caregivers in EU member states. Institute for Social Work and Social Education, Frankfurt. Available from: http://www.sociopolitical-observatory.eu
17. Jennings ET Jr (2007) Cele mai bune practici in administratia publica. Cum le recunoastem? Cum le putem folosi?
18. Kröger T, Yeandle S (2013) Reconciling work and care: an international analysis. In: Kröger T, Yeandle S (eds) Combining paid work and family care. Policies and experiences in international perspective. Policy Press, Bristol, pp 3–22
19. Křížová E, Janečková H, Běláček J (2016) Family carers' perspectives on integrated community care in the Czech Republic. Cent Eur J Public Health 24(4):289–296. Available from: https://cejph.szu.cz/pdfs/cjp/2016/04/07.pdf
20. De Meijer C, Koopmanschap M, Uva TB, van Doorslaer E (2011) Determinants of long term care spending: age, time to death or disability? J Health Econ 30(2):425–438
21. Montgomery RJV, Rowe JM, Kosloski K (2007) Family caregiving. In: Blackburn JA, Dulmus C (eds) Handbook of gerontology. Evidence-based approaches to theory, practice, and policy. Wiley, Hoboken, pp 426–446
22. OECD (2011) Providing and paying for long-term care. Available from: https://www.oecd.org/els/health-systems/47884865.pdf
23. OECD (2017) Long-term care expenditure. In: Health at a glance. OECD Indicators, OECD Publishing, Paris. Available from: https://www.oecd-ilibrary.org/docserver/health_glance-2017-81-en.pdf?expires=1538680281&id=id&accname=guest&checksum=A6B6553A51CC6FC95A2CB817EBC7A031
24. O'Connor DL (2007) Self-identifying as a caregiver: exploring the positioning process. J Aging Stud 21(2):165–174
25. Peterson E (2017) Eldercare in Sweden: an overview. Revista Derecho Social y Empresa, nr. 8. Available from: https://www.dykinson.com/cart/download/articulos/8280/
26. Rodrigues R, Huber M, Lamura G (2012). Facts and figures on healthy ageing and long-term care. European Centre for Social Welfare Policy and Research. Available from: https://www.euro.centre.org/downloads/detail/3059

27. Sardadvar K, Mairhuber I (2018) Employed family carers in Austria. The interplays of paid and unpaid work—beyond "reconciliation". Österreich Z Soziol **43**:61–72. Available from: https://link.springer.com/content/pdf/10.1007%2Fs11614-018-0283-0.pdf

28. Sowa A (2010) The long-term care system for the elderly in the Czech Republic. Available from: http://www.ancien-longtermcare.eu/sites/default/files/ENEPRI%20RR%20No%2072%20ANCIEN%20Czech%20Republic.pdf

29. Stelpstra M (2017) Fulltime work and informal care. The role of awareness and work related care arrangements. Universitatea din Utrecht. Lucrare de master. Available from: https://dspace.library.uu.nl/bitstream/handle/1874/338580/Master%20thesis%20Marinda%20Stelpstra%20-5625173.pdf?sequence=2&isAllowed=y

30. Svarcova I, Hoskova-Mayerova S, Navratil J (2016) Crisis management and education in health. In: The european proceedings of social and behavioural sciences EpSBS. **XVI**:255–261. http://dx.doi.org/10.15405/epsbs.2016.11.26

31. Szebehely M, Trydegård GB (2011) Home care in Sweden: a universal model in transition. Health Soc Care Community 20(3):300–309

32. Ulmanen P, Szebehely M (2015) From the state to the family or to the market? Consequences of reduced residential eldercare in Sweden. Int J Soc Welf 24(1):81–92

33. Werk and Mantelzorg (2016) CAO inspirator. Available from: https://docs.google.com/viewer?url=http://www.werkenmantelzorg.nl/%2FUserFiles%2Ffiles%2Falgemeen%2FWM_CAO_INSPIRATOR_juni_2015.pdf

Perceptions and Attitudes Concerning Ageing and Older People: A Qualitative Approach

Mihaela Ghenţa and Aniela Matei

Abstract The research objectives were to identify opinions, misconceptions and stereotypes about older people, as well as public policy measures that could help to overcome negative stereotypes and to promote a positive image of older people. Data collection included a qualitative research with representatives of public and private, social service providers for the elderly and representatives of organizations promoting the interests of the elderly. Findings highlighted that there is a major interest for the elderly and for its image in Romanian society.

Keywords Ageing · Older people · Qualitative methodology

1 Introduction

This chapter aims to contribute to the understanding of the perceptions and attitudes about ageing and older persons and it examines the opinions expressed by the main actors involved in the process delivery of social services for elderly. The chapter is organised as follows: the first part present theoretical aspects concerning older people, prejudices and stereotypes based on the age of a person, and it continues with an analysis of the data concerning older persons at national level, as well as of data collected through European Social Survey, Wave 4 in 2008. The last part is dedicated to the qualitative research applied in order to depict the current image of the older persons in Romanian society.

M. Ghenţa (✉) · A. Matei
National Scientific Research Institute for Labour and Social Protection (INCSMPS), Bucharest, Romania
e-mail: ghenta@incsmps.ro

A. Matei
e-mail: aalexandrescu@incsmps.ro

© The Author(s), under exclusive license to Springer Nature Switzerland AG 2021   217
D. Soitu et al. (eds.), *Decisions and Trends in Social Systems*,
Lecture Notes in Networks and Systems 189,
https://doi.org/10.1007/978-3-030-69094-6_18

2   Older Persons, Prejudices and Stereotypes Based on Age

Older people are a category of population that faces an increased risk of social exclusion. Most research focus on labor market integration [8] and on the exclusion of older workers, low-income earners, children and young people [9]. Such an approach often ignores the vulnerable position of the older persons in society, with a general lack of research concerning the relationship between social exclusion and ageing of the population.

Despite recent contributions [3, 12, 14], attempts have been made to review the existing evidence concerning the exclusion of elderly people from the process of social exclusion. At the individual level, ageing is associated with an increased probability of leaving the labor market. In many situations, such a decision is associated with a decrease in income. But retirees may also be at increased risk of social isolation because of the loss of social work-related contacts—which in turn could lead to accelerated cognitive decline. According to Ehlers et al. [4], the risk factors for social exclusion in older ages are:

- the social security policies, the physical environment issues, the health and life expectancy, the social networks;
- the stereotypes, prejudices and age discrimination;
- the barriers to the lack of affordable and age-appropriate living conditions;
- the access to transport systems that could reduce the isolation of older people;
- an increased the participation of elderly people in cultural and recreational activities and voluntary activities respectively.

Age can be a prejudice on the basis of which a person can be devalued [1, 11]. Prejudices are the expression of a negative attitude, sentiment towards a person or group as a result of the characteristics of the group to which the person to whom they are exposed belongs. Stereotypes are fixed, generalized or simplified beliefs about the characteristics of a particular group of people that are then applied to all members of the group. Stereotypes may be negative or positive and the discrimination is the differentiated behavior towards a person belonging to a particular group, behavior that is based on certain stereotypes or prejudices [13]. Although there are studies [15] demonstrating their contribution to society, most of the times older people are perceived as a burden or as dependents, unrealistic, needing attention persons, consumers of significant public financial resources. Unlike other forms of discrimination (based on gender or ethnicity) age discrimination is generally socially acceptable and widespread, so it can lead to negative actions and expressions towards older persons, even from those members of the society that are well-intentioned [2, 10].

3　Attitudes Towards Ageing, Discrimination and Stereotypes Associated with Ageing at National Level

Romania is facing an ageing process of population. On January 1, 2017, the population aged 65 and over represented 89.21% of the total population aged under 18 and the ageing index was 114.3% for the same year. This is the result of the growth in the population aged 65 and over in recent years. During 2011–2017 the ageing phenomenon continued, with the share of the elderly population increasing from 16.12% in 2011 to 17.79% in 2017 [6]. The average length of life expresses, in a synthetic way, the state of health of the population. The value of this indicator has increased almost continuously for both sexes. The highest life expectancy was recorded in 2017: for men, the average life expectancy was 72.28 years and for women 79.24 years. The rate of material deprivation and severe material deprivation at the national level remains high and very high among the population aged 65 and over. Differences from European averages are widening for persons aged 75 years and over, reaching a maximum of 26% points in 2011, Eurostat [6]. Romania ranks last among the EU states in terms of the participation of Romanian women and men in voluntary activities.

At national level,[1] the Romanians perceive the beginning of old age at the age of 61 (the age most often indicated by the respondents as marking the barrier between the moment when a person is considered elderly, being 60 years old) [7]. More than a half (56%) consider that 20-year-olds and 70-year-olds are two separate groups belonging to the same community and only 8.7% consider that the representatives of the two age groups form one group. 38.3% of the respondents consider that the age discrimination phenomenon is "quite serious" and 24.5% think it is "very serious", and among them, 15.8% are people aged 65 and over. Older people in the rural area consider, compared with the urban population of the same age, that the phenomenon of age discrimination is a "fairly serious" and "very serious" phenomenon (59.9% of the elderly in the rural area, compared to 41.1% of the elderly living in the urban area) [7].

Prejudices and discriminatory treatment on grounds of age were the most often experienced forms of discrimination in the last 12 months: 41.3% of respondents, regardless of age, experienced at least once a form of discrimination on the grounds of age and 8.6% were "often" and "very often" the subject of age-related prejudice. People aged 65 and over declare that they have experienced discriminatory treatment and "often" and "very often" prejudices in the last 12 months represent 24.1%, leading us to the idea that these people are aged more prone to inequality. Regarding the different types of discriminatory treatment, people aged 65 years and over mostly mention lack of respect and inappropriate treatment [7].

A low percentage of respondents considers that social status in old age is very high (2.2%). The standard of living of pensioners is perceived differently: 15.5% of active people aged 25–44 years old consider that the standard of living of pensioners

---

[1]The analysis of stereotypes, perceptions and age-based discrimination from Sect. 2 is based on the European Social Survey, Wave 4 [5], conducted in 31 European countries.

is very low compared to 23.7% in the case of those aged 45–64 years. In the case of people aged 65 and over, the negative perception is even more accentuated, about one-third (27.2%) of the respondents who were of those ages considering that the standard of living of the pensioners is very low. As regards the role of older people in society, people aged 65 and over perceive themselves as a burden for the public health system in a larger extent compared with respondents of other ages (11.4% compared to 4, 6% for those under the age of 65). In contrast, respondents aged under 65 believe that people aged 70 years and over have a low contribution to the country's economic life (15.2% compared to 12.3% for people aged 65 years and over) [7].

The extent to which people belonging to different age groups perceive themselves as similar (share common goals, interests, and concerns) provides an idea of the possibility that they interact with each other [1]. From the perspective of spontaneous intergenerational interactions between people from different age groups, the data collected show that only 25% of respondents had no friends under 30 years old, other than family members, compared with about a half (48.9%) of respondents who had no friends over the age of 70 years old, apart from their family members. Therefore, younger and older people are more likely to develop friendships with people of the same age [7].

4 Perceptions and Attitudes About Ageing and Older Persons

4.1 Framework Methodology

In order to identify perceptions and attitudes among main actors involved in the process delivery of social services for elderly, a qualitative research approach was developed. The main objectives were:

- Objective 1: Collecting opinions and experiences about the image of the elderly in society;
- Objective 2: Identifying the most common misconceptions and stereotypes about elderly people;
- Objective 3: Identifying measures that can help to overcome negative stereotypes and promote a positive image of older people;
- Objective 4: Identifying ways to encourage the participation of elderly people in social, civic, cultural, sports activities.

Data collection included focus group discussions. The project team developed a focus group guide centred on the following main topics of discussion:

- The image of the elderly person in the contemporary Romanian society;
- Misconceptions and stereotypes about the elderly;

- Promoting a positive image of the elderly in society;
- Identifying ways to encourage the participation of older people in social, civic, cultural, sporting activities.

The target group participating in the focus group was composed of: (1) representatives of public and private, social service providers for the elderly; (2) representatives of organizations promoting the interests of the elderly.

The qualitative research was conducted in 20 November 2018 and the focus group was audio recorded.

5 Results and Discussion

The main results of the focus group are organized starting from the main topics of discussion of the qualitative research.

5.1 The Image of the Older Persons in the Contemporary Romanian Society

There is a reluctance to indicate a certain age beyond which a person is considered elderly. Drawing on experience with older people, participants felt that old age could not be reduced to a number, being determined mainly in accordance with the person's mood and the activities carried out by each person. In Romanian society, including through media broadcasts, the perception is that old age is set at a much lower age, at 55, even 50 years old in some cases. Gender differences were highlighted, women being frequently associated with old age, starting with 45 years old, especially in relation to the labour market.

In general, the age associated with old age is the age at which the shift from occupational status to retired status occurs. Often, this moment can generate or accentuate the isolation, marginalization, social exclusion of the elderly person. Life course before retirement also determines the person's behaviour after retirement. The involvement of the elderly in activities is a direct consequence of functional capacity, health and motivation. Most of the older persons want to be active, as the feeling of belonging to the age group is, in some cases, reduced, the perception of age being different from the real biological age. Lifelong attitude is important and the behaviour of the individual should be oriented towards achieving a balance between professional life and family life, which implies changes in mentality across society. The image of the older people is also determined by the attitude of each person.

The image of the elderly in society also includes positive and negative aspects, and the elderly have an important role in shaping and perpetuating this image. This image of the elderly is built from lifetime, even during the professional activity. Part

POSITIVE ASPECTS	NEGATIV ASPECTS
„a resource person"	„an inactive person"
„a persons with professional and personal competences"	„an isolated person"
„a punctual person" „a tolerant person"	„a marginalised person" „a person that has nothing to offer" „a reluctant person"

Fig. 1 The image of the elderly in society

of the negative aspects of the image of the elderly is developed in the workplace, mainly as some aspects of performance at work are related to the employee's age (Fig. 1).

5.2 Misconceptions and Stereotypes About the Elderly

The loss of motivation for active participation in community life is generally influenced by individual factors (values, interests), but also by physical, cultural, history of life factors. It is the family that, in some cases, draws the roles of the elderly, starting from the occurrence of the retirement status: the older person—the grandfather/grandmother, the older person—the caretaker.

Focus group participants [7] identified different types of elderly starting from the two main criteria: (1) the degree of social participation and (2) the residence environment of the elderly.

Depending on the degree of social participation two types of elderly have been identified:

- The elderly with a passive attitude, with little involvement in social, cultural, sporting, volunteer activities, focused on satisfying its basic needs. This elderly is the one who has a low level of income and was, most likely, dependent on benefits and social assistance system during his active life.
- The older persons with an active attitude and involvement in social, cultural and leisure activities. This is a person who has been active throughout his life, with a living standard above average, diverse concerns and social relationships, with access to a social support network (friends, acquaintances) and interests in the field of volunteering and leisure activities.

Depending on the residence environment of the person, two types of elderly have been identified:

- The older persons from urban area, as a person being in a more advantageous situation, as he/she has increased access to a variety of services. Like elderly

POSITIVE	NEGATIVE
„very good mentors" „serious persons" „patient persons" „volunteers" „pillars of the family"	„an inactive, depended persons" „ a person separated from the realities of the labour market" „people who do not accept the change" „people reluctant to technical progress"

Fig. 2 Stereotypes regarding the elderly

from the rural areas, they face loneliness, but the cause is the lack of consistency of social relations;

- The older persons in the rural area who are in a more advantageous situation compared with those from urban areas, from the point of view of communication and community integration. In the rural area, the elderly enjoys more respect, it maintains its functional capacity for a longer period of time, mainly due to the activities carried out in agriculture activities. But due to the depopulation of many rural areas, the elderly from these areas are isolated.

In some situations, elderly people themselves choose to participate as observers in the life of society for medical reasons or because they do not find their place and role in the community. The diminishing of the social circle leads to isolation, to the inability to initiate new relationships, contacts, activities. It creates a rift between generations and this rift is difficult to recover. The elderly withdraws from the social life and this retreat is explained as an "embarrassment" the elderly feels.

Focus group participants consider that the elderly choose to reduce these physical contacts, choose the isolation, in other words, this is an individual decision. The choice of the elderly can be the result of a personal decision, as it may be the result of the subtle, insidious or even direct forms of discrimination to which the elderly person is subject in public places in relation to younger generations. These forms of discrimination can take different forms: verbal (reverential language), paraverbal (aggressive or less friendly tone), non—verbal (attitudes, gestures etc.) (Fig. 2).

Overcoming negative stereotypes requires measures and interventions in public policy. The public policy proposals mentioned by the participants of the focus groups envisage an integrated policy approach (social assistance, labour market, health, education) throughout elderly life:

- It is appreciated that measures aimed at overcoming negative stereotypes are related to the human resources working in the social services system provided to the elderly. There is a need for training courses and an increase in the number of people employed in social services, especially those addressed to the dependent elderly.
- Home care services could have a valuable contribution in reducing stereotypes, despite insufficient public budget support for such services.
- Volunteer activities give to elderly the opportunity to regain the sense of utility.

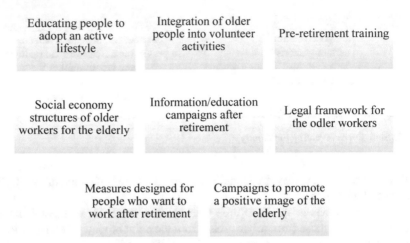

Fig. 3 Measures and interventions designed to overcome negative stereotypes

- Actions for people with dementia are required to overcome stereotypes related to the context in which these people (residential or at home) have to be cared for, with positive effects on their quality of life and the quality of life of the families they belong to. These actions should go from awareness campaigns concerning the incidence of this disease and the impact it produces on the person and the family, to training provided for family members, support measures for carers, social and socio-medical services for people affected by this disease.
- Preparatory measures for retirement should be correlated with those intended for older workers wishing to continue working (pre-retirement integrated services: reduction of worktime, teleworking, counselling services).
- As negative stereotypes develop through active life, the development of a legal framework for the older worker was considered necessary (Fig. 3).

5.3 Promoting a Positive Image of the Elderly in Society

This topic of discussion requires measures to be targeted at the same extent to a dependent and also to an independent elderly person. In the case of the dependent elderly it is necessary to train the staff involved in the provision of social services, as the lack of staff influences the quality of care and affects the image of the dependent elderly person. Elderly people are not considered to be people of interest in the media, which is why negative stereotypes are predominantly promoted regarding this category of population.

5.4 Encouraging Participation of Elderly People in Social Activities

Encouraging the participation of elderly people in volunteer activities is related to the existence of sufficient individual incomes, as practice shows that volunteer activities are suitable for those older people who have already covered their basic needs. This implies the development of networks involving active older persons, social service providers, cultural, sporting and church services. In rural areas, elderly people are involved in volunteer activities, especially related to church activity and the potential in this area must be developed by expanding this type of volunteering to other types of activities. Another form of encouraging the participation of the elderly in society is the development of social economy structures designed to promote traditions and customs in rural areas and to attract elderly people into such activities.

6 Conclusions

The interest for the elderly, for its image in society, and for the increase in social participation of the elderly is high from the perspective of different social service providers and social partners (trade unions, employers), older workers, and elderly themselves. This interest is reflected in the implementation of projects targeting elderly people, how they are perceived by the other members of society, promoting their active social participation, encouraging intergenerational volunteering. Life course before retirement also determines the person's behaviour after retirement. The involvement of the elderly in activities is determined by their functional capacity, health and personal potential. The image of the elderly is also determined by the attitude of the elderly. The loss of motivation for active participation in community life is influenced by individual factors (values, interests), but also by physical, cultural, life history factors.

Acknowledgements The work presented in this chapter has been supported by the Ministry of Labour and Social Justice (Research project "Măsuri destinate creşterii incluziunii sociale a persoanelor vârstnice", contract no. 3210 from 20.08.2018).

References

1. Abrahams D, Swift H (2012) Experiences and expressions of ageism: topline results (UK) from round 4 of the European Social Survey [online 20-02-2020]. Available online at: https://www.europeansocialsurvey.org/docs/findings/ESS4_gb_toplines_expe riences_and_expressions_of_ageism.pdf
2. Ayalon L (2014) Perceived age, gender, and racial/ethnic discrimination in Europe: results from the European social survey. Educ Gerontol 40(7):499–517

3. Börsch-Supan A, Kneip T, Litwin H, Myck M, Weber G (2015) SHARE: a European policy device for inclusive ageing societies. In: Börsch-Supan A, Kneip T, Litwin H, Myck M, Weber G (eds) Ageing in Europe—supporting policies for an inclusive society. de Gruyter, Boston, pp 1–22. ISBN 978-3-11-044441-4

4. Ehlers A, Naegele G, Reichert M (2011) Volunteering by older people in the EU [online 10-01-2020]. Available online at: http://hdl.handle.net/10147/299047

5. European Social Survey (2008) SS4—2008 documentation report [online 20-10-2019]. Available online at: https://www.europeansocialsurvey.org/docs/round4/survey/ESS4_data_docume ntation_report_e05_4.pdf

6. Eurostat (2018) [online 20-08-2019]. Available online at: https://ec.europa.eu/eurostat

7. Ghenţa M, Matei A, Mladen-Macovei L, Stroe C (2018) Măsuri destinate creşterii incluziunii sociale a persoanelor vârstnice [online 20-02-2020]. Available online at: http://www.mmssf. ro/j33/images/Documente/MMJS/Transparenta/2019/01022019_Raport_stiint_masuri_inc luz_pers_varstnice.pdf

8. Madanipour A (2011) Social exclusion and space. [online 05-12-2019]. Available online at: https://www.researchgate.net/publication/306203723_Social_exclusion_and_space

9. Moffatt S, Glasgow N (2009) How useful is the concept of social exclusion when applied to rural older people in the United Kingdom and the United States? Reg Stud 43:1291–1303

10. Officer A, Schneiders ML, Wu D, Nash P, Thiyagarajan JA, Beard J (2016) Valuing older people: time for a global campaign to combat ageism. Bull World Health Organ 94(10):710–710A

11. Officer A, de la Fuente-Núñeza V (2018) A global campaign to combat ageism. Bull World Health Organ 96(4):295–296

12. Scharf T, Keating N (2012) Social exclusion in later life: a global challenge. In: Scharf T, Keating N (eds) From exclusion to inclusion in old age: a global challenge. The Policy Press, Bristol, pp 1–16

13. Steele J, Choi YS, Ambady N (2007) Stereotyping, prejudice, and discrimination [online 05-12-2019]. Available online at: https://ambadylab.stanford.edu/pubs/2004Steele.pdf

14. Warburton J, Shardlow SM (2013) Social inclusion in an ageing world: introduction to the special issue. Ageing Soc 33:1–15

15. World Health Organisation (2015) World report on ageing and health [online 05-12-2019]. Available online at: http://apps.who.int/iris/bitstream/handle/10665/186463/978924069 4811_eng.pdf;jsessionid=F401803DFC065F7AC6D57BFF3988612F?sequence=1

Accreditation and Quality of Long-Term Care for Elderly: Social Provider's Perceptions and Representations

Adina Daniela Rebeleanu and Paula Cristina Nicoară

Abstract Generally speaking, accreditation is the appreciation made by an outside body to the institution in order to guarantee the quality of the service provided. This appreciation is the compliance of the institution with a set of explicit criteria and, especially, known by the two parties—service provider and beneficiary. Accreditation is not limited to a "verdict". Is accreditation a sufficient and necessary approach for quality of elderly care? Using the thematic analysis, from critical approach, the current research focuses the social providers' perceptions concerning the quality of long-term care for elderly, from accreditation perspective.

Keywords Care · Accreditation · Quality · Need · Providers

1 Introduction

In today's society, we often hear about the aging of population, dependency, but also about the extension of the active age, social participation, active aging. Population aging and changing family patterns are realities worldwide. The challenges that these developments have on the evolutions of social security schemes are current and often give rise to divergent views that oscillate between the tax burden that the inactive population and the need for social protection imply and the fact that the same population considered as inactive or dependent can become a real resource for the future.

For the decision-makers of social policy and practitioners (social workers, doctors, etc.), it is important to recognize that the process of awarding an age-based dependency status is a social construct rather than a biological one. There is no necessary

A. D. Rebeleanu
Faculty of Sociology and Social Work, Babeş-Bolyai University Cluj-Napoca, Cluj-Napoca, Romania
e-mail: adina.rebeleanu@socasis.ubbcluj.ro

P. C. Nicoară (✉)
County Agency for Payments and Social Inspection Cluj, Cluj-Napoca, Romania
e-mail: nicoara.paula@mmanpis.ro

© The Author(s), under exclusive license to Springer Nature Switzerland AG 2021
D. Soitu et al. (eds.), *Decisions and Trends in Social Systems*,
Lecture Notes in Networks and Systems 189,
https://doi.org/10.1007/978-3-030-69094-6_19

relationship between chronological age and need or dependence. Moreover, if we think in terms of transfers, the insured transfers from the active generation to the inactive population, some aspects often ignored are added—downstream flows, especially financial (donations, inheritances, financial aid) from the elderly (considered inactive) to their adult children and their grandchildren (which some authors call "family returns") (Masson [1]: 290–314).

The aspect of aging is one that has generated and will generate research in various aspects and from different perspectives. With the aging, *the need for long-term care* tends to grow significantly. This is a consequence of the elderly's dependence, which obviously requires a wide range of services, not just social, but rather social-medical, as the state of dependency is strongly correlated with the disease. Thus, personal care services, as specified by the law on social work, are addressed to dependent persons that require significant help to carry out the basic and instrumental activities of daily life. Dependency is therefore a loss of functional autonomy, having physical, psychological or mental causes and advertising personal care services.

When this care is offered/needed for a period longer than 60 days, we talk about *long-term care (see for conceptualization and standards in national context Law no. 292 of 2011, Law no. 17 of 2000, Government Decision no. 566 of 2015, Order No. 2126 of 2014, Government Decision No. 867 of 2015, Government Decision No. 978 of 2015, 2011, Recommendation 2011/413/EU of 2011, Long, Active and Forceful Life. Promotion of Active Aging in Romania (June 2014)).*

Elderly care is a social service before it becomes an institutionalized activity. Without denying the role of different types of care for the third age, in accordance with national and international regulations, this chapter focuses on long-term care in elderly's homes. Issues concerning the quality of elderly care in these institutions are addressed from the perspective of accreditation/licensing, in the representation and perception of social services providers.

2 Argument

The perspective of accreditation and its impact on the quality of elderly care in residential institutions requires a holistic approach, which cannot miss the perceptions of professionals—social and medical service providers. The normative dimension of accreditation has been analyzed by authors in previous studies [2–5]. These studies have revealed the fact that, although dizilable, the legal framework only creates the premises for the obligation to meet and maintain quality standards in the elderly homes. It is a necessary step, but not enough, to guarantee the quality of long-term care services [4]. An important role in ensuring the quality of the care act lies with the care staff.

Essentially, quality is an attribute or set of attributes that describe, build, give shape to a product or service designed to meet the needs and goals of some customers. To be appropriate, elderly services, in an integrated approach, must be adapted to the needs and goals of the elderly.

"A better definition of the quality of social services also responds to the need to address a demand for social services that is becoming more and more complex and diverse, as well as the need to protect those who are part of the social services consumers, vulnerable groups members as well as to improve the results of social services for users and other stakeholders" is mentioned in the European Voluntary Quality Framework for Social Services (SPC/2010/10/8).

National legislation establishes criteria for quality standards for services for the elderly (Order No. 2126/2014). National regulations become a reference for professionals working in care centers, but who have the opportunity to assess, in their interaction with the beneficiaries, the degree of suitability of institutionally pre-established criteria to the perceived and expressed needs of the elderly. Furthermore, professionals may assess to what extent standards allow the exercise of professional competencies in accordance with the responsibilities established by the clearance and professional recognition institutions.

For these reasons, in the research process, we started from the taxonomy of the needs proposed by Bradshaw [6]. The author distinguishes between the perceived, expressed, comparative and normative needs. The normative dimension of the need is represented in this study by current regulations on quality standards imposed on the process of accreditation/licensing of services. The experienced and expressed needs are those formulated by research participants (incorporating on the one hand institutional requirements, professional practice, and also of the beneficiaries—institutionalized elderly, as perceived in practitioner-client interactions).

3 Methodological Approach

The purpose of this research is to analyze the perceptions of social services providers on aspects related to elderly care quality in elderly homes.

The study was explorative. The data were gathered through two focus groups with medical, social and social-medical services providers, carried out in Cluj County, Romania, between April and May 2017. We used the thematic analysis of the data from critical perspective. The processing and analysis of the obtained data allowed the following thematic units and sub-themes to be outlined. These are presented in the following Table 1.

14 professionals from the public and private system, both in accredited centers and in institutions that operate without accreditation, participated in the two focus groups of. The length of service of the participating social workers is different (from one month to 12 years) and their average age is 31.5 years. Among the participants, there were also two managers of care centers. Suppliers that are active in both urban and rural areas were selected.

Table 1 Analyze the perceptions of social services providers (table based on ownresearch)

Thematic units of analysis	Details of thematic aspects
1. Social services quality	• Factors • Principle • Criteria
2. Accreditation of providers and licensing of social services	• Difficulties in the accreditation process • Utility of accreditation undertaking • Improvement proposals • Impact on elderly care quality
3. Elderly care in residential system	• Difficulties in the activities of the staff involved • Necessary qualities • Activities carried out • Interaction with the elderly
4. Activities in which social workers are involved and aspects on labor satisfaction	• Normative responsibilities • Responsibilities according to the profession deontology • Sources of satisfaction • Sources of insatisfaction

4 Results and Discussions

We will synthetically present some of the most important aspects outlined in the analytical approach.

1. *Quality in social services*

The participants in research believe that quality assurance in social services contributes to the qualification of staff in the field of social work. Besides, the existence of a social worker in the staff structure of a long-term care provider is also a normative requirement for accreditation. It is also necessary to guarantee the quality of service provided and continuous improvement. Another aspect that contributes to the quality of the services provided, in the respondents' opinion, is the approach of the elderly from a different perspective than the biomedical one, as well as the valorisation of his/her life (psychological-social and spiritual aspects) post-institutionalization.

> The standards also help us a lot, but there is a risk that the services will be more oriented towards the medical part if there is not qualified personnel in the social field - social work, psychology, I think that this is not something wrong, but the elder does not have only needs only for medical services, but he/she also needs social services; we are doing much on our medical side. (C.I., 34 years)

> The elder had a lifestyle before being institutionalized; it is important to take this into account after institutionalization. (I.D., 23 years)

> If I have a qualification, a profession does not necessarily mean I'm good in that profession, but the qualification gives me a luggage of knowledge that helps me carry out quality work. (L.H., 42 years)

First of all, qualified staff in working with elderly people; the mere fact that you're a registered nurse, nurse, social worker does not mean you know how to behave with the elderly; something extra is needed; there should be special courses for this. (A.B., 31 years).

The importance of the multidisciplinary team; in our home this works a bit harder, I collaborate with the registered nurse, with the manager, but we do not have weekly meetings for discussions, to make a report on what it was and go on, this is a very important aspect; this is a shortcoming... (L.H., 42)

Criteria that ensure the quality of services for the elderly are identified with the need for an integrated approach, taking into account the four perspectives of intervention on the beneficiary, namely biological, psychological, sociological and spiritual. The integrated approach, a condition for quality of care, is also mentioned and reiterated in the international documents governing social services (Biennial Report on social services of general interest SEC(2008)2179 and SEC(2010)1284), the Commission Recommendation of 3 October 2008 on the active inclusion of people excluded from the labour market (C(2008)5737), the Report "A Voluntary European Quality Framework for Social Services" [7] and, last but not least, the Report on the future of social services of general interest adopted by the Committee on Employment and Social Affairs of the European Parliament (2009/2222(INI) (the De Rossa Report). The local research also points to the importance of a comprehensive, holistic approach, integrated long-term care for the elderly [8–10].

We, as social workers, must look at the beneficiary as a whole, be receptive to all his/her needs; the nurse takes care only of the medical part; it is up to the social worker to identify the needs of the beneficiary, to feel and meet these needs. In our case (social workers) there are considered several aspects: medical, spiritual, social, but we have very little time to stand and talk with the beneficiaries, we have a lot of office work, files ... one social worker for 80-90 persons is not enough, you do not get to talk with everyone often; we organize groups of occupational activities, but ... it is not enough. (L.H., 42)

The complexity of the social worker's role in an elderly care center was frequently mentioned in discussions. On the one hand, in the context of accreditation, on the other hand, as a *sine qua non* condition of the quality of care, in the context of the integrated approach. It also becomes a source of satisfaction, but also of dissatisfaction for the professional work of social workers. The need to recognize the role of the social worker in a home for elderly people was frequently mentioned in the context of multidisciplinary team work. It is a fact that in the centers where the activity is especially medical, such situations are encountered—confusions between medical and social-medical services.

Here, there is this role confusion - I often find myself in the situation of being told that we have a "patient"; we have beneficiaries who have speech problems after a stroke and I am told "could not you do therapy with him, speech therapy?" ... but then I'm no longer a social worker ... (CI, 34 years)

2. *Provider accreditation and licensing of social services*

We would like to remind that the present research has started from within the accreditation activity, which is an assessment of the care services provided to elderly people in elderly homes. Accreditation is a complex process whereby the social service provider demonstrates its own functional, organizational and administrative capacity in providing social services, provided the quality standards in place are complied with and the state recognizes its competence to provide social services. In the context of Romania, the current accreditation process in the social field targets both providers and social services. The assessment for the accreditation of residential social services/homes for the elderly to obtain the operating license is based on the Minimum Quality Standards, regulated by *Order no. 2126 of November 5, 2014.* Most of the time, the preparation of the file for the provider's accreditation/licensing of social services is placed under the responsibility of the social worker. The perceptions of social workers on accreditation in fact reflect the correspondence between the normative dimension of accreditation and the needs expressed by beneficiaries. The professional level at which accreditation is achieved and perceived may contribute to the improvement of services, as well as the organization and functioning. On the other hand, it can be a basis for developing relevant public policies in the social field (a desirable response to what Bradshaw calls comparative needs).

The following thematic sub-categories have been identified: difficulties in the accreditation process, the usefulness of the accreditation process, effects on the quality of care, and proposals to improve the current regulations on quality standards in elderly care centers.

If the provider's accreditation is perceived to be simple from the procedural point of view and the requirements to be met, the licensing of social services is associated with administrative difficulties, which do not necessarily depend on the accredited supplier (e.g., the difficult procedure for obtaining the security clearance for fire fighting) or lack of qualified personnel (e.g., the need to have a physiotherapist, according to quality standards). In fact, the licensing of social services is a cumulation of pre-existing conditions that are required to be met in order to provide quality social services for the beneficiary. And these pre-conditions are considered as difficulties in obtaining the license.

> I, personally, was in the following situation ... the manager asked me to write the procedure manual - not what strictly relates to my profession, but what is happening throughout the center - it was very difficult for me ... almost two months, I did not know who to ask what a procedure manual looked like, what a procedural manual was ... Finally, I started writing it ... but it's very hard because I have to write including how to change the pampers, for example, or how to wash the window, how to disinfect all kinds of stuff; Did you know that the door is washed from top to bottom ... or the window? I did not know ... it would be good to have a draft, what it should contain, how to think about writing it; it's very difficult. This is my only problem. (M.T., 28)

> As related to the procedure manual, we should know why we need to have procedures; because that's what we find upon audits; it is always something new for which we have no procedure; there should be a list and then we would know. (E.K., head of the center, 52 years)

The discussions about the need and usefulness of social service licensing converge to its acceptance but, at the same time, prefigures the emergence of a specific assessment to obtain quality grades (an aspect that may be found in the legislation but without procedures, yet).

I believe that accreditation/licensing is a necessary and useful approach; you're covered. (M.T., 28 years old)

I think it is necessary - this is the difference between a home where specialized services are offered and a person who cares for the elderly at home. (I.D., 23 years)

From the point of view of legal regulations, it is OK, but I still have a question: why are we not, however, unitary services, why do not we have the same standard. (C.I., 34)

... why if we have the same accreditation, the same law to comply with, why services are not the same ... (C.I., 34 years)

The aspects presented by the participants predict the competitiveness of the services provided by C.P.V., also provided by the legislator, and the methodology for the classification in quality classes will be regulated. Law no. 197 of 2012 on *quality assurance in the field of social services* specifies the possibility of three quality classes, class III corresponding to the fulfillment of the minimum quality standards specifically regulated for each social service and under which the operating license/temporary operating license is granted. For the inclusion in quality class II and class I, detailed indicators will be developed, with reference to the level specified in the minimum quality standards, which will be approved by a subsequent normative instrument. Or, the classification in higher quality classes below the minimum level will constitute, according to the legal provisions (Law No. 197/2012, art. 16), a criterion for the level of the beneficiary's contribution, a contribution that is correlated with the degree of performance and services offered by nursing homes for the elderly (NHE).

Social workers believe that the process of accreditation/licensing of social services is necessary and beneficial for the provider, but which ultimately positively impacts the beneficiary. It also provides adequate work conditions for the staff in centers, which is an important factor in the quality of care.

The license contributes to quality ... if it was not for the standards, we did not care to develop, make some changes, it's easier not to do ... .but if the law forces you ... for example if the law provides a certain number of sq.m./beneficiary, you no longer have 3-4 beneficiaries in a room, but only 2, for the beneficiaries is much better; or the bathrooms, if there is a bathroom in each room, it is a great feature for the beneficiaries; another positive thing is the obligation of a physiotherapy room ... the beneficiaries move very little, we have work to do ... (L.H., 42 years)

I think the activity in the field of elderly care is a bit more professional – there is the social worker - with an important role, and the multidisciplinary team. (CI, 34)

Here, for staff, it was a step forward, because we were forced to make separate rooms for staff; we made a medical office, we did not have one until licensing, we did this according to the standards, it was good ... but it took our space; the beneficiaries do not know; we had done and offered the best for the beneficiaries even before the standards. (E.K., Head of the Center, 52 years)

3. *Elderly Care in Residential System*

Among the difficulties encountered in the professional activity, the social workers mentioned the lack of supervision, the management style of the manager, as well as contradictory values of the staff and the beneficiaries. The lack of specialized staff that provides services complementary to social and medical ones, is another factor that affects the quality of elderly care.

The supervisory institution is legally regulated, but its implementation arrangements are less professional, and this situation is acutely felt by social workers.

> ... but supervision only exists on paper ... unfortunately, we are not supervised ,,, and we would need it; there is no formal supervision; you can report to anyone, but we need confirmation if we did a good intervention or not ... or how we should have done it... (I.D., 23 years old)

> We would need supervisors ... we are forced to have supervision ... but supervision has to be paid, the organization cannot afford to pay ... and then we intercede ... between us ... it is needed; once, I was supervised for about 11 h ... the supervisor is a qualified person, who knows how to listen to you, to guide you, to answer the weirdest questions ... we need supervision ... because otherwise we get to burn out and look for another job. (CI, 23 years old)

The values and beliefs of the beneficiaries and the staff can be contradictory and can induce uncertainty for professionals: *"I think it depends very much on our own values and those of our profession ... for example, I had an atheist beneficiary who was in the terminal phase ... he wrote to me on a piece of paper that there is no future, that there will be nothing after death; his last days, he did not want to sleep any more because he was afraid the inevitable would happen and was not sure whether there would be anything after or not; honestly, this made me think ... my mind was just thinking about that ... then I would have liked to talk to someone, to tell him/her what my mood was and he/she to tell me what to do ...." (ID, 23 years).*

Conflicts can come from the encounter of different life philosophies, in which youth's confidence meets with experiences of illness and fear of death. Such clashes of values may create confusion in the minds of professionals who need guidance from superiors or supervisors. In the absence of formal supervision, each social worker finds his or her own ways of solving the situations in which he or he is engaged in "interviewing". Supervision will need to be institutionalized in the future and backed up with the necessary infrastructure.

The leadership style of the person who manages the social service is a variable that ultimately influences the beneficiary, the quality of the services offered to him/her:

> *"If I did not have a communicative manager, with whom I can talk, I think my job would have been less pleasant; she made me love social care, thanks to her. If the manager is ok with the employee, the employee is much more open, friendly to the beneficiaries." (M.T., 28).* At the same time, however, the manager must *"give you some freedom ..." (L.H., 42 years),* be *"flexible" (M.T., 34 years)* and *"prompt" (O.R., 27 years).* The qualities required for a coordinator/chief were also mentioned by the participants *"to be a leader, a professional, to come from the social-human field, to be able to understand what it is about." (C.I., 34 years); "to be determined, ... responsible." (E.K., Head of the Center, 52 years).*

The lack of specialized staff is another problem identified by social workers in the homes for the elderly:

> We have problems finding good registered nurses ... the good ones do not come to the homes, we face problems and finding involved caregivers... graduating from a nursing school is not a guarantee. (E.K., Head of the Center, 52 years)

I see the nurses - they have things to do daily; their work is enough and they have no time to interact with the elderly to be beneficial to the elderly. (L.H., 42)

Staff training is another important aspect for quality care by focus group participants:

Here, the lack of knowledge in the work with the elderly; for a nurse it would be important to know that there is a certain type of communication with the elderly beneficiary ... (C.I., 34 years)

Care staff has neither the time, nor the necessary training, nor the skill to communicate with the dementia beneficiary ... (L.H., 42)

The qualities needed for long-term care staff are *"calmness" (AR, 24 years)*, *"patience" (OR, 27)*, *"not to be touchy, not to take things personally." (EC, 24 years)*. Working with the elderly involves a certain existential situation, without which the results and, finally, the satisfaction of the work will not be right.

4.   *Activities involving social workers as well as aspects of work satisfaction*

By definition, social work is oriented towards providing support to individuals who are temporarily unable to work. It is considered the condition of a democratic society, and professional values are found in the values promoted by social policy. The compliance with the right to self-determination, the compliance with the decision-making freedom, right and access to quality services are just some of the ethical principles under which social workers carry out their work. By the nature of the profession, the emphasis is on absolute priority given to the person in difficulty and his/her right to obtain competent support. Optimal conditions for practicing the profession, appropriate workplace conditions, investment in continuous training of human capital are inherent conditions for the quality of social services provided to beneficiaries [7].

The social worker in a N.H.E. is involved in socializing and leisure activities. Sometimes he/she is involved in rehabilitation activities (occupational therapies).

Another beneficial activity is occupational therapy, but unfortunately few people can do activities - refuse or cannot, or prefer individual activities." (L.H., 42 years); we do "recreational activities: remy, puzzle, chess." (O.R., 27 years); "We have musical therapy, human-animal interaction therapy, especially for dementia beneficiaries; and laughter therapy, ... we also bring social animators; beneficiaries learn quite hard to communicate with each other; we encourage them; we also organize birthday parties where we also invite their families. (C.I., 34)

Leisure activities are more frequent and conducted with a certain rhythm in licensed N.H.E. but these aspects need to be improved both in licensed N.H.E. as well as in unlicensed N.H.E. (currently). For this purpose, it is preferable to identify an efficient external evaluation method that will effectively support the beneficiary, which will also ensure the quality of services from this point of view.

By showing an interest in the professions whose particularity consists in providing support to people in difficulty, Sommers-Flanagan and Sommers-Flanagan [11] identified a number of motivational issues for guidance in such professions; it is about: the desire to help others; genuine curiosity about people's lives; the need for status or professional fulfillment; a higher degree of self-understanding and self-healing. Beyond these elements, the authors emphasize that financial motivation is not on this list, the base for choosing a profession that assumes support—such as social worker—are predominantly intrinsic and non-material motivations. to the sources of satisfaction of the participants in the research, they focus on the quality of interaction with the elderly and the professional activity—the successes recorded in each case, the successes achieved through the introduction of positive changes in social cases with which they interact professionally.

> My satisfaction is that people like to talk to me. (C.I., 34 years);
>
> After 6 months, a beneficiary with severe dementia, who is not aware about himself, called me by my name! (M.T., 28 years);
>
> Gratitude in all its forms; I have a routine, I go every day and greet them … when I'm very busy or on vacation they miss me. (O.R., 27);
>
> When you see that, especially the persons who cannot move, have the desire to communicate with you, they open to me … then, the involvement of the beneficiaries in the activities that we carry out … (L.H., 42 years);
>
> My greatest satisfaction is that they like to come to the activities and, when I do not organize activities, they ask me why … and they get involved. (E.C., 24)

5 Conclusions

By synthesizing, the perceptions of long-term social services providers converge towards the need to approach elderly care from the four perspectives of intervention (biological, psychological, sociological, spiritual). In the respondents' opinion, the lack of an integral and integrated approach of the elderly beneficiary is felt as a frustration in the work of social workers. It becomes an expressed need that should be integrated not only in the normative dimension of accreditation of services but also in actual practice.

It is mentioned the necessity of continuous improvement of the involved personnel, the quality and the complexity of the elderly care, requiring the professional development. The work of the social worker is considered to be insufficiently valued, and sometimes his/her responsibilities are not known, responsibilities that translate into practice would contribute to achieve an integrated approach to elderly care. It seems to be the most strongly felt and expressed need by social services providers. The legal standardization of responsibilities through regulations on the quality of long-term care is not sufficient. The accreditation process has brought the work of the social worker to itself, but there is still some role confusion, the social worker being perceived as a person who can be approached in any situation.

One recognizes and reaffirms the importance of supplier accreditation and licensing of social services with all pre-requisites necessary from a legislative point of view, steps that, although procedurally demanding, are important because of the positive effect on the life of the beneficiary. Accreditation and licensing are desirable, but also a guarantor of quality of care.

The problems encountered in the professional activity can be located on two levels: administrative problems (regarding the various documents to be drafted by CPV and/or the public social care services and mutual communication between social services providers) and problems related to the professional approach of the elderly care (including the missing of professional supervision). If this last aspect could be solved through continuous training courses for staff serving long-term care centers, for administrative problems, the existence of a mechanism of collaboration between the local institutions involved in the delivery of various opinions/authorizations would be useful. This would avoid situations where a N.H.E. starts working without having all the necessary documents. Obtaining authorizations is also due to the fact that the changes/adaptations of the space and the related investments must be made by the N.H.E. manager.

The client is the one who decides on quality, but the quality of care provided to the institutionalized elderly is rather the result of multidisciplinary teamwork.

Work satisfaction for social services providers in elderly care centers is generated by the quality of interaction with the elderly and by the professional activity. It is confirmed that orientation towards the social worker profession is based on psychological-emotional and moral motivations, rather than economic, material [11].

References

1. Masson A (2007) Les avatar de l'altruisme parental. In: Paugam S (ed) Repenser la solidarité. L'apport des sciences socials. Presses Universitaires de France, pp 289–314
2. Nicoară PC (2014) Aspecte legislative privind îngrijirea de lungă durată în căminele pentru persoane vârstnice în România. Revista de Asistenţă Socială, nr.1/2014, Editura Polirom, Iaşi, pp 113–124
3. Nicoară P, Rebeleanu A (2013) Cost and quality standards implemented in the residential centres for elderly. In: Runcan, Rata, Cojocaru (eds) Applied social sciences: social work. Cambridge Scholars Publishing, UK, pp 225–239
4. Nicoară P, Rebeleanu A (2018) Accreditations of long-term care centers for elderly. Sci Ann Al.I.Cuza, Univ Iasi Sociol Soc Work Sect XI-th 11(1)
5. Rebeleanu A (2011) Cadrul legislativ în asistenţa socială din România. Prezent şi perspective. Presa Universitară Clujeană, Cluj-Napoca
6. Cookson R, Sainsbury R, Glendinning C (eds) (2013) Jonathan Bradshaw on social policy. selected writings 1972–2011. York Publishing Services Ltd, online 23.03.2017 available at https://www.york.ac.uk/inst/spru/pubs/pdf/JRB.pdf
7. The Social Protection Commitee (2010) A voluntary european quality framework for social services, SPC2010/10/8 final, available https://ec.europa.eu/social/BlobServlet?docId=6140& langId=en
8. Şoitu D (2018) The integrated social design of future long-term care. Sci. Ann. Al.I.Cuza Univ Iasi Sociol Soc Work Sect XI-th 11(1)
9. Şoitu D (2015) Resilience and vulnerability: competing social paradigms. Sci Ann Al.I.Cuza Univ Iasi Sociol Soc Work Sect XI-th 8(1)
10. Soitu D, Rebeleanu A, Gavrilovici C, Oprea L (2013) Client vulnerability within the relationship with social and healthcare services. In: Soitu D, Gavriluta C, Maturo A (eds) Recent trends in social sciences: qualitative theories and quantitative models. Universitatii Al. I, Cuza, pp 29–47

11. Sommers-Flanagan R, Sommers-Flanagan J (2007) Becoming an ethical helping profes-
    sional—cultural and philosophical foundations. Wiley, Hoboken

The Mourning in Nursing Homes and the Nursing Homes Workers Necessities

Ana Vallejo Andrada⬤, José Luis Sarasola Sanchez-Serrano⬤, and Evaristo Barrera Algarín⬤

Abstract In Spain as in the rest of Europe elderly group which has been increasing during the last few decades, our society has created new answers for this group necessities, as it could be nursing homes. Nursing Homes in many cases constitute a separate social sphere, which differ from family homes in many aspects, one of them is the morning process The aim of this research is to analyse the similarities and differences of the necessities of each professional group in their work with the residents morning process, to fulfil this porpoise we have created a survey which have been send to fifty-eight nursing homes workers in order to identify the barriers and facility factors which can interfered in their work with the residents mourning.

Keywords Elderly · Mourning · Grief · Nursing home · Nursing home workers · Senior citizens · Residents and Spain

1 Introduction

As a consequence of the increase in life expectancy, in many countries elderly people have become one of the most representative groups of our society. As a answer of that, our society has created new specific resources especially designed for them, whereas in some occasions this resources do not fulfil their purpose correctly.

One of these resources are the nursing homes. Nursing homes in the way there have created, constituted their own social sphere, with its own rules, timetables and relations.

A. Vallejo Andrada (✉) · J. L. Sarasola Sanchez-Serrano · E. Barrera Algarín
Departament of Social Work and Social Services, Pablo de Olavide University, Utrera Road, 1, 41013 Seville, Spain
e-mail: avaland@upo.es

J. L. Sarasola Sanchez-Serrano
e-mail: jlsarsan@upo.es

E. Barrera Algarín
e-mail: ebaralg@upo.es

© The Author(s), under exclusive license to Springer Nature Switzerland AG 2021
D. Soitu et al. (eds.), *Decisions and Trends in Social Systems*,
Lecture Notes in Networks and Systems 189,
https://doi.org/10.1007/978-3-030-69094-6_20

The aging process is not univocal, direct or simple. There is no consensus in defining it, but to consider that people express the physical, psychological and social defects over time, and manifested a highly asynchronous differential mode [1], p. 42. As we can see in Martín's definition, the ageing process have certain characteristic, that change the way that elderly people live their lives,

In nursing homes these factors get link to the nursing homes own characteristics, creating a new ambit of relations, between residents, family, workers and institution. In this new ambit relations ships affect each other's, creating its own ecosystem, completely differentiated from the outside world.

For this reason, is vital, that we analyse all the social factors that compose it. We have already talk about one of them: The residents, the elderly people who live in them. But what about the workers? Throughout this chapter we are going to get focused in the professional view and how the intuitional aspects can interfered in their work in one specific topic: The Mourning.

2   The Mourning and the Professional Intervention

As it is described by White, mourning is a natural process in human being, which initially should not be treated in a pathological manner, as long as this process is not complicated [2], p. 179.

Following this line Díaz et al. tells us that grief is a social process, and that as such, the fact of feeling support and being able to verbalize and share the experience will be fundamental for its resolution. However, the current society can exert its influence in the opposite direction, that is, forcing the sufferer to be well immediately, to be distracted, to avoid contact with pain or tears, causing the process to be inhibited and complicated [3], p. 11.

Making the professional intervention in the grieving process, something indispensable in our society, especially in those places where people are in greater contact with death. As the nursing homes.

In Nursey Homes, we can find a great variety of professions, specialized in the care of the elderly, which created together the assistance network that this group needs. Being essential networking and collaboration between each of them.

- *Care Workers*: Its general competence in providing auxiliary care in institutional settings to elderly people who have some degree of dependency or who present a situation of need (physical, mental or social health), applying the most appropriate strategies and procedures to maintain their health and improve their autonomy personal and social [4], p. 49.
- *Directors*: Its general competence is to manage the resources available, human and material, so that the institution provides the services for which it is intended. Using the best strategies of organization and teamwork that optimize the performance of all staff and ensuring that the best methods and procedures for quality care are applied [4], p. 49.

- *Occupational therapist*: Its general competences are based on organizing the human and material resources available to organize the department or Occupational Therapy Area whose objective is to maximize the functionality and support the level of occupation of the elderly, offering adaptation-compensation techniques through the simplification of tasks, the modification of the environment and the application of techniques of the discipline [4], p. 49.
- *Physiotherapist*: Its functions are to organize and develop the area of physiotherapy seeking the rehabilitation of the person, or what is the same, to diminish the effect of the injuries that produce disability and handicaps, to control the symptoms by physical means and to allow the eldest to [4], p. 49.
- *Psychologists*: Its general competences are to manage the resources available, human and material, so that the institution provides the services for which it is intended. Doing psychological and neuropsychological evaluations, diagnoses and developing stimulation and re-education programs more appropriate for each case. Also, assisting in a therapeutic way in those cases and in specific programs that require it [4], p. 50.
- *Social Workers*: Its main function is to organize the functioning of the social area so that it provides a service adapted to the needs of the users, of an integral and quality character. Sometimes, in the absence of other professionals, can be responsible for the area of socio-cultural animation [4], p. 51.

In order to promote a proper attention to the grief of our residents, we need to provide our professionals with factors, which will not only affect the treatment of grief, but their entire work in the centre.

As well we have previously expressed, elderly people in a residential context, do not have the same stages of grief that elderly people outside this context. Assuming that in some stages of grief, the centre's policy and actions of professionals are key to overcoming grief. How can we help residents better?

Despite it is true there is a clear heterogeneity, among different professional profiles and their functions, we can find some common elements that could facilitate their work with grief (Table 1).

Table 1 Recommendable attitudes to mourning work

Recommendable attitudes to mourning work	
Accompany the affected resident	Provide space and time for the person
Facilitate feeling expression	Give him or her individualise attention
Create help groups	Have coalify workers to deal with mourning
Create activities that attract the person and distract him or her	Talk about mourning with naturality
Facilitate participation in bereavement rituals	Use correct language
Create work teams	Create a communication line between all the professionals who work with the affected patient

Source Prepared from the material obtained by Rodríguez and Fernández [5], p. 279

3   The Mourning Protocol in Spanish Nursing Homes

Once it is specified the importance of bereavement care, the characteristics of mourning in the homes of elderly and professionals involved, we need to define the Spanish protocols for his process.

Regarding the bereavement care protocols, we can define 2 types of protocols:

- Those centres in which there is no specific protocol for grief attention, being it included in a section of the death protocol, as one of the steps that it is followed when a resident died in the centre.
- Those centres that do have a specific protocol for grief attention, assuming this is a differentiated section of the death protocol.

Then we will describe more clearly each of the two types, by providing protocols from different nursery homes.

- *Attention to grief included in the death protocols*: This is the example of the ICERON and LARES foundation, in which this section is included as part of the death protocols.

    In the first case this section appears after the preparation of the corpse and it differences between family's, resident's and residential worker's care [6], p. 6.

    In the second example, bereavement care, is not listed as a section by itself, but as two hyphens in the sequence of actions to perform at the death of a resident, the first referred to the family accompaniment, after death notification and the agreement to transfer to the funeral centre and the second once the process is completed, as an accompaniment by the psychologist or another professional stipulated to the roommate of the deceased.
- *Specific protocol for bereavement*: This is the protocol described by the Council of Social Affairs of the Community of Madrid, in which the grief, attention appears as a specific section.

    However, duel already appears in a previous section called "Protocols of Care Management: Care Death" as an introduction definition before the beginning of the chapter. After this, within the guide, we find a specific chapter dedicated to this phenomenon, entitled "Protocols of Care Management: Bereavement Care". This chapter begins with the specification of the users to whom it is addressed and with various definitions, including the one of the griefs, the types of grief and the phases. Then talks about the professionals involved in the grieving process, description of the procedure, technical standards and place and custody records.

Although in the last example of Grief Protocols in nursing homes, the process is more specified, we can see an absence of differentiation, between the type of relationship the resident has with the deceased and some reference if the deceased was not a nursing home resident but affects one of the elderlies who lives in the nursing home. Because our centre is our residents' home, we can find ourselves with the situation that someone external to it dies and it affect our residents.

4 Methodology

This research is part of an extended investigation about mourning in nursing homes. On this occasion, we have decided to focus on the necessities of the different professionals who work in the nursing homes. We have created a survey which have been send to fifty-eight nursing homes workers: Nursing Assistants (24.60%), Social Workers (21.19%), Nurses (22.8%), Occupational therapists (7%), Physiotherapists (7%), Psychologists (5.3%) and Sociocultural Motivators (5.3%).

Having these hypotheses:

1. There is a difference between the professional role and the necessities each of them must work with the residents mourning.
2. The ratio of workers per nursing home is the most select option as problem to work with the residents mourning.
3. The most of the surveyed consider the social worker and the psychologist the two professionals who work with the residents mourning process.

Despite our first questionnaire having sixteen questions, for this section we have used eleven questions especially related with the protocol's topic, four of them related with general information and eight with more specific questions which have been answered by 58 Spanish nursing home workers.

With ages between 16 years old and 55 years old, people surveyed ages are divided in four groups: 16–25 years old (25.60%), 26–35 years old (59.60%), 36–45 years old (7%), 46–55 years old (8.88%).

In relation with gender, 80.7% are women and 19.3% are men. They are from different cities of Andalucía and Extremadura (Spain). And the nursing homes where they work are private with public funds (52.6%), private (33.3%) and public (21.11%). With relation to the nursing home character, there are non-religious (68.4%) and religious (31.6%) centres.

About the number of residents, we have divided them in six groups: Less than 20 residents (3.40%) 21–40 residents (28.10%), 41–60 (21.10%), 61–100 (14%), 101–120 (5,30%) and more than 120 (28.10%).

Our variables are (Table 2):

5 Results

In this section, we are going to focus on the results of our research and the possible relationships between the factors which could be interesting for our conclusions.

- Facilitating elements and barriers in mourning work.

In Table 3 we can appreciate that the most common barrier for all the workers is the working time per patient, follow by the mourning formation. Being having a face to face contact with the residents the most common facilitating aspect.

Table 2 Methodology variables

Code	Variables	Question
VDPM0001	Age	1
VDPM0002	Sex	2
VDPM0003	Centre location	3
VDPM0004	Centre ownership	4
VDPM0005	Centre character	5
VDPM0006	Number of residents	6
VDPM0007	Professional role	7
VDPM0010	Existence of professional work with mourning	10
VDPM0012	Do not have a specific area to work with mourning	12
VDPM0013	Do not have a face to face contact with the residents	12
VDPM0014	Low economical and material resources	12
VDPM0015	Scarce ratio	12
VDPM0016	Do not have enough time to work with each patient	12
VDPM0017	The inexistence of a multidisciplinary work	12
VDPM0018	The inexistence of communication with the other professionals	12
VDPM0019	Lack of organizational flexibility	12
VDPM0020	Nonexistence of formation about mourning	12
VDPM0021	Have a specific area to work with mourning	13
VDPM0022	Have a face to face contact with the residents	13
VDPM0023	Economical and material resources	13
VDPM0024	The ratio	13
VDPM0025	Have enough time to work with each patient	13
VDPM0026	The existence of multidisciplinary work	13
VDPM0027	The communication with other professionals	13
VDPM0028	The organization flexibility	13
VDPM0029	The existence of formation about mourning	13
VDPM0030	Multidisciplinary work with mourning	14
VDPM0031	Professional categories who work with the patient mourning	15

Source Author compilation

- Facilitating elements and barriers in mourning work social animators' answers

Regarding with the social animator answers we can notice that the multiciliary work is as facilitating aspect as well as a barrier, being the working time per patient and have a face to face contact with the residents the other two most vote facilitating aspects and the organization flexibility and mourning formation the barriers (Table 4).

Table 3 Facilitating elements and barriers in mourning work

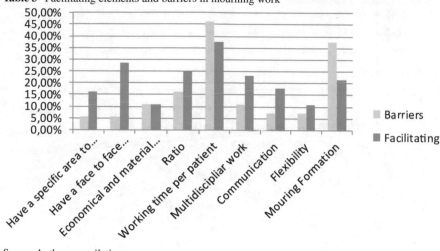

Source Author compilation

Table 4 Facilitating elements and barriers in mourning work social animators answers

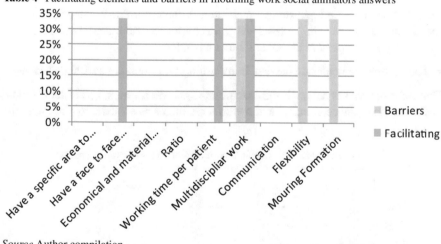

Source Author compilation

- Facilitating elements and barriers in mourning work nurses' answers

Most of nurses interviewed pointed the mourning formation follow by the working time per patient as the two principal barriers and have a face to face contact with the elderly as the most important facilitating element (Table 5).

Table 5 Facilitating elements and barriers in mourning work nurses' answers

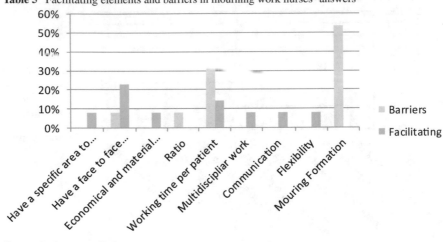

Source Author compilation

- Facilitating elements and barriers in mourning work physiotherapies answers

As it is show in Table 6 physiotherapies identify the mourning formation, having a specific area to work with the residents and economical and material resources as a facilitating element and at the same time as a barrier.

- Facilitating elements and barriers in mourning work nursing assistants answers

The working time per patient is the most pointed answer for nursing assistants follow by the ratio and the economical and material resources as a barrier to work with the

Table 6 Facilitating elements and barriers in mourning work physiotherapies answers

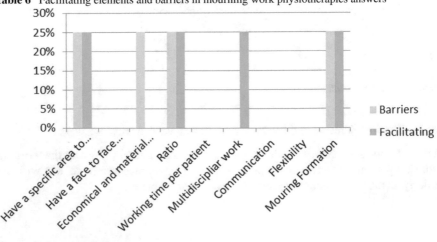

Source Author compilation

Table 7 Facilitating elements and barriers in mourning work nursing assistants answers

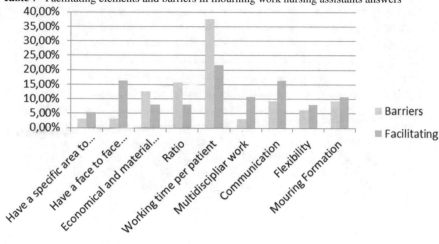

Source Author compilation

patient mourning. Being the working time per patient, the communication between the different professionals and have a face to face contact with the elderly the most pointed facilitating elements (Table 7).

- Facilitating elements and barriers in mourning work psychologist answers

In relation with the facilitating elements and barriers identified by psychologist, this collective has mentioned have a specific area to work with patient's mourning and multiciliary work as a possible barrier and a facilitating element (Table 8).

Table 8 Facilitating elements and barriers in mourning work psychologist answers

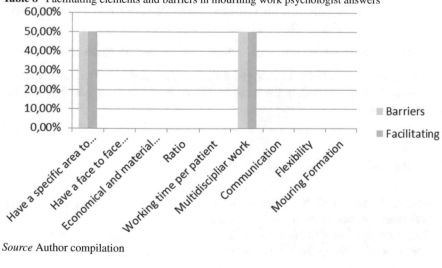

Source Author compilation

Table 9 Facilitating elements and barriers in mourning work occupational therapist answers

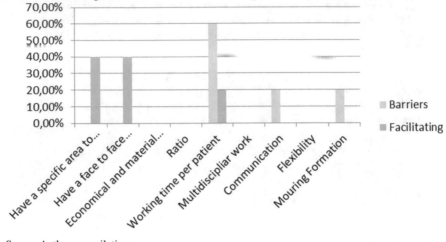

Source Author compilation

- Facilitating elements and barriers in mourning work occupational therapist answers

As it is show in Table 9 occupational therapist feel the working time per patient, communication between the different professionals and the mourning formation are the most remarkable barriers for their work with the residents mourning and have a specific area to work and have a face to face contact with residents as the most remarkable facilitating elements.

- Facilitating elements and barriers in mourning work social workers answers

Most of the social workers interviewed pointed the working time per patient and the mourning formation as the principal barriers, being also these two options the most facilizing element selected (Table 10).

- Who is the reference professional in the mourning work in your nursing home?

Following the results of the bellow graphic, the psychologist is the nursing home professional more selected by the workers follow by the nursing assistant and the social workers.

6 Discussion

After a deep results and theory analysis, we can remark some similarities between the previous mentioned theory and our results.

The first notable resemblance is between the recommendable attitudes to mourning work, among the list we have some recommendations like the creation

Table 10 Facilitating elements and barriers in mourning work social workers answers

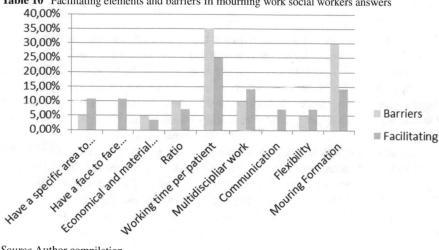

Source Author compilation

of work teams, a communication line between all the professionals who work with the affected patient and coalify professionals [5], p. 279.

Our surveyed have pointed: have mourning formation, multidisciplinary work and face to face contact with the elderly as barriers and facilitating aspect of the mourning work, being this a cohesion between the theory and results.

The second remarkable coincidence is related with the referent profession for the mourning work in the nursing homes, as it is pointed in the cinceron protocol the roommate's deceased must accompaniment by the psychologist [6], p. 6. And as it showed in the Table 11 our surveyed have pointed the physiologist as the reference professional with a 21.6%.

Table 11 The reference professional in the mourning work

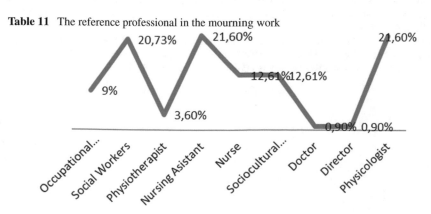

Source Author compilation

7 Conclusion

To conclude this chapter in relation with our hypothesis we can point that our results show us that some of our hypothesis are contrasted whereas others no.

Despite our first hypothesis is contrasted, as well as the third one. The second hypothesis it is not, the ratio of workers is not considering for the surveyed as the biggest complicating element in their work with the residents mourning process, but it is the working time per patient. Something which would be higher with a more adequate ratio.

It is remarkable how despite the nursing assistant are one of the workers with low study level, it is one of the selected professionals, as a reference in the mourning work, and how specially this collective demand some type of formation in this file. This could be because of the time they spend with the elderly. That show us the importance of educate the entire working equip in this area and the necessity of insist in the face to face contact with the residents specially in a duel.

References

1. Martín M (2003) Trabajo social en gerontología (ISBN: 9788497561235)
2. White R (2015) Dialectic of mourning. J Theor Humanit 20(4):179–192
3. Díaz P, Losantos S, Pastor P (2014) Guía de Duelo en Adultos para profesionales socio-sanitarios. [online 17-05-2017]. Available online at: https://www.fundacionmlc.org/noticia/guia-duelo-adulto/
4. IMSERSO (2015) Marco de actuación para las personas mayores: propuestas y recomendaciones para favorecer el ejercicio de los derechos de las personas mayores. [online 03-09-2017]. Available online at: http://www.fundacionpilares.org/modeloyambiente/materiales-utiles/publicacion/marco-de-actuacion-para-las-personas-mayores-propuestas-y-medidas-para-fortalecer-el-ejercicio-de-los-derechos-de-las-personas-mayores/
5. Rodríguez B, Fernández A (2002) Intervenciones sobre problemas relacionados con el duelo para profesionales de Atención Primaria (II): intervenciones desde Atención Primaria de salud. MEDIFAM 12(4):276–284
6. Ortega M (2013 5 Febrero) Protocolo de Exitus. Asocición Ciceron. [online 03-02-2017]. Available online at: file:///G:/Universidad/Master/TFM/Documentos/Protocolos%20de%20duelo/Protocolo%20de%20%C3%A9xitus%20-%20Asociaci%C3%B3n%20Cicer%C3%B3n.htm

The Mourning Influence in the Burnout Syndrome Among Nursing Home Workers in Spain

Ana Vallejo Andrada⊙, José Luis Sarasola Sánchez-Serrano⊙,
Evaristo Barrera Algarín⊙, and Francisco Caravaca Sánchez⊙

Abstract Nowadays they are a weight amount of mourning and grief studies and definitions, most of them related to death and focused in family and friends of the deceased. However just a few authors have research and define the mourning process the professionals (nurses, social workers, psychologist, …) who has worked directly with the deceased, have to get throw. That kind of mourning has been calling disenfranchised grief. With the aim to study that phenomenon, we have explored the burnout symptom and disenfranchised grief among 90 nursing home workers in Spain, using a questionnaire composed by The Maslach Burnout Inventory; Grief Support in Health Care Scale; Disenfranchised Grief making an univariate, bivariate, and multivariate data analyses were conducted using SPSS statistical software version 22. Having as a result the association between these three scales and some demographic variables.

Keywords Nursing homes · Grief · Burnout · Disenfranchised grief · Elderly · Spain

A. Vallejo Andrada (✉)
Pablo de Olavide University, Utrera Road 1, 41013 Seville, Spain
e-mail: avaland@upo.es

J. L. Sarasola Sánchez-Serrano
Jaen University, Campus Las Lagunillas S/N, 23071 Jaén, Spain
e-mail: jlsarsan@upo.es

E. Barrera Algarín
Department of Social Work and Social Services, Pablo de Olavide University, Seville, Spain
e-mail: ebaralg@upo.es

F. Caravaca Sánchez
Department of Psychology, Social Work and Social Services Area, Jaén University, Jaén, Spain
e-mail: caravaca@ujaen.es

© The Author(s), under exclusive license to Springer Nature Switzerland AG 2021   251
D. Soitu et al. (eds.), *Decisions and Trends in Social Systems*,
Lecture Notes in Networks and Systems 189,
https://doi.org/10.1007/978-3-030-69094-6_21

1 Introduction

Because of the acceleration of the average life expectancy and related growth of the elder population, addressing long-term care (LTC) needs has become an increasingly vital issue. Yet, the current LTC workforce faces major barriers in meeting these needs, including insufficient training and support mechanisms for staff, [15] high levels of burnout, [5] and high turnover rates in particular among direct care staff who provide the bulk of hands-on care [8].

Who work directly with residents in long-term care homes, death, dying, and grief are usual experiences in their day-to-day working life [24].

Palliative/end-of-life care literature identifies that specialized training, skills, and education requirements are needed for the delivery of palliative care [6]. Support for grief, loss, and bereavement is acknowledged to be an important component of holistic palliative care.

Among this chapter we are going to link the burnout symptomatology and its relation with grief, especially with the most common type of grief among workers, which is disenfranchised grief.

With the aim of the better understanding of these two topics, we are going to start describing both problematic (burnout and disenfranchised grief) and showing previous researchers finding in the nursing homes areas related with them.

1.1 Burnout

The World Health Organisation (WHO) declared burnout as a "occupational phenomenon" in International Classification of Diseases 11th revision (ICD-11), stating that burnout is a syndrome resulting from "chronic workspace stress that has not been successfully managed" [23]. This recent declaration recognized burnout as a serious health issue, and WHO is planning to develop "evidence based" guidelines for mental well-being in workplace.

Theoretically, Burnout depends on individual susceptibility and on the environment in which the professional is inserted, including the daily workplace and the society to which the individual belongs [21].

As the numbers of elderly people requiring nursing or residential care increase the wellbeing of nurses and care workers employed in this setting grows in importance. Nursing is inevitably a demanding and stressful job in a complex organizational setting, and it has been widely regarded as one of the most stressful occupations, associated with high levels of staff turnover, absenteeism and levels of burnout [7].

Nursing home RNs report higher rates of burnout and job dissatisfaction than employed in any other setting, including hospitals, and are often unable to complete necessary care due to insufficient time and resources [22].

Burnout and job dissatisfaction are both key drivers of staff turnover a significant problem in nursing homes that has been consistently linked to worse care quality [11].

Factors that contribute to burnout among human-service workers in nursey homes has been conceptualized according to several theoretical models, we have decide to use the Maslach Burnout Inventory (MBI) developed by Maslach and Jackson [13].

This scale has been previously used by Martínez et al. [12] in The Person Centered approach in Gerontology: New validity evidence of the Staff Assessment Person-directed Care Questionnaire, where the researchers prove correlations between perceptions of person-centred care delivery frequency and burnout.

By Shinan-Altman et al. [19] in the connection between illness representations of Alzheimer's disease and burnout among social workers and nurses in nursing homes and hospitals: a mixed-methods investigation where the researchers measure the staff perception of AD and associations with burnout.

And by Kubicek and Korunka [10], in Does job complexity mitigate the negative effect of emotion-rule dissonance on employee burnout? Where the writers approach Role of emotion rule dissonance in burnout. Between others.

1.2 Disenfranchised Grief

Most grief researchers and therapists indicate that most grievers have adequate internal resources and social support to readjust to life after a loved one's death without needing significant intervention [25].

However, there are factors that can complicate grief in a minority of individuals and lead to poorer outcomes. For instances, studies suggest that caregivers who are caring under both difficult and rewarding circumstances are at greater risk for complicated grief [17].

About the grief experiences of professional caregivers in long term care settings. Previous studies of this population that have used qualitative methods to explore staff responses to patients' deaths suggest that staff may experience symptoms of grief [9, 14].

Rickerson and colleagues [16] in their research surveyed 203 care-workers including five sections, First, staff provided demographic information about themselves and described their past healthcare work experience. Next, staff estimated how many of their patients had died in the past month, and then identified (from a checklist) the symptoms that they experienced as a result of those deaths [16].

Having as a conclusion that staff responded to the deaths of patients in very different ways and described a wide variety of grief-related symptoms. Some staff reported experiencing several symptoms, others reported few or none. Experience does not appear to "protect" staff from the experience of grief. On the contrary, there may be a cumulative component of staff grief that has not been previously examined in these settings [16].

Largely, the topic of death has been muted and grief experienced by front-line staff has been recognized as one form of "disenfranchised grief." Considering the lack of attention to how patient death affects direct care staff, an important question is how well staff members can be prepared to deal with death and related grief [8].

Following this new perspective, we have researchers like Anderson and Gaugler one, title the grief experiences of certified nursing assistants: personal growth and complicated grief. Whereas they have focused their study in nursing assistants and ours compile all the working categories in nursing homes, we can define this study like an important antecedent to ours, as the disenfranchised grief variable is proved in health care workers.

For their study, Anderson and Gaugler, created a questionnaire composed by: The Death Attitude Profile-Revised (DAP-R); General Self-Efficacy Scale (GSE); Scale of Perceived Social Support (MSPSS); The 12-item Personal Growth (PG) subscale from the Hogan Grief Reaction Checklist (HGRC); The 19-item Inventory of Complicated Grief (ICG) and Disenfranchised Grief Scale (DGS) [2].

With a sample of 136 nursing assistants, the get the following conclusion: Disenfranchised grief was found to be a strong and significant predictor of personal growth, participants who experienced higher levels of disenfranchised grief and who worked in urban facilities reported lower levels of personal growth in relation to the deaths of nursing home residents in their care. These findings partially support that higher levels of disenfranchised grief would be associated with less personal growth [2].

1.3 Grief and Burnout Previous Association Experiences

Whereas the association between disenfranchised grief and burnout, is something quite new, there are several researchers that prove the association between grief and burnout in the nursing home context, an example of it is the Boerner, Gleason and Jopp research [3]. Factors, Texas Revised Inventory of Grief and Burnout and Maslach Burnout Inventory, for a sample of 143 nursing home workers, of which they get among others the conclusion of: The new insight that grief avoidance relative to grief symptoms was more influential in predicting depersonalization. It suggests that those who made efforts to avoid their grief may have been more likely to report instances of depersonalization [3].

The strong focus on controlling emotions as a way of dealing with staff reactions to patient death may deserve some further reflection as it could come with a cost in terms of the staff members' ability to care for and about patients, which is not desirable in the long time care context [3].

Based in previous literature, our current research is aimed to explore the association between grief and burnout among nursing home workers in Spain.

2 Method

2.1 Participants and Procedure

Ninety nursing home workers from the province of Seville were contacted by mail for the current research between December 2019 to March 2020. Self-report questionnaires were distributed by mail to the nursing home workers that voluntarily agreed to participate through Google Form. Prior to complete the questionnaires, participants received information with each questionnaire that introduced the research team, the structure of the questionnaire, the aims of the study and guaranteed participant anonymity and voluntarily. Participants did not receive any benefit from participation. The study was approved by the Ethics Committee of the Pablo Olavide University (Spain), and by the board of executives of two facilities randomly selected from the province of Seville where the research were conducted. The final sample was composed by 90 nursing home workers being mostly women (84.4%), additional information about demographic and work characteristics are shown in Table 2.

2.2 Measures

The questionnaire was order as follows: demographic and work variables, Burnout and grief variables. It took participants approximately 10–20 min to complete the questionnaire. Completed questionnaires were returned directly to research staff via Google Form.

A demographic and work data questionnaire was used. It included workers' demographic variables such as age (coded as a continuous variable) and gender (coded as $0 =$ male and $1 =$ female), years of nursing home experience (coded as continuous variable), and duration of employment at the present nursing home (coded as continuous variable), position staff (as $0 =$ staff, $1 =$ social worker and $2 =$ supervisor), current working sector ($0 =$ private company and $1 =$ public administration) and working hours ($0 =$ working full time and $1 =$ working part time).

Burnout was assessed with the Maslach Burnout Inventory–Human Service Survey (MBI–HSS) [13], using the Spanish version developed by De las Cuevas [4]. The MBI is a twenty-two items scale including three dimensions: (1) Emotional Exhaustion (EE; composed by nine items), Depersonalization (DP; composed by five items) and Personal Accomplishment (PA; composed by eight items). Using a seven-point Likert scale ranging from 1 ("completely disagree") to 7 ("completely agree") participants were asked to rate how often they felt the feelings described for each item. For the current research, the cut-off points proposed by Seisdedos [18] in order to classified severity of burnout of each of the dimensions (Table 1).

Variables on grief were measured using two scales. First, the fifteen-item version of Grief Support in Health Care Scale (GSHCS) [1] were used adapted and validate to Spanish by Vega and colleagues [20]. It consists in an instrument divided into three

Table 1 Cut-off points for MBI dimensions

Dimensions	Low	Moderate	High
EE	<14	15–24	>25
DP	<3	4–9	>10
PA	<32	33–39	>40

five-item subscales: (1) recognition of the relationship, (2) acknowledgement of the loss and (3) inclusion of the griever. Using a seven-point Likert scale ranging from 1 ("strongly disagree") to 5 ("strongly agree"), a higher score on the GSHCS indicate higher levels of grief specific support. Secondly, Disenfranchised Grief was measured using the instrument developed by Anderson and Gaugler [2]. The Disenfranchised Grief instrument is composed by fifteen items answered each item by a 5-point Likert scales ranging from 1 ("strongly disagree") to 5 ("strongly agree"). The Grief Symptoms Scale was designed to determine the presence and strength of factors that may disenfranchise the grief.

2.3 Data Analysis

Univariate, bivariate, and multivariate data analyses were conducted using SPSS statistical software version 22 with a minimum significant level set at 95% ($p \leq 0.05$). Data analysis were conducted in X stages: First, descriptive statistics were computed across the study sample to understand demographic and work variables. Second, mean scores (M) and Standard Deviation (SD) of each of the MBI dimensions were calculated for the current sample and compared with the normative sample enrolled in the validation conducted by Seisdedos [18]. Third, according to Seisdedos [18] cut-off points proposed by Seisdedos [18] MBI dimensions (EE, DP and PA) were calculated for each dimension (low, moderate and high) according to severity. Next, mean scores (M), Confidence Interval to 95% (95% CI) and Standard Deviation (SD) were calculated for the two scales measuring grief: (1) GSHCS and (2) Disenfranchised Grief. Three models were calculated using hierarchical linear regression using MBI subscales as a reference to explore demographic, work and grief variables association.

3 Results

Table 2 shows participant demographic and work characteristics. Participants' mean age was 34.4 years (SD = 8.5), relatively experienced in nursing home work (M = 7.8; SD = 6.7 years) and with a mean of 4.6 years (SD = 4.3) working in current position. Of the sample, most participants were women (84.4%; CI 95%: 76.7–91.1).

Table 2 Participants characteristics ($N = 90$)

	M	SD
Age	34.4	8.5
Years worked in nursing homes	7.8	6.7
Years in current position	4.6	4.3
Gender	%	95% CI
Female	84.4	76.7–91.1
Male	15.6	8.9–23.3
Position		
Staff	61.1	51.1–72.2
Social worker	27.8	18.9–37.8
Supervisor	11.1	5.6–17.8
Work sector		
Companies	82.2	74.4–90.0
Public administration	17.8	10.0–25.6
Working hours		
Working full-time	68.9	58.9–77.8
Working part-time	31.1	22.2–41.1

Note SD: standard deviation; CI: confidence interval

Regarding job characteristics, approximately two out three (61.1%; CI 95%: 51.1–72.2) were staff workers and four out five (82.2%; CI 95%: 74.4–90.0) were working in private companies. Moreover, 68.9% (ICI 95%: 58.9–77.8) were working full-time versus 31.1% (CI 95%: 22.2–41.1) of working part-time.

In terms of MBI dimensions scores, differences between current sample (N = 90) and normative sample from the Spanish validation conducted by Seisdedos [18] are shown in Table 3. Overall, the average scores for the EE dimension (M = 19.0; SD = 10.2 and M = 20.3; SD = 11.3) were higher among current sample compared with the normative sample. However, there is a difference as regards the EE and DP dimension, showing current sample a higher average score of DP (M = 10.2; SD =

Table 3 Comparison of burnout scores of current sample with scores for the normative sample

Group	MBI subscales		
Current sample (N= 90)	EE	DP	PA
Mean	19.0	10.2	31.4
Standard deviation	14.3	5.6	9.2
Spanish norms for MBI (N = 1138)	EE	DP	PA
Mean	20.3	7.6	35.7
Standard deviation	11.3	5.0	8.0

Note EE: Emotional Exhaustion; DP: Depersonalisation; PA: Personal Accomplishment

Table 4 Incidence of low, moderate and high levels of burnout for current sample ($N = 90$)

	MBI subscales						Burnout[a]	
	Low		Moderate		High			
	%	95% CI	%	95% CI	%	95% CI	%	95%CI
EE	47.8	37.8–57.8	22.2	13.3–31.1	**30.0**	21.1–40.0		
DP	12.2	6.7–18.9	37.8	28.9–47.8	**50.0**	38.9–60.0		
PA	**40.0**	30.6–49.8	38.9	28.4–48.6	21.1	13.3–29.1		
Burnout							**17.8**	10.0–24.6

Note EE: Emotional Exhaustion; DP: Depersonalisation; PA: Personal Accomplishment. CI: confidence interval[a]Burnout is categorized as high EE and DP and low PA; EE-high: $\geq$25, moderate: 24–15, low: $\leq$14; DP-high: $\geq$10, moderate: 9–4, low $\leq$3; PA-high: $\geq$40, moderate: 39–33, low: $\leq$32

5.6 and M = 7.6; SD = 5.0) and a lower average score in the dimension of PA (M = 31.4; SD = 9.2 and M = 35.7; SD = 8.0) than the normative sample. Results from the severity of MBI dimensions according to cut-off points proposed by Seisdedos [18] are reported in Table 4. Almost one out three (30.0%) reported high EE levels, half of the sample (50.0%) showed high DP attitudes and 40.0% of participants reported feeling so low for PA. As shown Table 4, almost one out five of the sample (17.8%) reported burnout, characterized by high EE and DP and low PA.

As regards grief subscales, the average scores for current sample are display in Table 5. The average score for GSHCS subscales were: recognition of the relationship (M = 20.1; CI 95%: 18.8–20.7), acknowledgement of the loss (M = 17.0; CI 95%: 15.9–17.8) and inclusion of the griever (M = 13.0; CI 95%: 11.9–13.9). Regarding disenfranchised Grief, the average score were 48.7 (CI95%: 46.8–50.3).

As may be observed in Table 6, three models were conducted for explore association between MBI dimensions: EE (Model I), DP (Model II) and PA (Model III) with the demographic, working and grief subscales variables. In Model I, significant associations were found between EE and the following variables: years working in nursing homes (p = 0.046), working part time (p = 0.041), acknowledgement of the loss (p = 0.015) and inclusion of the griever (p = 0.023). Exploring association between DP and the variables explored (Model II), exclusively inclusion of the griever (p = 0.014) were statically associated. Model III shown that grief subscales,

Table 5 Descriptive statistics for Grief subscales for current sample ($N = 90$)

	Grief subscales		
	M	95% CI	SD
Recognition of the relationship	20.0	18.8–20.7	4.2
Acknowledgement of the loss	17.0	15.9–17.8	4.5
Inclusion of the griever	13.0	11.9–13.9	4.7
Disenfranchised Grief	48.7	46.8–50.3	8.5

Note CI: confidence interval; SD: standard deviation

Table 6 Hierarchical linear regression models for MBI subscales

	MBI subscales									Burnout (Model III)		
	EE (Model I)			DP (Model II)			PA (Model III)					
	B	SE	p	B	SE	p	B	SE	p	B	SE	p
Demographic and work variables												
Age	−0.005	0.225	0.983	0.021	0.094	0.821	0.086	0.172	0.618	0.005	0.066	0.295
Years worked in nursing homes	0.728	0.359	**0.046**	0.054	0.150	0.719	−0.259	0.274	0.348	0.016	0.009	0.079
Years in current position	−0.354	0.431	0.413	0.055	0.180	0.761	0.175	0.330	0.597	−0.006	0.012	0.611
Male[a]	−0.230	4.272	0.957	0.624	1.788	0.728	−2.346	3.268	0.475	0.009	0.114	0.938
Social Worker[b]	−4.194	2.333	0.076	−1.443	0.976	0.143	−0.914	1.784	0.610	−0.101	0.060	0.099
Companies[c]	−1.962	4.235	0.644	0.701	1.772	0.693	1.523	3.239	0.640	−0.024	0.111	0.832
Working part-time[d]	6.575	3.022	**0.041**	0.453	1.464	0.758	−4.265	2.676	0.115	−0.006	0.050	0.910
Grief variables												
Recognition of the relationship	−0.538	0.396	0.178	−0.064	0.172	0.709	0.673	0.284	**0.020**	−0.028	0.011	**0.013**
Acknowledgement of the loss	0.875	0.353	**0.015**	0.180	0.153	0.244	−0.298	0.253	0.242	0.021	0.010	**0.038**
Inclusion of the griever	−0.828	0.358	**0.023**	0.454	0.112	**0.014**	−0.237	0.256	0.359	−0.002	0.010	0.813
Disenfranchised grief	−0.347	0.220	0.118	−0.024	0.095	0.800	0.454	0.157	**0.005**	−0.008	0.006	0.184

Note EE: Emotional Exhaustion; DP: Depersonalisation; PA: Personal Accomplishment. SE: Standard Error
Bold is for variables with p-value of <0.05
[a]Omitted variable: Female; [b]Omitted variable: Supervisor; [c]Omitted variable: Public administration; [d]Omitted variable: Working full-time

specifically recognition of the relationship (p = 0.020) and disenfranchised grief (p = 0.005) were statically associated with lower levels of PA. Finally, Model IV explore the demographic, work and grief variables associated to burnout (categorized as high EE and DP and low PA), none of the demographic and work variables were statically associated, in terms of grief, burnout were associated with recognition of the relationship (p = 0.013) and acknowledgement of the loss (p − 0.038)

4 Discussion and Conclusion

Current research provided important insights into the nursing homes perspective, as it associate the burnout and the disenfranchised grief in an innovative way, as to our knowledge this is one of the first study which combine Maslach Burnout Inventory–Human Service Survey (MBI–HSS) [13], Grief Support in Health Care Scale (GSHCS) [1] and Disenfranchised Grief [2] among nursey home workers in Spain.

In which concern to our objective: to explore the association between grief and burnout among nursing home workers in Spain, we can affirm that it has been fulfil.

In concordance with previous researches [7, 22], the level of burnout of our sample are considerably high 17.8%, being Emotional Exhaustion and Depersonalisation the most remarkable subscales of the MIB. In relation with Disenfranchised Gried and GSHCS scales, as other authors affirm [2], the Disenfranchised Grief of our sample has a media of 48.7 which can be considerer a representative number of workers, and in relation with the GSHCS the subscale with a high media is recognition of the relationship, being this las date extremely related with Boerner, Burack, Jopp and Mock, conclusions [3].

About Disenfranchised Grief Scale, GSHCS and MBI, as other authors have already done, [3] we can confirm there is an association between them. Despite the MBI scale just have a direct association with the GSHCS variables: Recognition of the relationship and Acknowledgement of the loss. There are also associations in MBI variable Emotional Exhaustion and GSHCS variables: Acknowledgement of the loss and Inclusion of the griever; MBI variable: Depersonalisation and GSHCS variable: Inclusion of the griever; MBI variable: Personal Accomplishment and GSHCS variables: Recognition of the relationship and the Disenfranchised Grief Scale.

Regarding the demographic variables, as well as other author researches [19], there is an association between years worker in nursing homes and Emotional Exhaustion variable of the MBI scale. Whereas we have also found an association between working part time in nursing home and this variable which not appear in the previous mentioned study.

Following this data, we can conclude that there is an association between the three scales, as each MBI scale it is as least associated with one GSHCS scale or with the Disenfranchised Grief Scale. Which means that Disenfranchised Grief among Nursey home workers have a direct effect in their levels of Burnout, and their levels of Burnout in their Disenfranchised Grief levels.

That conclusion may change the Burnout and Grief perspective among Nursey Home Workers, and it could help to visibly the importance of giving correct grief strategies to workers or adapt grief protocols in nursery homes including the staff grief in other to improve to prevent burnout and as a consequence of it improve productivity and quality of service for the residents.

References

1. Anderson KA, Ewen HH, Miles EA (2010) The grief support in healthcare scale: development and testing. Nurs Res 59(6):372–379
2. Anderson KA, Gaugler JE (2007) The grief experiences of certified nursing assistants: personal growth and complicated grief. OMEGA J Death Dying 54(4):301–318
3. Boerner K, Gleason, Jopp D (2017) Burnout after patient death: challenges for direct care workers. J Pain Symptom Manag 54(3):317–325
4. De las Cuevas C (1994) El desgaste profesional en atención primaria: presencia y distribución del síndrome de burnout. Laboratorios Servi, Madrid
5. Dill JS, Cagle J (2010) Caregiving in a patient's place of residence: turnover of direct care workers in home care and hospice agencies. J Aging Health 22(1):713–733
6. Froggatt KA (2001) Palliative care in nursing homes: where next? Palliat Med 15(1):42–48
7. Jamal M, Baba VV (1992) Shiftwork and department-type related to job stress, work attitudes and behavioral intentions: a study of nurses. J Organ Behav 13(5):449–4664
8. Kathrin B, Burack O, Jopp D, Mock E (2015) Grief after patient death: direct care staff in nursing homes and homecare. J Pain Symptom Manag 49(2):214–222
9. Katz J, Sidell M, Komaromy C (2000) Death in homes: bereavement needs of residents, relatives and staff. Int J Pall Nurs 6(6):274–279
10. Kubicek B, Korunka C (2015) Does job complexity mitigate the negative effect of emotion-rule dissonance on employee burnout?Work Stress 29(4):379–400
11. Laschinger HKS, Leiter MP (2006) The impact of nursing work environments on patientsafety outcomes the mediating role of burnout/engagement. J Nurs Adm 36(5):259–267
12. Martínez T, Suárez-Álvarez J, Yanguas J, Muñiz J (2015) The person centered approach in gerontology: new validity evidence of the staff assessment person-directed care questionnaire. Int J Clin Health Psychol 16(2):175–185
13. Maslach C, Jackson SE (1981) The Maslach burnout inventory. Consulting Psychologist Press, Palo Alto, CA
14. Moss AH (1993) Discussing resuscitation status with patients and families. J Clin Ethics 4(2):180–182
15. PHI (2016) U.S. home care workers: key facts. 2016. [online 08-04-2020]. Available online at http://phinational.org/sites/phinational.org/files/phi-home-care-workers-key-facts.pdf
16. Rickerson E, Somers C, Allen C, Lewis B, Strumpf, Casarett D (2005) How well are we caring for caregivers? Prevalence of grief-related symptoms and need for bereavement support among long-term care staff. J Pain Symptom Manag 3(3):227–233
17. Schulz R, Boerner K, Shear K, Zhang S, Gitlin LN (2006) Predictors of complicated grief among dementia caregivers: a prospective study of bereavement. Am J Geriatr Psychiatry 14(1):650–658
18. Seisdedos N (1997) MBI: Inventario de Burnout de Maslach ['Maslach Burnout Inventory']. TEA, Madrid
19. Shinan-Altman S, Werner P, Cohen M (2016) The connection between illness representations of Alzheimer's disease and burnout among social workers and nurses in nursing homes and hospitals: a mixed-methods investigation. Aging Mental Health 20(4):352–361

20. Vega P, Melo J, González R (2015) Validación de escala de apoyo en duelo en atención de salud para población hispanoparlante. Psicooncologia 12(2–3):355–366
21. Vega NV, Sanabria A, Domínguez LC, Osorio C, Bejarano M (2009) Síndrome de desgaste profesional. Revista Colombiana de Cirugia 24(3):138–146
22. White EM, Aiken LH, McHugh MD (2019) Registered nurse burnout, job dissatisfaction, and missed care in nursing homes. J Am Geriatr Soc 67(10).2066 2071
23. World Health Organisation (2019) Burn-out an "occupational phenomenon": international classification of Diseases. [online 10-04-2020]. Available online at https://www.who.int/mental_health/evidence/burn-out/en/
24. Wowchuk SM, McClement S, Bond J (2007) The challenge of providing palliative care in nursing homes, part 2: internal factors. Int J Palliat Nurs 13(1):345–350
25. Zisook S, Shear K (2009) Grief and bereavement: what psychiatrists need to know. World Psychiatry 8(1):67–74

Social Systems—Policies and Educational Prerequisites

European Union—Crises and Challenges in Twenty-first Century

Bogdan Ștefanachi

Abstract The European Union is passing through an endogenous poli-crises which is emphasized by exogenous features originated in the new geostrategic architecture of contemporary world. The economic crises overlaid on the necessity of a new social contract needed to re-legitimize European Union among its own citizens should be managed in a context of a recalibration of the international balance of power. Issues and events, like the conflicts in the neighborhood of the community, the impact of immigration or redefining relations with traditional partners or negotiations with emerging powers under full resurrection of nationalism is proposing a more variable geometry of the European Union. The chapter is analyzing these major challenges as a starting point of rediscovering a successful story that offers a new political identity to the European integration project.

Keywords European union · Crises · Challenges · Legitimacy · Nationalism · Reform

1 Introduction

The connecting element of the whole tradition of the European continent unification lies in trying to identify a scenario for avoiding conflicts between these states. As a matter of fact, security and well-being have been preserved over time as the main driving forces behind European integration. But the substance of European integration, which is nothing but the essence of the idea of the United Europe, was built on the idea of cultural and intellectual unity of the European continent. Therefore, European unification is essentially a cultural and institutional process, the institutions being nothing more than a form of objectified culture. But in order to move from the idea of unification to its implementation in specific institutional projects, it took time to transform the identity of Europe into a successful integrationist strategy. The historical perspective provided the image of "a bellicose past once entangled all

B. Ștefanachi (✉)
Faculty of Philosophy and Social-Political Sciences, Alexandru Ioan Cuza University of Iasi, Carol I Avenue no. 11, 700506 Iași, România

© The Author(s), under exclusive license to Springer Nature Switzerland AG 2021

D. Soitu et al. (eds.), *Decisions and Trends in Social Systems*,
Lecture Notes in Networks and Systems 189,
https://doi.org/10.1007/978-3-030-69094-6_22

European nations in bloody conflicts. They drew a conclusion from that military and spiritual mobilization against one another: the imperative of developing new, supranational forms of cooperation after the Second World War. The successful history of the European Union may have confirmed Europeans in their belief that the domestication of state power demands a mutual limitation of sovereignty, on the global as well as the national-state level" [1].

Today, the European Union seems to be farther from these desires as well as from the therapy for which it was initiated by the founding fathers. The integration infrastructure transformed into a traditional way of overcoming economic difficulties and political malfunctions, transposed into *the ever closer Union* formula, seems to have been replaced by a logic increasingly centered on centrifugal trends. The European Union is passing through an endogenous poli-crises which is emphasized by exogenous features originated in the new geostrategic architecture of contemporary world. The economic crises overlaid on the necessity of a new social contract needed to re-legitimize European Union among its own citizens should be managed in a context of a recalibration of the international balance of power. Issues and events, like the conflicts in the neighborhood of the community, the impact of immigration or redefining relations with traditional partners or negotiations with emerging powers under full resurrection of nationalism is proposing a more variable geometry of the European Union. In this context, the country-specific approaches in the various bi-or multilateral formulas are relevant, but more important is how these formulas could be translated into supranational policy approaches that reflect the common interest of the Union.

Sixty years after the entry into force of the Treaties of Rome, the European Union is being more and more conceived between *reform and decline* [2]. If the remedy is obvious, the reform attempted without much success since the mid-1980 s is closer to the image of a Sisyphean effort. To the extent that the European Union remains consistent in promoting the underlying fundamental objectives—made war impossible and generated unprecedented prosperity on the continent, and to the extent that the twenty-first century will not become the century of European decline, then the reform of the European Union becomes the only solution to overcome the current impasse.

However, the evidence of the remedy does not facilitate its administration, since the reform cannot be limited to a simple aggregation of preferences based on a predefined procedure, but the optimal procedure itself must be selected. It is only that the choice of procedure has to go from accepting the fact that as an *unidentified political object*, the European Union has a hybrid structure that attempts to engage intergovernmental elements with federalist aspects, to mediate between a minimalist vision in terms of a system of integration and cooperation economic relations between independent governments, and a maximalist vision, which, in terms of economic and political union, is a kind of United States of Europe. The tension between these two perspectives as well as the way to balance them generated an institutional project lacking clarity [2]; and the lack of clarity is reflected in institutional impediments, which, in turn, constitute obstacles to institutional reforms; the lack of clarity is reinforced by different views on reform, as a result of divergences between Member

States. On the other hand, a highly discrete foreign policy of the European Union deepens these divergences and, at the same time, made community space vulnerable in the context of new international realities.

Consequently, contrary to the background noise that give consistency of a political solution structured on the negative valences of protectionism, the economic and political decline of the European Union can only be avoided by a positive reform which needs to mitigate protectionism and thus, to consolidate the EU position on *the big chase table* of contemporary geopolitics.

2 European Union Endogenous Poli-Crisis

The shock wave generated by the decision of British electorate to leave European Union brought again to foreground not only the fragility and ebb probability of the European integration, but, more important, the imminence of real actions that reiterate the viability of this project. The wave of populisms and Euroscepticism that accompanied and survived the Brexit erodes the inner structure of EU and complicates even more the process of identifying solutions. No doubt the arguments for saving this Europe public good—represented by EU—are in large number (but they can be easily compressed under the shape of 2 fundamental dimensions—peace and prosperity) but, to complete them a radiography of crisis and their implications upon EU is compulsory.

Thus, the economic crisis started in 2008 in the USA determined a strong shock in the EU raising more and more question marks regarding the stability of the Euro aria. The Greek economic crisis, the economic problems in the Iberian Peninsula and those in Italy or Ireland (the PIIGS labeled states), the monetary fragmentation of the communitarian aria between the Euro aria and the aria of national currencies or the deep internal fragmentation of the Euro aria bring to foreground the gravity of the economic issues in EU. But, the same gravity wear the identified solutions under the shape of exaggerated austerity policies adopted at the EU level denoting a sort of "autism" of the public policies neglecting almost entirely the economic and social cohesion. More, on the Brexit ground, different points of view that question even the highest stage of European integration begin to shape namely the monetary union.

However, it should not be neglected that the rate of unemployment in EU went under 8% in February 2017, compared to 8.9% in the same period in 2016 or that Great Britain registered in 2016 one of the lowest unemployment rate, namely 4.6%. But, great economic disparities quantified under this point of view are still registered among states like Germany (3.9%) and Greece (23.1%) or Spain (18%), maintain a significant deficit of economic convergence of whose centrifuge tendencies should not be neglected. But, beyond the Euro crisis there should be underlined that the problems that called such solutions, firstly political and secondly economic, are still real: the necessity of a new impulse to unity, maintain Germany in the European institutional logics and taking Europe to the status and scope of the USA. Considering an error, the *politicization* of Euro, Euro renunciation as technical solution to even more obvious political issues it could lead to a similar great tactic error.

A different level at which we can identify the acting of the communitarian aria is the social one. Thus, from the demographic point of view, the population of the EU has increased extremely slow (from 406 7 million in 1960 to 506.8 million in 2014) fact that, connected to life expectancy mark a sensitive tendency of aging of the population (European working population, aged between 20 and 64 years old, is declining 0.4 percent/year, anticipated tendency by year 2040) [3]. Also, the migration waves, increasingly difficult to control, as a result of the transformed *Arab Spring* (2010) into an apparent endless winter affects deeper and deeper the solidarity of member states and also the possibility of unitary management of this social pressure.

The terrorist attacks in the EU aria committed by citizens of Muslim religion that are of second or third generation in Europe complete the social crisis with a series of question marks regarding the limits of the multicultural pattern, launching new challenges to cultural integration within the social European space. Also, other effect of the Brexit is seen as the comeback of national pride to the detriment of European patriotism which, conceived by the European intellectuals after the pattern of the American exceptionalism, it proves to be more and more an artificial patriotism [4]. The European patriotism should have represented the cure for the different exacerbated nationalist behaviors under the shape of chauvinism or xenophobia, but "this state changed as the EU enlarged, meaning that if the political and geostrategic reasons were in favor of the enlargement direction, the economic reasons and especially the cultural ones did not certify the fast pace of the implementation of this huge action" [4]. Thus, added the economic with the social aspects they feed the resurgence of the Eurosceptic nationalism totally opposed with the fundamental principles the Union was conceived and build, taking the European pattern away from the ideal of an *ever closer union.*

But, more important, the economic and social crisis turn into a deep *existential crisis* of political nature. The comeback of nationalism only points out the dysfunctional characteristics of the excessive hybridization of the EU underlying the split between the federal pattern supporters (Germany) and the states that sustain a Europe of nations (France and the states of the Vise grad group). The uncertainties regarding the optimum organizational formula come from a series of breaks of legitimacy in the conditions in which it has not been fully achieved the loyalty transfer between the elites and the member states' citizens towards the communitarian institutional assembly [5]. The triggering of the European integration process has assumed from the beginning the democratic values as fundamental for the success of the project and then, by defining the *Copenhagen criteria* transformed them into express conditionals that have to be internalized by the states willing to follow the European project. But, the apparent consensus regarding this axiological set is eroded "by the lack of a social and cultural European plan" [6] that rather supports an European "negative consensus: advocating for the fundamental democratic values is not justified by the cognitive, emotional and estimative assumption of them, but of the fact that the values of the European past proved to be wrong. In this manner "the hope that democratization will transform the young into good European citizens" [6] who will refuse nationalist attractions is at least illusory the more so as the post war memory seems to have vanished and, its place is taken by a first generation that lacks the communist memory. If EU has represented a true solution to the dramatic results of the two conflictual ideologies—communism and Nazism—today, the European citizens, alienated from the lessons of the past are more and more inclined to consider peace/stability as a given and to jeopardize "the virtuous circle" of integration. Assuming this false premise it becomes more and more evident that "Europe without democracy would go back to the Empire (an enlarged, technologically updated version of the Holy Romano-Germanic Empire) and that the sovereign democracies without EU would go back to the confrontation of nation-states" [7]. *The end of history* [8] as postulated by Fukuyama has been, at its turn, wrongly accepted as a given and thus, it has been neglected the danger of the excessive technocratisation of the politic act of which its main effect was the obliteration, and then antagonizing the citizen in relation with the politic decider [9]. Translating EU into sets of impersonal rules or approaching it exclusively economic terms has limited the understanding of the deep implications of the European integration. Placing right "at the heart of European integration" the man, Pope Francis was perfectly right when counter argued that "Europe is not a conglomeration of rules to obey, or a manual of protocols and procedures to follow. It is a way of life, a way of understanding man based on his transcendent and inalienable dignity, as something more than simply a sum of rights to defend or claims to advance". *Losing citizens* on the road (reflected in the constant increase of the democratic deficit) eliminated the possibility of an emotional approach to EU—increasing the difference between EU and citizens—and thus, it created a space to the appearance "of the angry electorate" who, unsatisfied with the way the benefits of globalization are redistributed, is prone to radicalize. Moreover, the most accentuated massification of society offers the necessary props to adopt simplest solutions of xenophobe or populist type; speculating real disappointments of European citizens, the representatives of the populist parties identify/transform the fight against EU into the perfect antidote for their own lack of political vision and performance. Therefore, the politic crisis for which a right litmus is the number and the increasing weight of the populist parties can only be overcome through a new social agreement, an agreement to (re-)legitimize EU in the consciousness of its own citizens. The new legitimizing speech should temperate the divergent views within the EU space, that even though maintain their common point peace and prosperity seem to embrace radically different semantics of those. Beyond the compatibilization of politic views, the new social agreement must not view in Rousseau-ist manner only the relation between the sovereign and general will but will have to solve one of the great challenges of all contemporary democracies—reestablishing the social agreement between generations [10]. If, as Donald Tusk claimed, the Brexit has also good part reflected in catalyzing the energy of 27-member states to reinforce unity, then, for sure, a new legitimate narration is necessary to generate durability to this process.

3   The (Exogenous) Crises of European Geostrategic Project

The debates generated by Brexit underline not only the cleavage between different visions of the national governments regarding the future of UE or the break between the European politic elites and citizens but bring to foreground also the aspects that belong to the statute and position of EU in the dynamics of the international relations. No doubt, EU is integrated part of the global geopolitical equilibrium. Being a "geopolitical project of Kantian inspiration" [11] *pax europeana* developed in the subsidiarity of EU proposes, through Copenhagen criteria a new geopolitical pattern, based on different set of values than of the brutal power balance [11]. Beyond these alternative valences of the European power Parag Khanna considered that the world has 3 relatively equal centers of influence: Washington, Brussels and Beijing, the planet being simultaneously Americanized, Europeanized and Chinesed [12]. In a somehow similar way Zbigniew Brzezinski placed Europe next to China and Russia in the *geostrategic triad* together with whom the USA have to live only that the national security advisor of President Carter also said "Europe, despite its economic power, of the significant economic and financial integration and of its solid transatlantic friendship—is, de facto, a military United States protectorate" [13]. In accordance to this view and having as a pretext the Golf War the Belgian Foreign Affairs Minister Mark Eyskens plastically defined EU as being an "economic giant, a political dwarf and a military worm".

Questioning, in a mainly geostrategic keynote, the future of EU, Stephen Walt [14], found out that there can't be identified too many reasons of optimism. Thus, beside the economic and socio-political crisis, a first challenge is the overextension/over-enlargement that transforms EU into the victim of its own success. To the extent that as Arnold Toybee claimed, the empires are susceptible of the "mirage of immortality" another malign component of their behavior is the overextension itself; differently said, the expansion of Europe "respects the Newtonian law of inertia: an object in motion remains in motion" [12]. The enlargement of the communitarian aria has raised the heterogeneous level, mainly as the result of the fact that each member state maintained its own domestic political arrangements and, later, favored by the lack of a functional common foreign policy (supranational that would imply real sovereignty transfer) member states have defined different interests from the common ones. Such discrepancies in views make more and more difficult to complete now what the Belgian Prime Minister Leo Tindemans from the mid 70' claimed "EU will not be completed until it won't have a common defense policy". The disappearance of Soviet Union, beside the numerous positive implications, eliminated one of the most important motivations of the deepening of the European integration: the disappearance of the Soviet military threat offered solid arguments to nationalize the foreign policy concerns and decreased (to elimination) the engagement of a *true common foreign and security policy* (the incoherent European response regarding the Ukraine crisis is a proof in this regard). Also, the implications of the state of a "political dwarf" type of EU are visible in the way it has been managed the decreasing of the security environment in the Mediterranean Sea pool, in the way EU didn't succeed to engaging in a convincing manner in managing the crisis generated by the failure of some states from its South periphery or in the way in which the refugee crisis didn't receive viable solutions. Thus, a wavering foreign policy in the context in which a redefinition of the transatlantic partnership is more and more evident accentuates the profound crisis of EU. The geostrategic "irrelevance" of EU must become integral part of the concerns to identify a new legitimacy speech even more, if the strategy of overcoming the crisis built on *an ever closer union* continues to be functional, then the complete political integration becomes natural level of European integration consolidation.

4 Conclusions

Along with the end of the Cold War, a series of transformations of quantitative nature but especially of qualitative one, that influence fundamentally international interactions, structured under the form of three main characteristics: globalization, liberalization and innovation ([15], pp. 243–253). Initially, the communitarian space, through study and later, through repeated enlargements, subsumed to the principles of the unique market has become one of the main actors in eliminating barriers that ingrained foreign exchange and introducing the Euro currency on 1st of January 1999—that proved to be a "historical innovation" ([15], p. 252)—constituted an important catalyst for this process. However, the economic crisis from the end of the first decade of the twenty-first century implied a profound crisis of the Euro-aria that, at its turn, underlined more and more clearly the structural deficiencies (economical but mainly political) of the European integration project, placing the EU into a dysfunctional triangle defined in terms of national policies, European politics and global markets.

In the United States of America, the debt crisis in the eighteenth century had transformed a confederation into a federation as a direct consequence of pooling sovereignty. To duplicate this model, the European Union needs more than a financial crisis (which is already had), needs a united political society; or, precisely, Brexit itself is the proof that national identity is still prevalent on the continent and, moreover, it is still the only support for democracy (the low turnout for European elections is an argument for the absence of a European demos that to support a functioning European democracy). But, on the other hand, the transfer of sovereignty and the profound tenets of the European federalist project cannot be canceled so far neither by Brexit nor by the increasingly radical Euroscepticism. What is real, however, is that the project of European integration cannot be carried out by continuing excessive compromises or anemic political reactions, without removing the feeling of exclusion, without eliminating the coexistence between deepening European integration and excessive austerity as a mean of overcoming the economic crisis, or the autistic way to manage immigration when the very mechanisms that should protect the European space are endangered.

In this regard, Jean Claude Juncker was perfectly right in saying that "too much Europe could kill Europe" and, therefore, it is necessary to return to a series of pragmatic coordinates of integration that provide solutions and thus to re-legitimize the European Union project. Therefore, at least as long as Europe is a more peaceful, prosperous and more influential place than it would have been without the European Union the return to the nation-state avatar could prove a hasty and misleading solution. Moreover, taking into account the dangers of fragmented Europe before 1945 and those of the divided Europe during the Cold War, the EU is less looking like a bureaucratic monster and resembles more as a miraculous achievement: miraculous by relative efficiency, and not only by its very existence. But the refusal of the fact that the *European dream has come to an end* must entail the scenario at the center of which the principle of more integration must be complemented by a complementary quality, namely better integration (and sometimes in terms in which European Commission synthesizes its own path to reform, but which can also be applied to the Union as a whole, this may mean doing less, but doing better). European integration is not necessarily by itself the root cause of obvious popular skepticism and dissatisfaction, but rather, it has largely "failed in the eyes of many Europeans to provide a credible model of managing globalization rather than simply going along with it; it has failed to be part of the solution and not just part of the problem" [16].

In the absence of an obvious alternative solution, lacking the ingredients of radicalism and populism, repositioning EU in the new global landscape must consider the fact that globalization remains the fundamental process that shapes the international space, and avoiding the marginalization within global governance holds only on the extend to which the *iron law* of integration will prove to be operational again: *an ever closer union* that will lead EU out of crisis more mature and stronger. In the keynote speech at opening ceremony of 1st China International Import Expo, Chinese president Xi Jinping asserted that Chinese economy is like an ocean and big winds and storms are only to be expected. Without them, the ocean wouldn't be what it is! In the *State of the Union Address* 2017, Jean Claude Juncker believes that believe: the wind is back in Europe's sails, but meantime the wind got the intensity of a storm. It remains to be seen if the European ship will come out of the storm as it always does with the ocean, and under the conditions in which Europe is (still) our common future [17] a new legitimacy speech for the European project is absolutely necessary to project and to sustain the solidarity and convergence of interests.

References

1. Habermas J, Derrida J (2005) February 15, or, what binds european together: plea for a common foreign policy, beginning in core Europe. Levy (eds) Old Europe, new Europe, core Europe. Transatlantic Relations after the Iraq War, ISBN 1-84467-520-3
2. Alesina A, Giavazzi F (2008) The future of Europe: reform or decline. ISBN 978-02-625-1204-6
3. Cioculescu Ş (2017) Mergând pe gheață cu frâna de mână trasă: Uniunea Europeană și deficitul cronic de gândire și acțiune strategică, Naumescu (coord.). Criza globală și ordinea globală în era Trump. ISBN 978-606-719-902-4

4. Jucan M. (2017) Criza europeană: clivajul dintre politică şi cultură, Naumescu (coord.). Criza globală şi ordinea globală în era Trump. ISBN 978-606-719-902-4
5. Haas EB (1958) The uniting of Europe: political, social and economic forces 1950–1957. ISBN 0-268-04346-9
6. Hoffman S (2003) Sisiful european. Studii despre Europa (1964–1994). ISBN 973-8356-92-X
7. Naumescu V (2017) Criza Uniunii Europen şi noua ordine globală. Şapte perspective. Naumescu (coord.). Criza globală şi ordinea globală în era Trump. ISBN 978-606-719-902-4
8. Fukuyama F (1992) The end of history and the last man. ISBN 9780029109755
9. Liberalism after Brexit. The Politics of Anger (2016) The Economist [online 15.01.2020]. Available online at https://www.economist.com/leaders/2016/07/02/the-politics-of-anger
10. Ferguson N (2014) Marele declin. Cum decad instituţiile şi economiile. ISBN 978-973-46-4297-7
11. Kahn S (2008) Geopolitica Uniunii Europene. ISBN 9789975794688
12. Khanna P (2008) The second world: empires and influence in the new global order. ISBN 9781400065080
13. Brzezinski Z (2000) The geostrategic triad—living with China, Europe and Russia. ISBN 9780892063840
14. Walt SM (2016) Does Europe have a future, foreign policy [online 15.01.2020]. Available online at https://foreignpolicy.com/2015/07/16/does-europe-have-a-future-stephen-walt-testimony-house-foreign-affairs-committee/
15. Cassis Y (2007) Capitals of capital: a history of International Financial Centres, 1780–2005. ISBN-13 978-0-521-84535-9
16. Tsoukalis L (2016) In defence of Europe: can the European project be saved? ISBN 978-0198755319
17. The Rome Declaration (2017) [online 15.01.2020]. Available online at https://www.consilium.europa.eu/en/press/press-releases/2017/03/25/rome-declaration/pdf

An Analysis of the Relative Importance of Social, Educational and Environmental Expenditures on Life Expectancy at Birth. Evidence from Europe

Pedro Antonio Martín Cervantes, Nuria Rueda López, and Salvador Cruz Rambaud

Abstract The framework of this chapter is the analysis of health expenditure as a variable which, from an intuitive point of view, must be positively correlated with life expectancy at birth. Traditionally, health outcomes have been studied from a macroeconomic perspective without taking into account new categories of public expenditures related to health such as social, educational and environmental protection expenditures. In consequence, the objective of this chapter is to investigate the relative importance of these public expenditures on health outcomes. The employed methodology is based on the main metrics which analyze the relative importance, by using the data of twenty-five European countries for the period 1995–2017. This study has found that social expenditure has the greatest relative importance on explaining life expectancy. On the other hand, our results show that public educational expenditure has the least relative importance. Finally, there is no evidence demonstrating the extent to which environmental expenditure contributes to improve population health.

Keywords Life expectancy at birth · Relative importance · Health outcome · Public expenditures

1 Introduction

A growing body of literature has explored the determinants of health status of the population since Auster et al. [2] by following two different approaches. On the one hand, a *micro-data analysis* investigated health outcomes from an individual perspective based on epidemiological studies (see, for example, [23, 36, 37, 47,

P. A. Martín Cervantes · N. Rueda López · S. Cruz Rambaud (✉)
Department of Economics and Business, University of Almería, Almería, Spain
e-mail: scruz@ual.es

P. A. Martín Cervantes
e-mail: pmc552@ual.es

N. Rueda López
e-mail: nrueda@ual.es

© The Author(s), under exclusive license to Springer Nature Switzerland AG 2021
D. Soitu et al. (eds.), *Decisions and Trends in Social Systems*,
Lecture Notes in Networks and Systems 189,
https://doi.org/10.1007/978-3-030-69094-6_23

50, 78]). On the other hand, a *macro-data approach* allowed researchers to analyze health outcomes at a country level -by using state, regional or local data- and to obtain suitable economic policy implications. For this reason, in this work we prefer to follow the second point of view.

Some studies referred to the determinants of health outcomes have been focused on developing countries and have employed explanatory variables different from those analyzed in the case of developed countries, and more related with the degree of economic and social development, such as the access to drinking water, the degree of literacy, belonging to a specific ethnic group, the degree of urbanization, the geographic location, the access to health and the fertility (see, for example, [16, 18, 33]).

On the other hand, most previous literature which follows a macroeconomic approach are referred to OECD countries [4, 6, 21, 26, 42, 51, 54, 55, 61, 65, 79] or to a mix of OECD or non-OECD countries with a middle level of development [5, 10, 13, 24, 45, 52, 56, 62–64]. To a lesser extent, the analyses are referred to a specific OECD country, such as the United States [2, 25, 35, 39, 40, 72]; Canada [11, 12]; Spain [46, 71]; Turkey [22]; or Italy [58]. A second group of studies, less numerous, are limited to a group of European countries [14, 48, 53, 75]. Finally, other contributions refer to Brazil [68] and, more recently, have been extended to Asian countries [28, 66]. This group of researches -focused on developed countries- considers that the health status is determined by a combination of health care resources, lifestyle and socio-economic factors.

Paying attention to health resources, some of the previous works have considered the health care expenditures financed by public sector. It should be noted that, recently, there has been a move towards including new categories of public expenditures related to health such as social and educational expenditures [70], but there has so far been no inclusion of environmental protection expenditures. In addition, this type of literature has ignored the analysis of the level of contribution of these expenditures in explaining or not the population health gains.

In consequence, the objective of this chapter is to investigate the relative importance of these components of public budget on health outcomes, by including environmental expenditures. For this purpose, we employ the methodology based on the main metrics which analyze the relative importance of these public expenditures. To do this, we use data of twenty-five European countries for the period 1995–2017.

The structure of this work is as follows. In Sect. 2, we include a review of the main empirical studies published to date on this topic and referred to developed countries. In Sect. 3, we present the relative importance metrics approach and the main results. Finally, Sect. 4 presents the main conclusions and offers future lines of research in this area.

2 Literature Review

Focused on developed countries, there is a wide consensus on classifying the determinants of health status of the population into three categories, specifically, health resources (health care expenditure, number of doctors and nurses, and number of hospital beds), socio-economic factors (gross domestic product (GDP), income *per capita*, education, income distribution, unemployment, poverty and pollution, among others) and lifestyle-related factors (such as consumption of alcohol and tobacco and some proxy for diet (for example, consumption of sugar, calories, vegetables and sugar)) [32].

With respect to health resources, most previous researches have included some expression of healthcare expenditure, in addition to other socio-economic and lifestyle variables. First, from a broader perspective, some researches consider total health expenditure. Auster et al. [2] were the first scholars to analyze the health status of the population through an aggregate production function by using data of health expenditure in United States in 1967. Wolfe and Gabay [79] also considered health expenditure in their study referred to twenty-two OECD countries in 1960, 1970 and 1980. Hitiris and Posnett [26] used *per capita* health expenditure to investigate the determinants of mortality in twenty-eight OECD countries during the period 1960–1987. Crémieux et al. [11] considered total healthcare expenditure (public and private) and its effect on life expectancy and mortality in ten Canadian provinces in the period 1978–1992. Thornton [72] included health expenditure in his research of United States in 1990 whilst Nixon and Ullmann [53] preferred to focus on fifteen European countries in the period 1980–1995 and considered total health expenditure *per capita* and also this parameter as a percentage of GDP. Halicioglu [22] used *per capita* health expenditure to investigate the determinants of life expectancy in Turkey in the period 1965–2005. Heijink et al. [24] evaluated the relationship between healthcare expenditure and avoidable mortality at a macro-level in fourteen western high-income countries between 1996 and 2006. Later, Jiang et al. [28] included *per capita* health expenditure, among other variables, to study the effect of the social development on life expectancy in thirty-one Chinese provinces in 2000 and 2010.

A second group of studies are referred to public health expenditure as a possible determinant of health outcomes. Cochrane et al. [10] considered the total health expenditure and the percentage of healthcare expenditure financed by the public sector in eighteen developed countries in 1970, 1969 or 1971. Peltzman [56] considered the requirements of medical prescription for the consumption of certain drugs and public expenditure on health as determinants of the mortality rate in twenty-two middle-income countries in the period 1970–1980. Additionally, Elola et al. [14] considered the two previous expressions of health expenditure, and analyzed the relationship between different health systems and health status of population in seventeen European countries. Or [54] considered total health expenditure in absolute terms and public expenditure in relative terms to study premature mortality in twenty-one OECD countries between 1970 and 1992. Or [55] extended the analysis to the period 1970–1995. Berger and Messer [4] analyzed mortality rates in twenty OECD countries during the period 1960–1992, taking into account total health expenditure and the percentage of public expenditure. Lichtenberg [40] considered public and private health expenditure and medical innovation in the United States in the period 1960–1997. Self and Grabowski [64] employed public and private *per capita* health expenditure data in a sample of 191 developed, middle income and less developed countries in 2000. Linden and Ray [42] investigated the relationship between life expectancy at birth and private and public health expenditures in the context of a panel of thirty-four OECD countries -that were grouped into three clusters depending on the size of their public health expenditure over the GDP- during the period 1970–2012. Toader et al. [73] assessed the impact of healthcare public financing on population health status, whilst controlling other non-financial health determinants by using a panel dataset regarding the European members of OECD from 1970 to 2014.

A third group of researches, more reduced, has included in the analysis a specific component of healthcare expenditure such as pharmaceutical expenditure. This was the case of Miller and Frech [51] in their research referred to eighteen OECD countries. Shaw et al. [65] considered an aggregate life expectancy production function and pharmaceutical consumption for a sample of nineteen OECD countries. Crémieux et al. [12] used public and private pharmaceutical expenditures in their analysis of the determinants of health status in a group of Canadian provinces during the period 1975–1998.

Whereas there is some consensus in international studies when classifying the categories of factors which determine the health status, there is more controversy when identifying the contribution of healthcare resources, in particular healthcare expenditures. In this sense, some authors have found a statistically significant positive relationship with total health expenditure [11, 22] whilst Nixon and Ullmann [53] and Hitiris and Posnett [26] identified a significant positive but small impact; finally, Thornton [72] concluded there is no significant effect on population health outcomes. On the other hand, the positive significant contribution on health outcomes is related to public healthcare expenditure [14, 40, 42, 55, 73], although in some cases no significant impact is detected [64] and, in other works, this effect is moderately adverse [4, 56].

All previous studies have explored the extent to which expenditures (total, private or public) on health sector are or not associated with population health outcomes. However, a growing body of literature has analyzed whether higher social expenditures are associated to health outcome gains. Stuckler et al. [69] examined whether there is a historical association between levels of social expenditure and population health outcomes in fifteen European countries for the period 1980–2005. Karim et al. [34] investigated whether the association between population health and welfare state regimes is still valid when the welfare states of East Asia are added to the analysis. Social and health expenditures as a percentage of GDP were examined in thirty welfare states, categorized into six different regimes (Scandinavian, Anglo-Saxon, Bismarckian, Southern, Eastern European and East Asian). Bradley et al. [6] examined the relationship between health and social expenditures and health outcomes in thirty OECD countries from 1995 to 2005. Five years later, Bradley et al. [7]

investigated this relationship by using data of the fifty USA states and the District of Columbia for the period 2000–2009. Vavken et al. [76] searched for associations between healthcare expenditure (total, public and private) and total social insurance expenditure, after adjusting potential confounding variables, by using aggregate data of Austria for the period 1997–2008. McCullough and Leider [49] examined the relationship between expenditure on social services, health factors and health outcomes at a county level in USA for the period 2010–2015. Beckfield et al. [3] found that increases or decreases in social investment by the national government are associated with changes in cause-specific mortality and years lived with disability between 2005 and 2010 in eleven European countries. Álvarez-Gálvez and Jaime-Castillo [1] used individual (micro) and country (macro) data to explore the direct and indirect effects of the welfare state on health inequalities in a sample of European countries.

Additionally, recent empirical studies have included the educational expenditures as a determinant of health outcome. Van den Heuvel and Olaroiu [75] analyzed the relationship between life expectancy at birth and expenditures on healthcare, social production, and education as a percentage of the GDP by taking into account the quantity and the quality of the healthcare system and several lifestyle indicators in thirty-one European countries by using data of 2013. Reynolds and Avendano [61] tested the hypothesis that greater social expenditure (included education expenditure and family, unemployment, incapacity, old age, and active labor market programs expenditures) are positively associated with life expectancy across twenty OECD countries between 1980 and 2010. Contrary to the results referred to the effect of health expenditure, in the case of social expenditure, most of the existing literature concluded that a greater expenditure is associated with a better health status (with the exception of the study by Karim et al. [34] and Álvarez-Gálvez and Jaime-Castillo [1]).

Whilst all previous literature has investigated the existence of a potential link between health (and non-health) expenditures and health outcomes, however few of them reported which types of public expenditures contribute, to a greater extent, to population health gains. Our study, therefore, is going to investigate the relative importance of the public budget composition -specifically the percentage of public expenditure on healthcare, education, social services and environment in GDP—on health status.

3 Methods

3.1 Current Prospects for Life Expectancy: A Very First Insight

The intuition seems to have spread that greater health expenditure must unequivocally lead to greater life expectancy at birth. However, this observation is partially true because the consolidation of welfare states in the first half of the last century determined that life expectancy at birth would increase in a quasi-linear manner whilst public expenditure increased. Very soon, it was noticed that, when time passes, this relationship gradually began to fade until reaching a point of *de facto* saturation [44] in which an increase in public health expenditure does not necessarily imply an increase in life expectancy.

Therefore, such a "saturation point" must be an essential element in the analysis of life expectancy. Castanheira et al. [9] verified this hypothesis by creating a Bayesian model according to which life expectancy describes a random walk with drift, by deducing that life expectancy increases until reaching a saturation point to subsequently decrease until it reaches an asymptotic rate increase, denoted by Z^c [60].

Obviously, each country individually follows its own guidelines regarding the evolution of the variables "life expectancy" (represented by LE) and "public health expenditure" (denoted by H). Therefore, the joint comparison given by the elasticity LE-H exemplifies, to what extent, an increase in public health expenditure does necessarily mean a longer life expectancy. Considering the joint data of LE and H (percentage increments) of twenty-five European countries for the period 1995–2017 (see Table 6, Dataset A, in Appendix), the following model has been implemented in order to calculate the elasticity following the original prototypical model by Schultz [67]. Let us start from the following model:

$$LE = \beta_0 + \beta_H H + \beta_t t + \varepsilon, \tag{1}$$

where $E(\varepsilon) = 0$. It is well known that the elasticity, denoted by $\eta_{LE,H}$, is the quotient between the percentage variations of both variables:

$$\eta_{LE,H} = \frac{\partial LE/LE}{\partial H/H} = \frac{\partial LE}{\partial H}\frac{H}{LE} = \frac{\%\text{ variation in } LE}{\%\text{ variation in } H}. \tag{2}$$

As $\frac{\partial LE}{\partial H} = \beta_H$ and $\frac{H}{LE} = \frac{H}{\beta_0 + \beta_H H + \beta_t t}$, expression (2) would remain as:

$$\eta_{LE,H} = \frac{\beta_H H}{\beta_0 + \beta_H H + \beta_t t}. \tag{3}$$

Finally, the results can be derived from Fig. 4 in Appendix in which a certain negative trend is observed in the elasticity of those European countries that, traditionally, have been investing more in public health expenditure (e.g., Australia, Denmark, Belgium, France or Sweden). In the same way, those countries with a greater positive elasticity are either "late starters" in terms of investment in public health expenditure or come from the area of the former Eastern European economies (e.g., Romania, Bulgaria, Hungary or Slovenia), or those countries which, during the analyzed period, carried out profound reforms in the financing of their health systems (e.g., United Kingdom in 1979 or The Netherlands in 2015).

These conclusions are supported by Fig. 5 in Appendix in which, by using the same database, the scatterplot of the annual increase in life expectancy is represented in relation with the annual increase in public health expenditure. Observe that the joint trend of both variables is slightly negative, describing two types of outliers: those countries in which an increase in public health expenditure determines an appreciable increase in life expectancy (e.g., Estonia, Lithuania or Hungary) and those others in which the effort of the public treasury in the health sector is not exactly linked to a significant increase in life expectancy (e.g., France, Germany, Belgium or Finland).

Marchetti [44] indicated that not only the variable "public health expenditure" is fundamental to explain life expectancy, but also many other conditioning factors such as biological components, improvements in behavioral habits related to health or technology. For Johnson [31], the statement that the increase of life expectancy is due only to the economic growth of nations (specifically, the subsequent public expenditure on health), without taking into account other variables such as education, health habits or the incidence of environment, would be a merely "holistic" interpretation.

In this paper, we are going to develop the perspective presented by Johnson [31] by introducing some of those other variables which, predictably, have something to say about the increase in the variable "life expectancy". Far from the usual analysis of life expectancy and based on simple regression models, we will use a methodology derived from it, the so-called *relative importance*. To the extent of our knowledge, this is the first work in which the analysis of life expectancy will be addressed from the conceptual framework of relative importance.

3.2　The Relative Importance Metrics Approach

There is neither an explicit definition nor an accurate origin of this methodology. According to Johnson and Lebreton [30], the roots of this methodology appeared in the 1930s [15], but its conceptual bases can only be found in the statistical analysis of Hoffman's cognitive processes [27], in which the term "relative weight" (that is to say, the percentage of variance explained by the linear model assigned to each individual predictor) was already introduced [77]. The definition by Johnson and Lebreton [30], increasingly rooted in the existing literature, highlights the concept of relative importance: "The proportionate contribution each predictor makes to R^2, considering both its direct effect (i.e., its correlation with the criterion) and its effect when combined with the other variables in the regression equation".

Due to the non-existence of an explicit definition of relative importance, we can only find more or less intuitive analytic definitions such as Wallard [77]'s which considers p random variables as predictors and y a response. Thus, it is possible to define a relative importance function as a function RI which associates to each predictor (defined by its index j in the set $P = \{1, 2, \ldots, p\}$) a value of importance:

$$RI : P \to \Re,$$

where $\mathfrak{R}$ is the set of real numbers. In the same way, there will not be just one, but several measures or "metrics" in the final determination of the relative importance of regressors used in the analysis (see [8, 17, 20, 29–31, 41, 43, 77]).

The main metrics used to analyze the relative importance are different specifications based on different starting hypotheses. According to Wallard [77], these would be classified, basically, according to two methods: methods for variance allocation and methods of variance decomposition. Next, we are going to briefly describe each one of the different metrics which will be used in the empirical part of this work:

(a)   Methods (or metrics) for variance allocation:

- Variance allocation "first". In this method, the measures are the covariances of the predictors given the response, that is to say,

$$\mathrm{First}(j) = \mathrm{cov}(y, X_j).$$

- Variance allocation "last". In this method, the relative importance of a predictor j is given by the increase in R^2 when predictor j is included in the model compared to the R^2 with the other $p - 1$ predictors.
- Variance allocation "betasquared". In this method the relative importance is the square of the standardized regression coefficient (like "first" and "last" metrics, these variance allocations are the sum of the measures corresponding to all (p) predictors).

In this measure, the standardized coefficients are introduced as scale-invariant versions of the coefficients, adjusted with estimated standard deviations according to the following formula:

$$\hat{\beta}_{k,\mathrm{standardized}} = \hat{\beta}_k \frac{S_{kk}}{S_{yy}},$$

where S_{kk} and S_{yy} are, respectively, the empirical variances of the regressor X_k and the response y.

(b)   Methods (or metrics) of variance decomposition:

- Decomposition Hoffman-Pratt. It assigns to a predictor j the product of the standardized multiple regression coefficient by the marginal correlation between the predictor and the response. In case of standardized predictors, one has:

$$\mathrm{Pratt} = \beta_j \rho_{yj}.$$

Taking into account the properties of the OLS regression, it can easily demonstrated that this measure leads to the following decomposition of R^2:

$$R^2 = \text{cov}\left(y, \sum_j \beta_j X_j\right) = \sum_j \beta_j r_{yj}.$$

- LMG, Shapley value or average. This measure representative of relative importance can be calculated by averaging, on all possible orderings of the p predictors, the increase in the R^2 when the predictor j is added to the model based on the other predictors entered before j in the model (see also Lindeman et al. [41]; Grömping [20]; Feldman [17]):

$$\text{LMG}(j) = \frac{1}{p!} \sum_r \Delta_j(r).$$

- PMVD (Proportional Marginal Variance Decomposition). This measure is computed similarly to $\text{LMG}(j)$, but including the corresponding weights attached to each single permutation:

$$\text{PMVD}(j) = \frac{1}{p!} p(r) \sum_r \Delta_j(r).$$

Within the variance decomposition metrics, there are two essential measures: Genizi and CAR (Correlation Adjusted marginal-coRelation) scores. In both cases, an orthogonal decomposition of the variance is performed according to the relative weights procedure by Johnson [31]. In the first case [19], relative weights are computed by using a specific set of orthogonal predictors which minimize the sum of the squares. In the second case [80], starting from a similar procedure, the square correlations between the response and a Z vector are computed, remaining as:

$$\text{CAR}(j) = \lambda j^2.$$

The procedure of relative weights [31], closely related to the previous measures, opens the doors to the research in the field of relative importance, because, according to this methodology, the regression coefficients are assigned to the uncorrelated variables instead of the correlated primary original predictors. In addition, Johnson [31] simplified this approach by combining two sets of squared regression coefficients. Figure 1 shows a graphic representation of the relative weighting procedure by Johnson [31], by considering the number of predictors (four) implemented in the empirical part of this document.

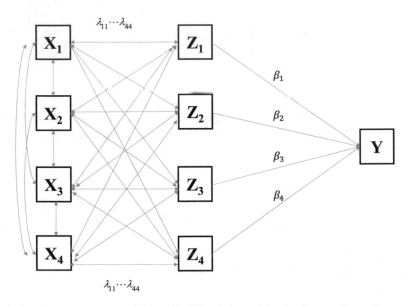

Fig. 1 Graphic representation of Johnson's [29] relative weights for four predictors *Source* Own elaboration based on Johnson [29]

3.3 Results

The use of multiple regression has usually constituted the main methodological approach to the analysis of life expectancy. However, and without denying representativeness to the results obtained through this technique, it would be necessary to find certain complementary types of analysis, by contributing new lines of research equally effective in the field of life expectancy, this being the primary objective of this work. In the words of Tonidandel and LeBreton [74]: "The goal of such analyses (relative importance) is to partition explained variance among multiple predictors to better understand the role played by each predictor in a regression equation" (see Table 1).

Table 1 Correlation matrix

	LIFE_EXP	Enviro	Health	Educa	Social
LIFE_EXP	1	0.25	0.41	0.2	0.55
Enviro	0.25	1	0.16	−0.24	0.05
Health	0.41	0.16	1	0.2	0.6
Educa	0.2	−0.24	0.2	1	0.37
Social	0.55	0.05	0.6	0.37	1

Source Own elaboration

Continuing our analysis, from an exploratory perspective, it is especially significant to highlight how the explained variable "life expectancy" exhibits variability substantially greater than explanatory variables. This phenomenon could be justified by the fact that the different analyzed nations, *grosso modo*, make relatively similar reimbursements in public expenditures (social, health, educational and environmental) whilst there are still tangible differences in the life expectancies of these countries (see Table 2).

The adjustment of the variable "life expectancy" to a simple linear regression model contemplating the four strands of *ad hoc* public expenditures included in our analysis (summarized in Table 3) determines especially a strong relationship with public social expenditures, a fact which we are going to confirm below by using different measures of relative importance.

Table 2 Descriptive statistics

	LIFE_EXP	Enviro	Health	Educa	Social
Mean	80.089	0.0065714	0.061	0.048571	0.16304
Median	81.35	0.006	0.0615	0.049	0.1605
Min	74.8	0.002	0.026	0.028	0.095
Max	83.4	0.014	0.084	0.068	0.249
S.D.	2.7505	0.0029367	0.01474	0.0096932	0.041835
C.V.	0.034343	0.44689	0.24163	0.19957	0.2566
Perc. 5%	74.845	0.00245	0.03005	0.03025	0.10265
Perc. 95%	83.265	0.01355	0.0831	0.06665	0.2463
IQR	4.325	0.005	0.024	0.016	0.06925

Source Own elaboration

Table 3 Simple regression coefficients: Life expectancy explained according to four regressors (Enviro, Health, Educa and Social)

	Estimate	Std. Error	t value	Pr(>|t|)	
(Intercept)	71.819	3.068	23.407	<2e-16	***
Enviro	212.476	164.789	1.289	0.2101	
Health	15.5	39.54	0.392	0.6987	
Educa	19.763	52.933	0.373	0.7123	
Social	30.473	14.494	2.102	0.0467	*
Signif. codes: 0 '***' 0.001 '**' 0.01 '*' 0.05 '.' 0.1 ' ' 1					
Residual standard error: 2.384 on 23 degrees of freedom					
Multiple R-squared: 0.3603, Adjusted R-squared: 0.2491					
F-statistic: 3.239 on 4 and 23 DF, p-value: 0.03019					

Source Own elaboration

Fig. 2 Percentage of
response variance explained
by each regressor. Method
LMG *Source* Own
elaboration

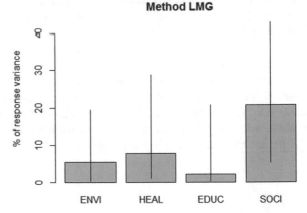

Relative importances for LIFE_EXP
with 95% bootstrap confidence intervals

Method LMG

$R^2 = 36.03\%$, metrics are not normalized.

Figure 2 displays the percentage of response variance of the variable "life expectancy" which has been calculated by taking into account the four aforementioned regressors. Thus, we would obtain the relative importance of each regressor by applying the LMG method [41].

As detailed in Table 4, the analysis has been amplified to the eight measures of relative importance which analyze the impact of the four regressors on the life expectancy according to different methodologies. For a greater degree of detail, Table 5 shows, in a synthetic way and as a diagnosis, the effect on the regression equation (see Table 3) when more predictors are added.

As a colophon, according to Table 4, Fig. 3 shows the relative importance in terms of the natural decomposition of R^2, starting from the contributions made by

Table 4 Eight relative importance selected metrics

	LMG	PMVD	Last	First	Betasq	Pratt	Genizi	CAR
ENVIRO	0.053775	0.049732	0.046238	0.061695	0.051465	0.056348	0.054489	0.054932
HEALTH	0.078306	0.009996	0.004274	0.170580	0.006899	0.034306	0.077347	0.061493
EDUCA	0.021669	0.004330	0.003877	0.041345	0.004851	0.014161	0.024704	0.016884
SOCIAL	0.206569	0.296261	0.122935	0.303895	0.214819	0.255504	0.203780	0.227011

Response variable: LIFE_EXP
Total response variance: 7.565437
Analysis based on 28 observations
Four regressors: ENVIRO HEALTH EDUCA SOCIAL
Percentage of variance explained by model: 36.03%
Metrics are not normalized

Source Own elaboration

Table 5 Average coefficients for different model sizes

	1X	2Xs	3Xs	4Xs
Enviro	232.638	224.9446	214.3658	212.4758
Health	77.07126	55.81662	34.02188	15.5001
Educa	57.69794	38.13387	25.23623	19.76266
Social	36.24404	34.3096	32.33105	30.47273

Source Own elaboration

each regressor and taking into account the eight metrics presented in this chapter. In summary, we could make the following assessment:

- According to all implemented metrics, the social expenditure (SOCIAL) is the most important predictor in the decomposition (explanation) of R^2.
- Oppositely, the least important predictor would be educational expenditure (EDUCA).
- There is no consensus among the different measures of relative importance to determine those in the second and third place on explaining life expectancy. Thus, HEALTH would be in second place according to the methods of LMG, First, Genizi and CAR, and in third place according to PMVD, Last, Beta and Pratt methods.

4 Discussion and Conclusions

Traditionally, most studies on the determinants of health status, from a macroeconomic point of view, have relied on some expressions of health resources such as healthcare expenditure. These researches have tried to answer the question whether increases in health expenditure actually result in increased healthcare outcomes, but the results have been rather inconsistent.

Health expenditure differs among European states, but the general trend is that it tends to rise with GDP *per capita*. Thus, higher expenditure could be associated with significant gains in health status of population, though this association is not applicable in high-expenditure countries [59]. These findings are reinforced by our results shown in Fig. 5 in Appendix. This fact suggests that other economic, social and lifestyle factors are also of importance [31, 34]. Moreover, Marchetti [44] concludes that there is a point of saturation in which increases in public health expenditure do not necessarily imply increases in life expectancy. This may merely reflect an inflated administration, expensive technologies, poor comparative effectiveness, or personal financial advantages for interest groups or individuals [76].

For an effective health policy, it is important not only to analyze how healthcare expenditures affect health status of population but also other public expenditures. This work, therefore, has investigated whether there is an association between life expectancy and the composition of the public budget, especially the proportion of

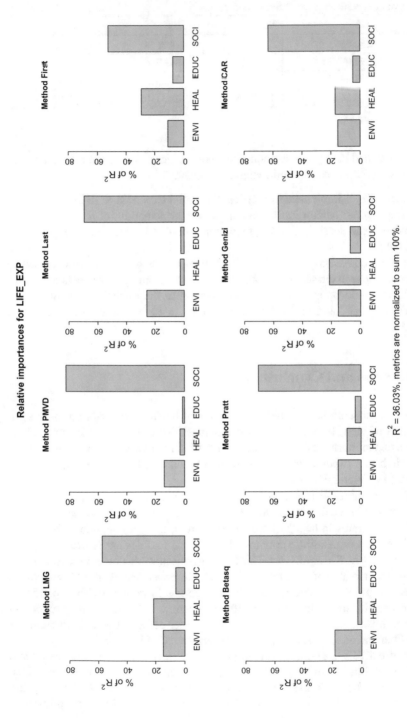

Fig. 3 Percentage of R^2 explained by each regressor, applying eight different methods *Source* Own elaboration

public expenditure on healthcare, social services, education and environment on GDP [38].

Comparative studies on population health between countries have increasingly started to look at the social policy instruments to explain the differences in health outcomes, especially in social and education expenditures. This study has found that social expenditure has the greatest relative importance on explaining life expectancy. In consequence, the ratio social expenditure to GDP may influence health outcomes beyond that resulting from health expenditure alone. Additionally, previous studies have confirmed the positive effect of social expenditure on health status [6, 7, 49, 69, 76]. On the other hand, our results show that public educational expenditure has the least relative importance. It does not mean that expenditure on education has no value in explaining health outcomes, but it has less influence compared to other expenditure items. In effect, other researches have included this type of expenditure as a determinant of health outcome and have concluded that investment in education contribute to gains in life expectancy [61, 75].

Whereas the impact of investing in this non-health sectors (social and educational programs) is well analyzed [49], there is no evidence so far demonstrating the extent to which environmental expenditure contributes to improved population health. Pollution -generally represented by emissions of polluting substances [52–55] or by an environmental quality indicator [45]—has been incorporated into this type of literature as a determinant of health status. However, this predictor is not derived from public environmental expenditure. Consequently, one of the main contributions of this study is the inclusion of expenditure on environmental protection in the analysis. We have found that, depending on the implemented metrics, environmental public expenditure is in the second or third place—in terms of relative importance—on explaining life expectancy.

Policy makers need to consider not only the direct impacts of allocating public funds but also whether these investments are likely to produce improvements in health outcomes [57]. By taking into account the results of our study, the main recommendation to public policy is rethinking the prioritization of public expenditure related to health outcomes.

Thus, this study seeks contributing to the progress of health status determinants research by adding additional public expenditures. Future research needs to incorporate wider indicators of government intervention such as environmental policy instruments.

Acknowledgements The work presented in this chapter has been supported by the Ministry of Economy and Competitiveness of Spain (research project No. DER2016-76053R: "La sostenibilidad del Sistema Nacional de Salud: reformas, estrategias y propuestas").

Appendix

See Figs. 4, 5 and Table 6.

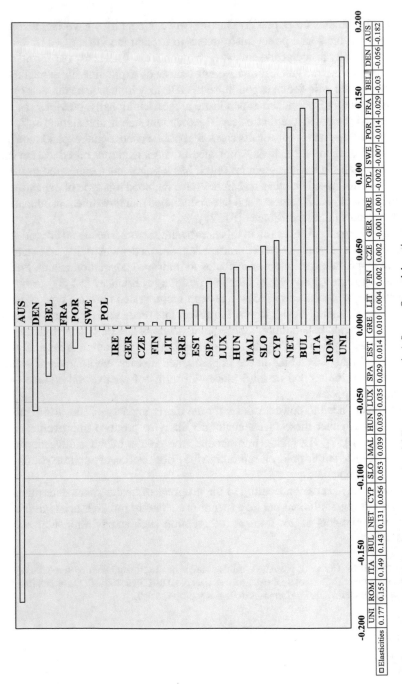

Fig. 4 Health expenditure—life expectancy elasticity (25 European countries) *Source* Own elaboration

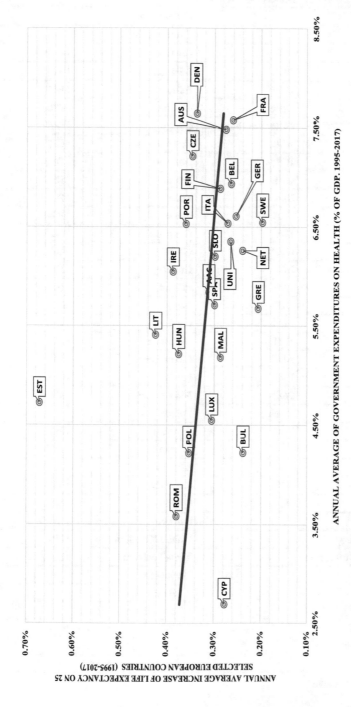

Fig. 5 Scatter plot of expenditures on health vs. life expectancy (25 European countries) *Source* Own elaboration

Table 6 Breakdown of the datasets used in this work

Dataset (A)		Dataset (B)	
Countries (25)		Countries (28)	
Belgium	BEL	Austria	AUS
Bulgaria	BUL	Belgium	BEL
Czech Republic	CZE	Bulgaria	BUL
Denmark	DEN	Croatia	CRO
Germany	GER	Cyprus	CYP
Estonia	EST	Czech Republic	CZE
Ireland	IRE	Denmark	DEN
Greece	GRE	Estonia	EST
Spain	SPA	Finland	FIN
France	FRA	France	FRA
Italy	ITA	Germany	GER
Cyprus	CYP	Greece	GRE
Lithuania	LIT	Hungary	HUN
Luxembourg	LUX	Ireland	IRE
Hungary	HUN	Italy	ITA
Malta	MAL	Latvia	LAT
Netherlands	NET	Lithuania	LIT
Austria	AUS	Luxembourg	LUX
Poland	POL	Malta	MAL
Portugal	POR	Netherlands	NET
Romania	ROM	Poland	POL
Slovakia	SLO	Portugal	POR
Finland	FIN	Romania	ROM
Sweden	SWE	Slovakia	SLO
United Kingdom	UNI	Slovenia	SLV
Time horizon:	1995–2017	Spain	SPA
Variables:		Sweden	SWE
• Life Expectancy		United Kingdom	UNI
• Expenditures on Health		**Time horizon:**	2017
		Variables:	
		• Life Expectancy	
		• Expenditures on Health	
		• Expenditures on Environment	
		• Expenditures on Education	
		• Social Expenditures	

Source Own Elaboration

References

1. Álvarez-Gálvez J, Jaime-Castillo AM (2018) The impact of social expenditure on health inequalities in Europe. Soc Sci Med 200:9–18. https://doi.org/10.1016/j.socscimed.2018. 01.006
2. Auster R, Levesoardln I, Sarachek S (1969) The production of health: an exploratory study, J Human Resour 4:411–436
3. Beckficld J, Morris KA, Bambra C (2017) How social policy contributes to the distribution of population health: the case of gender health equity. Scand J Public Health 46(1):6–17. https:// doi.org/10.1177/1403494817715954
4. Berger M, Messer J (2002) Public financing of health expenditure, insurance, and health outcomes. Appl Econ 34(17):2105–2113
5. Bergh A, Nilsson T (2010) Good for living? On the relationship between globalization and life expectancy. World Dev 38(9):1191–1203
6. Bradley EH, Elkins BR, Herrin J, Elbel B (2011) Health and social services expenditures: associations with health outcomes. BMJ Qual Saf 20(10):826–831
7. Bradley EH, Canavan M, Rogan E, Talbert-Slagle K, Ndumele C, Taylor L, Curry LA (2016) Variation in health outcomes: the role of spending on social services, public health, and health care. 2000–09. Health Aff 35(5):760–768
8. Budescu DV (1993) Dominance analysis: a new approach to the problem of relative importance of predictors in multiple regression. Psychol Bull 114(3):542–551
9. Castanheira HC, Pelletier F, Ribeiro I (2017) A sensitivity analysis of the Bayesian framework for projecting life expectancy at birth. UN Population Division, Technical Paper No. 7. United Nations, New York
10. Cochrane AL, St Ledger AS, Moore F (1978) Health service 'input' and mortality 'output' in developed countries. J Epidemiol Community Health 32:200–205
11. Crémieux PY, Ouellette P, Pilon C (1999) Health care spending as determinants of health outcomes. Health Econ 8:627–639
12. Crémieux PY, Mieilleur MC, Ouellette P, Petit P, Zelder P, Potvin K (2005) Public and private pharmaceutical spending as determinants of health outcomes in Canada. Health Econ 14:107–116
13. Cutler DM, Lleras–Muney A (2012) Education and health: International comparisons. NBERWorking Paper 17738, 1–30
14. Elola J, Daponte A, Navarro V (1995) Health indicators and the organisation of health care systems in Western Europe. Am J Public Health 85(10):1397–1401
15. Engelhart MD (1936) The technique of path coefficients. Psychometrika 1(4):287–293
16. Fayissa B, Gutema P (2005) Estimating a health production function for sub-Saharan Africa (SSA). Appl Econ 37:155–164
17. Feldman B (2005) Relative importance and value. Manuscript version 1.1, 2005-03-19. http:// www.prismanalytics.com/docs/RelativeImportance.pdf
18. Filmer D, Pritchett L (1997) Child mortality and public spending on health: how much does money matter? The World Bank, Policy Research Working Paper 1864
19. Genizi A (1993) Decomposition of R^2 in multiple regression with correlated regressors. Statistica Sinica 3
20. Grömping U (2007) Estimators of relative importance in linear regression based on variance decomposition. The American Statistician 61(2):139–147
21. Grubaugh SG, Rexford ES (1994) Comparing the performance of health–care systems: an alternative approach. South Econ J 60:1030–1042
22. Halicioglu F (2011) Modeling life expectancy in Turkey. Econ Model 28:2075–2082
23. Hansen CW, Strulik H (2017) Life expectancy and education: Evidence from the cardiovascular revolution. J Econ Growth 22:421–450
24. Heijink R, Koolman X, Westert GP (2013) Spending more money, saving more lives? The relationship between avoidable mortality and healthcare spending in 14 countries. Eur J Health Econ 14(5):527–538

25. Hill TD, Jorgenson A (2018) Bring out your dead!: A study of income inequality and life expectancy in the United States, 2000–2010. Health & Place 49:1–6
26. Hitiris T, Posnett J (1992) The determinants and effects of health expenditure in developed countries. Journal of Health Economics 11:173–181
27. Hoffman PJ (1960) The paramorphic representation of clinical judgment. Psychol Bull 57(2):116–131
28. Jiang J, Luo L, Xu P, Wang P (2018) How does social development influence life expectancy? A geographically weighted regression analysis in China, Public Health 163:95–104
29. Johnson JW (2000) A heuristic method for estimating the relative weight of predictor variables in multiple regression. Multivar Behav Res 7(3):238–257
30. Johnson JW, Lebreton JM (2004) History and use of relative importance indices in organizational research. Organizational Research Methods 35:1–19
31. Johnson SA (2011) Challenges in Health and Development: From Global to Community Perspectives. Springer Science+Business Media, Berlin
32. Joumard, I., André, C., Nicq, C., Chatal, O. (2008) Health status determinants: Lifestyle, environment, health care resources and efficiency. OECD Economics Department Working Papers, No. 627, OECD Publishing, Paris. http://doi.org/10.1787/240858500130
33. Kabir M (2008) Determinants of life expectancy in developing countries. The Journal of Developing Areas 41:185–204
34. Karim SA, Eikemo T, Bambra C (2010) Welfare state regimes and population health: Integrating the East Asian welfare states. Health Policy 94(1):45–53. https://doi.org/10.1016/j.healthpol.2009.08.003
35. Ketenci N, Vasudeva NRM (2018) Some determinants of life expectancy in the United States: results from cointegration tests under structural breaks. Journal of Economics and Finance 42:508–525
36. Khuder SA (2001) Effect of cigarette smoking on major histological types of lung cancer: A meta–analysis. Lung Cancer 31:139–148
37. Kiula O, Mieszowski P (2007) The effects of income, education and age on health. Health Econ 16:781–798
38. Kudlák A, Urban R, Hošková-Mayerová Š (2020) Determination of the Financial Minimum in a Municipal Budget to Deal with Crisis Situations. Soft Comput 24(12):8607–8616. https://doi.org/10.1007/s00500-019-04527-w
39. Laporte A (2004) Do economic cycles have a permanent effect on population health? Revisiting the Brener hypothesis. Health Econ 13:767–779
40. Lichtenberg, F. (2000) Sources of U.S. longevity increase, 1960–1997, NBER Working Paper No. 8755
41. Lindeman RH, Merenda PF, Gold RZ (1980) Introduction to bivariate and multivariate analysis. Scott Foresman & Co, Glenview
42. Linden M, Ray D (2017) Life expectancy effects of public and private health expenditures in OECD countries 1970–2012: Panel time series approach. Econ Anal Policy 56:101–113
43. Lipovetsky S, Conklin WM (2015) Predictor relative importance and matching regression parameters. Journal of Applied Statistics 42(5):1017–1031
44. Marchetti C (1997) Longevity and life expectancy. Technol Forecast Soc Chang 55:281–299
45. Mariani F, Pérez-Barahona A, Raffin N (2010) Life expectancy and the environment. Journal of Economic Dynamics and Control 34:798–815
46. Martín-Cervantes P, Rueda-López N, Cruz-Rambaud S (2020) Life expectancy at birth: a causal analysis of the healh sector. In: Flaut SD, Hoskova-Mayerova S, Ispas C, Maturo F, Flaut C (eds) Decision making in social sciences: between traditions and innovations, studies in systems, decision and control, Springer
47. Martínez-Sánchez E, Gutiérrez-Fisac JL, Gispert R, Regidor E (2001) Educational differences in health expectancy in Madrid and Barcelona. Health Policy 55:227–231
48. McAvinchey ID (1988) A comparison of unemployment, income and mortality interaction for five European countries. Appl Econ 20(4):453–471

49. McCullough JM, Leider JP (2016) Government Spending In Health And Nonhealth Sectors Associated With Improvement In County Health Rankings. Health Aff 35(11):2037–2043. https://doi.org/10.1377/hlthaff.2016.0708
50. Meara ER, Richards S, Cutler DM (2008) The gap gets bigger: changes in mortality and life expectancy by education, 1981–2000. Health Aff 27:350–360
51. Miller, R.D., Frech, T. (2002) The Productivity of Health Care and Pharmaceuticals: Quality of Life, Cause, University of California, Santa Barbara, UCSB Departmental Working Paper, No. 12–02
52. Monsef A, Mehrjardi AS (2015) Determinants of life expectancy: A panel data approach. Asian Economic and Financial Review 5(11):1251–1257
53. Nixon J, Ullmann P (2006) The relationship between health care expenditure and health outcomes – evidence and caveats for a causal link. European Journal of Health Economics 7(1):7–19
54. Or Z (2000a) Determinants of health outcomes in industrialised countries: A pooled, cross-country, time series analysis, OECD Economic Studies, No. 30, 2000/I
55. Or Z (2000b) Exploring the effects of health care on mortality across OECD countries. OECD Labour Market and Social Policy, Occasional Paper No. 46
56. Peltzman S (1987) Regulation and health: the case of mandatory prescriptions and an extension. Manag Decis Econ 8:41–46
57. Plough AL (2015) Building a culture of health: a critical role for public health services and systems research. Am J Public Health 105(Suppl 2):S150–S152
58. Porreca A, Cruz Rambaud S, Scozzari F, Di Nicola M (2019) A fuzzy approach for analysing equitable and sustainable well-being in Italian regions. International Journal of Public Health. https://doi.org/10.1007/s00038-019-01262-9
59. Poullier, J., Hernandez, P., Kawabata, K., Savedoff, W. D. (2002) Patterns of Global Health Expenditures: Results for 191 Countries. Discussion Paper 51. World Health Organization
60. Raftery, A.E., L. Chunn, J., Gerland, P., Ševčíková, H. (2013) Bayesian probabilistic projections of life expectancy for all countries. Demography, 50(3), 777–801
61. Reynolds MM, Avendano M (2017) Social policy expenditures and life expectancy in high-income countries. Am J Prev Med 54(1):72–79
62. Robalino, D.A., Oscar, F.P., Albertus, V. (2001) Does fiscal decentralization improve health outcomes? Evidence from a cross–country analysis. World Bank Policy Research Working Paper series 2565, 1–14
63. Rodgers GB (1979) Income and inequality as determinants of mortality: an international cross–sectional analysis. Population Studies 33:343–351
64. Self S, Grabowski R (2003) How effective is public health expenditure in improving overall health? A cross–country analysis. Appl Econ 35:835–845
65. Shaw JW, Horrace WC, Vogelf RJ (2005) The determinants of life expectancy: An analysis of the OECD health data. South Econ J 71(4):768–783
66. Shing-Jong L (2009) Economic fluctuations and health outcome: a panel analysis of Asia-Pacific countries. Appl Econ 41:519–530
67. Schultz H (1933) A comparison of elasticities of demand obtained by different methods. Econometrica 1:274–308
68. Soares RR (2007) Health and the evolution of welfare across Brazilian municipalities. J Dev Econ 84:590–608
69. Stuckler, D., Basu, S., McKee, M. (2010) Budget crises, health, and social welfare programmes. BMJ: British Medical Journal, 341 (7763), 77–79
70. Svarcova, I., Hoskova-Mayerova S., Navratil, J. (2016) Crisis Management and Education in Health, The European Proceedings of Social & Behavioural Sciences EpSBS, Volume XVI, pp. 255–261. https://doi.org/10.15405/epsbs.2016.11.26
71. Tapia-Granados JA (2005) Recessions and mortality in Spain, 1980–1997. European Journal of Population 21:393–422
72. Thornton J (2002) Estimating a Health Production Function for the US: Some New Evidence. Appl Econ 34(1):59–62

73. Toader E, Firtescu B, Oprea, F (2017) Determinants of health status and public policies implications –lessons for Romania. Transylvanian Review of Administrative Sciences, 52 E, 128–147
74. Tonidandel S, Lebreton JM (2011) Relative importance analysis: A useful supplement to regression analysis. J Bus Psychol 26(1):1–9
75. Van den Heuvel WJA, Olarolu M (2017) How important are health care expenditures for life expectancy? A comparative, European analysis. J Am Med Dir Assoc 18(3).276.e9–276 e12
76. Vavken P, Pagenstert G, Grimm C, Dototka R (2012) Does increased health care spending afford better health care outcomes? Evidence from Austrian health care expenditure since the initiations of DRGs. Swiss Medical Weekly 142:w13589
77. Wallard H (2015) Using explained variance allocation to analyse importance of predictors. In: 16th ASMDA conference proceedings, 30 June–4 July, Piraeus, Greece
78. Wise RA (1997) Changing smoking patterns and mortality from chronic obstructive pulmonary disease. Prev Med 26:418–421
79. Wolfe BL, Gabay M (1987) Health status and medical expenditures: more evidence of a link. Soc Sci Med 25:883–888
80. Zuber V, Strimmer K (2011) High-dimensional regression and variable selection using CAR scores. Stat Appl Gen Mole Biol 10(34):1–27

Who Is the Professional Pedagogist and How He Practices

Franco Blezza

Abstract The Italian Parliament approved (20–27 December 2017) the law establishing the figure of the professional Pedagogist (law 205/2017, art. 1 594–601) with university's enablement, thus recovering the serious Gap compared to the many more advanced countries. In this paper we summarize who is the professional Pedagogist and what is his professional profile, what are his conceptual and operative tools and what exemplary problems apply. Pedagogy is a social science, a science strictly speaking, and by its own nature a social profession, whose origins refer to classical Greece and to Latinity. "Pedagogist" is called, properly, the apical professional. We take in a synthetic examination some important examples of the rich and diversified "toolbox" the Pedagogist can use in its professional practice, in the context of a composite culture for a plurality of Inputs of various disciplinary origins, regarding the professional practising. Several social and psychological sciences give their contributions, like "educational sciences".

Keywords Pedagogist · Social professions · Professional pedagogy · Social pedagogy · Helping relationship · Educational sciences

1 Pedagogist: A Professional for Italy and for the Whole World

In Italy too there is officially the professional figure of the Pedagogist, who had been practicing for several decades without legal recognition [16].

The Italian Parliament approved definitively in December 2017 the law 205, which art. 1 comma 594–601 establishing the first discipline of this profession (with the Professional Educators), so finally aligning Italy with Europe at this specific regard, and giving to the pedagogical culture those professions recognized and ruled, as long existed just as happened for thousands of years in medicine surgery and jurisprudence,

F. Blezza (✉)
Department of Philosophical, Pedagogical and Economic-Quantitative Sciences, "Gabriele D'Annunzio" University, Chieti–Pescara, Italy
e-mail: franco.blezza@unich.it

© The Author(s), under exclusive license to Springer Nature Switzerland AG 2021   297
D. Soitu et al. (eds.), *Decisions and Trends in Social Systems*,
Lecture Notes in Networks and Systems 189,
https://doi.org/10.1007/978-3-030-69094-6_24

for centuries in architecture and for decades in psychology or psychoanalysis, for example [20].

The aim of this paper is to summarize some aspect of this profession, his toolbox and his specific professional practice, particularly on couple and family problems [4–6].

2 Pedagogy Is a Social Science and a Social Profession from the Beginning in the Occidental Civilisation

The Pedagogist is the professional of the highest and apical level expressed by the pedagogical culture and by the complex of educational subjects. Such a professional, like any other, is susceptible to specializations and hierarchical articulation.

Pedagogy is a profession of ancient history as those of the medical surgeon and of the jurist, about 2500 years in Western civilization. On the other hand, education is an essential human necessity as well as the health and the civil coexistence normed by positive law. The historians of pedagogy, and those of thought, know how and why in certain periods the integral fulfilment of such a need may not had the necessity of a specific professional contribution, even though the education always was the subject of study and reflection. The profound change in the time of the pedagogical professions should be placed in relation to the different educational paradigms that prevailed in each historical period, just as was the case for the professional contribution of the jurists. The social role of doctor surgeon, on the contrary, has had a history much more complex and was characterized by the constancy of a substantial presence and a strong social conspicuousness.

John Dewey (1859–1952) defined the Sophists *"the first body of professional educators in Europe"*. And it is of great interest the context of this affirmation, a particular perspective of the relationship between pedagogy and philosophy.

> The earlier history of philosophy, developed by the Greeks in Asia Minor and Italy, so far as its range of topics is concerned, is mainly a chapter in the history of science rather than of philosophy as that word is understood to-day. It had nature for its subject, and speculated as to how things are made and changed. Later the traveling teachers, known as the Sophists, began to apply the results and the methods of the natural philosophers to human conduct.
>
> When the Sophists [...] instructed the youth in virtue, the political arts, and the management of city and household, philosophy began to deal with the relation of the individual to the universal, to some comprehensive class, or to some group; the relation of man and nature, of tradition and reflection, of knowledge and action. Can virtue, approved excellence in any line, be learned, they asked? What is learning? It has to do with knowledge. What, then, is knowledge? How is it achieved? Through the senses, or by apprenticeship in some form of doing, or by reason that has undergone a preliminary logical discipline? Since learning is coming to know, it involves a passage from ignorance to wisdom, from privation to fullness from defect to perfection, from non-being to being, in the Greek way of putting it." [9].

It should be born in mind that in Dewey's English the term "pedagogy" and derivatives where unused, compared to the term "education", unlike for other western languages such as French, Italian and German; moreover, the English language is knowing its evolution also in this specific regard. And let us not forget that the fundamental work of Dewey of the [8], collection of five essays, was entitled *My pedagogic creed.*

3 The Nineteenth-Century Human Sciences that Where the Roots of Corresponding Professions in the Twentieth Century

Several professions that have gained great importance in the '900 have had their foundations in the context of the Middle-European world of German-speaking, or in its vicinity, in the previous century: particularly the professions of the psychological and psychoanalytic area, and the professions of sociological culture. The historical Modern Age had ended since the last decades of the '700 with the Enlightenment, the bourgeois revolutions, the industrial revolution and the deep social changes that have been achieved. It was starting a later historical age, which never find his technical denomination and was dominated by a particular "Bürgergeist" (bourgeois spirit) that had in the education an essential foundation and condition: an education carried out through the replication of prefixed models, and aimed at the construction of the genders in an extremely polarized sense as required by the "conjugal" [10, 11] or "nuclear" family [18] paradigm, a very important creation of that period. Such an educational investment was very strong but a-specific, did not require the educator to have a particular pedagogical culture, but only the previous adherence to those principles, which legitimized the idea that it did not require the contribution of specific professionals except in very special cases [2].

These ideas established themselves rapidly and deeply in Western Europe, that transition between two historical Ages was relatively short. *Die Wiener Kongress* with the pretended *Restaurierung* of the *ancien régime* appears to us an illusory attempt to let the clock of history go back. The social, cultural and economic reality was deeply changed and began to evolve, it would immediately take frantic rhythms as they were never seen before.

The exit from the Enlightenment was represented pedagogically by the fundamental figure of Johann H. Pestalozzi (1746–1827), pedagogist in the full sense since he has always been in direct and organic relationship with the reality of education, engaged in the agricultural colony and then for the orphans and as a teacher, but with an original theoretical synthesis that developed contextually, well documented by his scientific works.

Two were the main pedagogists who continued his work: Friedrich Fröbel (1782–1852), who devoted himself to *der Kindergarten*, i.e. to the childhood school, and Johann F. Herbart (1776–1841) who theorized pedagogy as a synthesis of ethics and psychology, a few years before psychology emerged from philosophy (λόγος on η ψυχή) and began to legitimately form part of the sciences in the strictest sense

Moreover, it's well known that all the sciences have been philosophies and have preserved one or more philosophies in their essence.

At that point, there were all the assumptions for the profession of pedagogist had a firm reference also in the same historical and cultural context of the Central Europe and particularly in the social pedagogy ([3], part. pp. 9–11). We are referring to the work of researchers pedagogists as Karl Mager (1810–1858), who first wrote of "sozial Pädagogik" in 1844 (in the "Pädagogische Revue" of which he was director from 1840 to 1848); he proposed this conceptuality as opposed to *Individualpäda-gogik* and alternatively to *Collectivpädagogik*; Friedrich A. W. Diesterweg (1790–1866), who dealt with the training of the teachers, and was critic of the notionism and favourable to a school for the people, and who first of all developed the concept of *learning by doing*; and Paul Natorp (1854–1924) who used first the term compound in the essay entitled precisely *"Sozialpädagogik"* (the subtitle evoked a theory of the desires that come from the basis of the community established in the common culture, or *Gemeinschaft* (Natorp 1899; on the web we can find several other articles, and short essays, by this author and many others ones).

We must bestow, or give back, a particular importance to the specifically pedagog-ical contribution of D. Émile Durkheim (1858–1917), let's not forget that the great scientist of the society began his academic career at the University of Bordeaux in 1887 with the call to the chair of *Social sciences and Pedagogy*, and was then called to the Sorbonne, where he entered in 1902, in 1906 as professor on the chair of *Pedagogy*, and that only in 1913 that chair would assume the name of *Pedagogy and Sociology*. He was not only a great social Pedagogist, he was one of the founders of Sociology, and this historical qualification perhaps leads to not giving the right importance to others; In fact, he was also an Anthropologist and a researcher on religions.

His pedagogical works were mostly gathered from articles and dispenses in posthumous volumes, from which it is evident its importance; the Web makes the right homage to a great Pedagogist and allows a specific study (Durkheim 1904/05 1918 1922). They were not, in short, minor works [11–13].

Among other things, we must take in account his birth in Alsace-Lorraine and the unmistakable origin of the surname (*das Heim* indicates in German the home, family, of origin, …) concur with that idea which we have mentioned, i.e. the funda-mental importance of the Central European culture for the foundation of a complex of sciences that would have given rise to a wide range of intellectual and social professions in the twentieth century.

Alsace Lorraine, after all, would have completed the German Kaiserreich only a few years after its birth (18 January 1871).

This is not an isolated case: Pedagogy is a science, a social science, fully fledged; and it's not just a philosophical science anymore.

4 *Professional* Pedagogy: A New Branch of *General* Pedagogy (Allgemeine Pädagogik)

We will call *professional Pedagogy* [6] a particular branch of *general* Pedagogy. This branch deals with the study, the proposal and the experimentation of principles, methods, techniques, procedures, specific vocabulary and the related organic arrangement for the specifically pedagogical professional practice. In a nutshell, everything constituting a necessary condition for the pedagogical profession and for the community of professional Pedagogists.

The essential contribution of professional Pedagogy is also indicated for initial and continuous training of intellectual professionals in the social and health sectors, in the School and in the psychological field.

We must widely explore the contribution it can offer to the professionalism of the university professors and researchers, regardless of the sector in which they are framed.

As for general Pedagogy, and for any branch of this social science, the development of professional Pedagogy must be in an organic relationship with the reality object of studies and applications, and must maintain it, also in order to obtain the necessary experiential feedback that, as is known from the Epistemology of the XIX and XX centuries. This observance of the experience that will follow, and Feedback consequent, is a condition of the scientificness of the discourse, and therefore, inter alia, of intersubjective transferability of every discourse, which in education is indispensable, and of openness and respect of the person. These are also indispensable conditions in human culture, towards a continual, a-teleological cultural evolution, endless and without τέλος.

Indeed, any idea, proposition, human creation, hypothesis, alternative we propose within the professional Pedagogy too, as in the general Pedagogy, is nevertheless subjected to what Pragmatists have taught us to call the "future experience". Karl R. Popper (1902–1994) resumed this conceptuality, and it is appropriate to read fully its statute of *das Abgrenzungskriterium*, which is well-known:

> Nun wollen wir aber doch nur ein solches System als empirisch anerkennen, das einer Nachprüfung durch die "Erfahrung" fähig ist. Diese Überlegung legt den Gedanken nahe, als Abgrenzungskriterium nicht die Verifizierbarkeit, sondern die Falsifizierbarkeit des Systems vorzuschlagen; mit anderen Worten: Wir fordern zwar nicht, daß das System auf empirisch-methodischem Wege endgültig positiv ausgezeichnet werden kann, aber wir fordern, daß es die logische Form des Systems ermöglicht, dieses auf dem als Abgrenzungskriterium Wege der methodischen Nachprüfung negativ auszuzeichnen: Ein empirisch-wissenschaftliches System muß an der Erfahrung scheitern können." [22], 6. Falsifizierbarkeit als Abgrenzungskriterium)

It follows that Pedagogy is a science (Wissenschaft stricto sensu) also according to this "Abgrenzungskriterium": a social science, as it develops through a constant position of problems and a continuous attempt to solve them through human creativity, practiced respecting the rules of inner coherence and external consistency, and any other typical rule of science. Professional Pedagogy is a branch in close relation, strongly synergic, with the branch of Social Pegagogy, also for the historical reasons that have been mentioned.

Particularly important is the rethinking of Popper about fifty years after the basic work just mentioned, following the formulation of Tichý's theorem [25, 26], on the question of verisimilitude [1]. We can't speak of some form of progress in the evolution of scientific knowledge and, more generally, of human knowledge. this concept is consistent with today's Pedagogy, which, moreover, not long ago, recognizes in cultural evolution an a-teleological process which has no fixed direction but only one verse, that of the time's arrow, that of the increase of global entropy, that of the "clock of history" whose hands, as is known and as already said, "can't go back".

5 How Practices His Profession the Pedagogist: An Experience

The fundamental aim of the remaining pages of this chapter will be to report synthetically and organically what has been studied, proposed and experienced in social and professional Pedagogy as professional Pedagogist. It is about thirty years of professional experience actually practiced, dealing mostly with family, couples, parenthood, generality problems, and connected decisions associated with these fundamental instances of sociality. But this experience offers contributions of obviously general validity. The following paragraphs that are, in fact, sub-paragraphs.

5.1 From Particular to General Cases, in Continuous Research: The Abduction (Ab-Ductio, απαγωγή)

First of all, no professions exist without adequate competence on the theories and general cases that the practicer has the task of treating: so, in the human physiology and pathology of the doctor surgeon, as for those of the animals of the veterinarian, as for the pharmacological principles of the pharmacist, or as for the science of constructions and for the other technical disciplines of the architect and civil engineer, et coetera.

The professional competence is highlighted in the specific modality in which the transition from the particular to the general is accomplished: the simplest and most immediate example is that of the medical doctor, who cures sick and does not cures sicknesses, each case with its singular and unfailing peculiarities, but could not cure them if it does not bring back the clinical frame of each individual sick to the general case of the sickness, such as diagnosis, prognosis and therapy. This necessary human mediation, which does not give certainty, from the logical point of view doesn't constitute a tautology but a form studied since classic Greece and Latinity: it can be called *abduction* or *retro-duction*. It is one of the underlying concepts of the composite contribution to the Pragmatism of Charles S. Peirce (1839–1914), to whom we must the first theorization of science as *fallibilism* [21].

But the speech is more general: most of the contributions, which are generally linked to the logic of the research of Popper and his students, we educators owe them to pragmatism classic, that is to say several decades before.

In the specific field of the social and professional Pedagogy, a passage of great importance consists in helping the interlocutors to bring to the explicit what of their respective life projects is long been implicit, or otherwise not expressed and not discussed, as deemed not in need of any analysis, as taken for granted, as it is assumed tacit and perpetual acceptance on the part of some proximal person, without any legitimacy, by the Partner particularly.

A typical casuistry for this specific matter concerns, indeed, a profound asymmetry between the two partners of a couple in the "outside" investment, work social relations and public roles, and the "inside", family intimacy and the domestic home: one of the two partners who, at the moment of the contraction of the bond, is invested with wide-ranging "outside", and the other "inside", even in full harmony and in agreement.

At a time not far away, the first person was necessarily the male and the second one the female; and for this determination would be given natural causes and motives. Today there are increasing reciprocal choices.

It is often noted that one of the two partners conceives this balance in diversity as definitive and pacifically accepted by the other one forever, and the other partner considers it only a temporary and conditioned step, for example, until the partner has achieved some social status or until the children have reached a certain age. At that point, that of the two partner who had accepted the pre-eminent investment "inside" begin to recapture its "outside" spaces, e.g. by rescuing its own study degrees and professional qualifications, opening free-trade or crafts or social cooperatives; to that, it happens that the other Partner refuses to rebalance the situation and considers this behaviour as a kind of breach of commitments, commitments that have never been taken.

Something similar also happens in the sex life: certain sexual behaviours can be held for years, with one of the two partners who decides to accept them only provisionally, while the other believes that the acceptance is definitive, forever, and without any exception. For example, one of the two accepts contraceptive behaviours in the other considering them provisional, while the other has not posed any term for these practices. Often this is not discussed unless after a long repetition of this sexual practice conditioned by a unilateral decision, until it is much more difficult to intervene if it is still possible.

Even about fidelity, there are those who believe that the other partner accepts unquestionably his treacheries without in turn betray: in this case too, it would be the male towards the female in past times, today happens the reciprocal also. In this and in all the other examples of casuistries that could be brought these beliefs that emerge only after many years, sometimes after decades, in their destructive character for the couple, were given to tacitly accepted and for not requiring any discussion from the beginning of the life of that couple. It's an original mistake.

We must add, for the professional experience that we have done, that where there is full and unreserved opening on the part of both partners cases like these are all solvable with mutual satisfaction and without residue, even when the behaviours of the one or of the other or of both have taken breaking or even dramatic features.

The mind goes immediately to the concept of μαιευτική (τέχνη), the second phase with the εἰρωνεία of the διάλογος as the great Athenian Σωκράτης proposed, a give birth to what has matured inside the interlocutor also making him face the difficulties connected. That can be as painful as childbirth pains. This juxtaposition is correct, but with the important difference of not having any claim to reach some ʽαλήθεια. On the other hand (here and elsewhere), this is accomplished without any involvement of the unconscious, being ideas perfectly conscious even if not discussed and implied or kept implicit, and without the slightest possibility that the professional intervention of the Pedagogist can be confused with a form of therapy. It is, on the contrary, a helping relationship that compared to the therapy is an exclusive alternative: not *to cure* someone or something, but *to care of* someone or something, or more directly and generally following *I care!* of Martin Luther King.

We are talking about a professional intervention with the word, but not a "therapy of the word": it's not a therapy. After pointing out all these concepts, we can outline some phases in logical and methodological sequence that effectively describe and scan the relevant professional practice of the Pedagogist.

6 Problem Posing

So much for the general features of a reflective experience. They are (i) perplexity, confusion, doubt, due to the fact that one is implicated in an incomplete situation whose full character is not yet determined; (ii) a conjectural anticipation—a tentative interpretation of the given elements, attributing to them a tendency to effect certain consequences; (iii) a careful survey (examination, inspection, exploration, analysis) of all attainable consideration which will define and clarify the problem in hand; (iv) a consequent elaboration of the tentative hypothesis to make it more precise and more consistent, because squaring with a wider range of facts; (v) taking one stand upon the projected hypothesis as a plan of action which is applied to the existing state of affairs: doing something overtly to bring about the anticipated result, and thereby testing the hypothesis. It is the extent and accuracy of steps three and four which mark off a distinctive reflective experience from one on the trial and error plane. They make thinking itself into an experience. Nevertheless, we never get wholly beyond the trial and error situation. Our most elaborate and rationally consistent thought must be tried in the world and thereby tried out. And since it can never take into account all the connections, it can never cover with perfect accuracy all the consequences." [9] cited work, Chapter Eleven: Experience and Thinking).

These words where written over a century ago. Living and operating "by problems" was already clearly ruled even through the consequent stages. For the continuation of the speech, we owe much to Popper regarding the philosophy of science (since *cited work* [22] that, we observe this choice carefully, did not bear the adjective *"wisseschaftilche"*, and this adjective will only appear in the English edition), although he preferred to call it "logic of research"; and to György Pólya (1887–1985), initially in the field of didactics of mathematics [23, 24] and then also of the teaching of physics and natural sciences.

Within the *Logik der Forschung*, the *problem* has its logical definition as *a contradiction between established assertions*, between two theories, or between a theory and an assertion describing a fact. Thus, the "problem" acquires an essential function for the evolution of the knowledge and for the human life: it is, in short, a positive factor, while the term has in the common language a negative inflection, as if we wanted a life without problems, that is a life impossible and without evolution. *Latine loqui, error felix culpa.*

Compared to these positions, and compared to other ones also, are at least *two* are the *changes* that are proposed today in social and professional pedagogy.

A *first change* concerns the precise distinction that must be established between the situation of imbalance, contradiction, conflict, difficulties that the living being encounters in its interaction with the environment and with the other living ones, which we would call rather "*problematic situations*", distinct from the "*problem*" properly said. Between the ones and the other there is the *human decision*, which is not automatic or mechanical, but on the contrary, is highly selective: only a very few problematic situations become problems, a small minority.

We would call then "*problem*" the *rationalization of a problematic situation*, namely the positive, constructive reaction of the man who intends to overcome what he always meets, ordinarily, commonly of non-harmonic in his interaction with the environment.

For instance, if a couple is in a highly problematic situation, we'll ask to the partners: are you willing to pose it a problem?

The *second change* warns that the human reaction in posing the problem doesn't have in itself any guarantee of success, as it would be for a naïvely optimistic reading of these messages, confused with simplistic assertions of some obscure and crude positivists in the XIX century. On the contrary, theories are human creations: all that is assumed and elaborated in an attempt to solve the problem is imperfect and fallible, like any human artefact.

In other words: what do you do when you have posed a problem? The correct answer is not "I solve it", but "I try to solve it", "I devise a solution hypothesis".

7 Theories and Human Creativity

After the problem has been posed, or rather already in the act of the position of the problem, it is then involved the very high human faculty that is *creativity*. During about two centuries, this general faculty was considered under the romantic and idealistic ide of "*genius and lack-of-rules*". I.e. it was denied to almost all human persons. On the contrary, any social instance is constantly a continual source of problematic situations, the most important of which must be posed as a problem and therefore is involved the practice of creativity.

This creativity, in turn, is not unregulated, its practice is always normed, and here too there would be heavy criticisms to address to the culture and education of the XIX and XX centuries.

In a nutshell, and referring to other works for the appropriate insights, the hypotheses created, devised and proposed in an attempt to solve human problems must be subjected to certain orders of rules of method, at least if we speak on peda gogy, on social sciences and on social service, on human relationship inter-personal, on civil and democratic coexistence.

8 Rules of Method

We synthetize these rule's orders below.

- Hypotheses are to be inserted into *broader thought environments*, otherwise they have neither Sinn, nor meaning nor applicability. This applicability corresponds to *die Anwendungsmöglichkeit*, the level of professional practice for the Pedagogist, neither theory nor praxis. This insertion is what one does with the scientific hypotheses which are inserted in laws, theories, branches and disciplines, both in the sciences of nature and in the human, social and of culture sciences (*Human- Sozial- Geistes-Wirtschafts-wissenschaften*).
- These hypotheses, with all the context in which they are inserted, must be subject to the rules of *inner coherence* i.e. the laws of Logic; for education we must highlight the rule of non-contradiction. Both Pragmatists and Popper, and much of the epistemology of the XX century, indicated to employ the so-called "classical" logic, Aristotelian logic systematized in the Middle Ages by Thomism. The Pragmatists and, among them, the physician and psychologist William James (1842–1910) believed that this logic was inscribed in the anatomy and physiology of the human brain [17]; Popper instead adopted it because binary, that is because it's the strongest one.
- The "inner" coherence is not enough, it comes sooner or later the moment in which it is necessary to test the advanced hypotheses, together with the whole system of thought within which they was developed, and thus is consequently required *the "external" coherence*; as it was clear to the Pragmatists. Popper theorized rigorously that there is no inductive method, there is no possibility of making "true" a hypothesis for how many positive confirmations it receives from experience, while only one empirical falsification is sufficient to logically affect the hypotheses and the whole system in which it was inserted. It is the idea of science as *fallibilism*, an idea proposed by Peirce (vol. 1), decades before Popper made it the basic idea of the 1900s that we have seen like *Abgrenzungskriterium*.

We could continue for a long time, but in summary we conclude this review with the application to human things of an authentic *historical and critical spirit*, whereby every human idea, by the fact that it is human, is subject to becoming historical, which does not represent an incoherence, and it is always susceptible to criticism from those who advanced it and from anyone else.

Ideas are for man, and never man for ideas. In family too, or better starting right from the couple and the family.

We respect, and we can also admire, those men who dedicate their lives to an idea, even when it is observed that the maintenance to the deeper end of that idea ends up butting or with the rules of inner coherence, and here we are at the Socratic εἰρωνεία, or with those of external coherence, reality contrasts falsifications of fact to those convictions, and in any case are violated in principle the rules of human historicity and of criticism.

What is not to be respected, and indeed whose condemnation does not allow exceptions, is any behaviour, in any place or entity, that tends to enslave other people or person to an idea, especially people who, for some reason, are in a situation of less strength or dependence: children compared to parents, elderly and disabled compared to people in the fullness of their resources, those who ask for help, those who need assistance, and so on. Here we must pay close attention, because in most cases the violation of this fundamental rule takes on lies and noble appearances: it's commonly replied that some people are enslaved to the ideas of others "for their own higher interest", "for the good of the family" or "to save the family", "for the good of the company", "for the work of everyone", "for the good of the players before the team", and so continuing, in short for a non-personal but more general, superior good or interest.

We must to not fall into any of these pitfalls or other comparable analogues, the answer can only be negative: the human person can never be made an instrument by others for any idea, however high the nobility and human valence are alleged about that idea.

Synthetically, the person can never be made an instrument for any purpose and for any condition: it is always and only a purpose to itself.

9 The Professional Pedagogist and Psychology as a Science of Education

The composite character of the general Pedagogy, and therefore also of the social and professional Pedagogy, allows to transpose within these domains the contributions coming from other disciplinary fields and from authors pertaining to them that they would never consider themselves Pedagogists; obviously, it must be done after reprocessing these contributions in order to integrate them with the complex of pedagogical-general and pedagogical-social and professional culture, and after turning to the specific aims of education that are not the aims of all those disciplines, or of others yet.

In this analysis we cannot go beyond an extremely concise and schematic scan: even for the bibliography, here we must remain on the general lines and we cannot detail the texts of individual authors or comprehensive manuals. Here we can, if anything, exemplify some of the general pedagogy "classical" manuals that offer the most suitable context for this task [19, 27].

We could and should begin with Sigmund Freud (1856–1939) and Psychoanalysis, except that for us Pedagogists the unconscious is a precluded domain: precluded by the awareness of not having the skills to enter the unconscious, before the laws that protect the exercise of certain professions or therapeutic arts, and anyway by the fact that we are not therapists.

We could continue talking about other Psychoanalysts, including the founders of other fundamental schools of Psychoanalysis: Carl Gustav Jung (1875–1961), Alfred Adler (1870–1937), Jacques Lacan (1901–1981).

Erich Fromm (1900–1980) offers to our profession a very important contribution, for example with the concepts of "paternal" love and "maternal" love that should be balanced and integrated, without which opens the way to one side for a whole series of pathologies that are not our competence, on the other for a whole series of educational shortcomings that instead directly involves our competences. But many others are the contributions that come to us from this important social psychologist, as, moreover, from the whole Frankfurter Schule to which he too referred: from criticism to civilisation and contemporary society, to the study of aggression, the destructiveness and the necrophilia, until the discourse on the modalities of having and being.

We understand how fundamental is the contribution of Carl Rogers (1902–1987), where we consider carefully also in this case that our competences are not therapeutic. We practice the helping relationship and, so working, we have very much to learn from him.

Instead, we can fully employ the conceptual tools proposed by Viktor Frankl (1905–1997), namely the search for *Lebenssin*, die *Dereflexion*, der *paradoxen Intention* and *der gemeinsame Nenner*.

In a similar way, it can be employed by the professional pedagogist *das Autogene Training* according to Jurgen H. Schultz (1884–1970) in its fullness, especially considering *die Formel*, the original fundamental propositional formula, and its subsequent literal translation "(Ich bin) ganz ruhig [ruhe]" and it is used with the appropriate adaptations and reformulations, rather than as a simple relaxation technique. For specifically pedagogical reasons and purposes we can propose an alternative propositional formula designed for the purpose, for example for maximum effort in study, work, sports, other activities, sexuality; in the search for maximum concentration; to overcome personal fears and obstacles; for release from addictions and risky behavior; and in practice in almost all the purposes that could be hypothesized in a pedagogical intervention as such.

We understand that in 2500 years and more of history of Pedagogy we have to analytically collect a huge collection of ideas, conceptual and operational tools, procedures, lexical forms for the profession of pedagogist. We could start with the Greeks, but without neglecting the Latins, beginning with their philosophies and literatures but not neglecting the ancient science and technique, first of all as methodology of interaction with the natural reality and with the social reality. These methodologies are widely misunderstood today, but they can teach us a lot with a direct access two millennia and more before Galileo, Gilbert, modern science i.e. the science of the modern Age.

On the other hand, Frankl himself considered direct descendants of the Socratic dialogue not only his own logoanalitic technique, but all the Psychoanalysis and any form of therapy by the word; to that same ancient classical source we draw us with the *pedagogical interlocution* [3–6] and other forms of *clinical colloquium* [7], as in the rest draw to the Greek-classical sources for a whole other complex of essential instruments and tools, from the ῾ρητορεία to the πολιτεία, from γνῶθι σεαυτόν to the sense of the human limit with the condemnation of the ῾ύβρις, and away continuing for a long time. The important choice, as has been said, is to not scotomize all the other immense contributions that we can derive from classical Latinity as well as from Greece, and the scientific and technical contributions in relation to the philosophical and literary ones, moreover, thus avoiding a rift in the classical culture that has no historical or scientific legitimacy.

However, what has been summarized and exemplified here can give a sufficiently substantial and organic idea of the profession of Pedagogist and of its culture to wide spectre, which we could also represent with the metaphor of a "toolbox" of a variety and richness that amazes even the most experienced scientists and professionals in the specific field of social sciences and professions.

10 Deliveries to the Professional Practice, Hand off the Baton in an Ideal Relay Race, Instead of a Conclusion

The undeniably nature of Pedagogy as an empirical science, which must be developed in an organic relationship with the educational experience and with the reality object of study and applications, is coherent with the etymon of the term designating this social science of man: a Latin etymon, straddling the Middle Age and the Modern Age: Paedagogia as "Paedagogi ars", the art of taking charge, of caring, the education in people who must be educated as social subjects.

The fact that, then, the Latin word "paedagogus" was a transposition of the greek term παιδαγωγός, and that this in turn derives from the classical Greek παίς— παιδός and ῾άγω, it's a different discourse. There existed in the classical Greek also the word παιδαγωγία which did not designate a science or a λόγος, but the activity of that particular subject who was instructed by the father to accompany the young people, the sons, on the social occasions more favourable for their education which were offered by the πόλις, εσπεχιαλλψ ατ ιτσ ἀγορά, πλαχε οφ πριμαρψ σοχιαλιτψ.

Anymore, it is just like in a relay racing, to hand off the baton transition to those who have the task of continuing the discourse in research and professional practice. Both have no end.

Not by chance, we find this suffix "-agogy" only in pedagogy, and in its subsidiary disciplines "andragogy" and "geragogy", these lasts are words rarely used in italian and in other languages, a real ἅπαξ λεγόμενον among the denominations of sciences and professional disciplines. As known, the main choice goes for the suffix "-logy". The doctors surgeons for their specialties prefer the suffix "-iatry" i.e. therapy, even if they don't miss the suffixes "-logy" (audiology, gynaecology, cardiology, …) and other different choices still (optics, radio diagnostics, childcare, obstetrics, …) as well as various compound words and phrases referring to surgery. This speech is of particular interest, because it explains how Pedagogy cannot be a simple "-logy" i.e. a speech without commitment and application, and in no case is it a "-iatry" i.e. a therapy.

11 Discussion

As for any empirical science of nature, or for any social science of culture, no study can close with a conclusion, but it must pass the discourse to the live of the "future experience", generally speaking: the professonal pedagogical practice [14, 15]. For this author, in our case a professional practice experience, of a helping relationship towards subjects for being in a couple or in a family in problematic situations [4–6].

References

1. Bartley WW III (ed) (1983) Postscript to the logic of scientific discovery. In: Popper KR, Vol. I Realism and the aim of science, Vol. II The open universe, Vol. III Quantum theory and the schism in physics, London. 0091514509
2. Blezza F (2005) Studiamo l'educazione oggi – ça Pedagogia generale del nuovo evo. Osanna, Venosa – PZ. 9788881672400
3. Blezza F (2010) La pedagogia sociale – Che cos'è, di che cosa si occupa, quali strumenti impiega, Liguori, Napoli. 9788820738136
4. Blezza F (2011) Pedagogia della vita quotidiana dodici anni dopo, Luigi Pellegrini editore, Cosenza. 9788881017379
5. Blezza F (2017) Il debito coniugale e altri dialoghi pedagogici, Libreria Universitaria, Limena – PD 2017. 9788862928243
6. Blezza F (2018) Pedagogia professionale. Che cos'è, quali strumenti impiega e come si esercita. Libreria Universitaria. Limena - PD. 9788833590608
7. Crispiani P (2001) Pedagogia clinica – La pedagogia sul campo, tra scienza e professione, Azzano San Paolo BG, Junior, 2001. 9788884340771
8. Dewey J (1897) My pedagogic creed, public domain edition on the Web

9. Dewey J (1916) Democracy and Education, public domain edition on the Web, Chapter Twenty-four: Philosophy of Education
10. Durkheim DÉ (1888) «Introduction à la sociologie de la famille» Extrait des Annales de la Faculté des lettres de Bordeaux, 10, pp. 257 à 281. Disponible in ligne, public domain edition on the Web
11. Durkheim DÉ (1921) «La famille conjugale» Extrait de la Revue philosophique, 90, 2 à 14. Disponible in ligne, public domain edition on the Web
12. Durkheim DÉ (1938) L'évolution pédagogique en France, Cours pour les candidats à l'Agrégation dispensé en.: Paris; 1 re édition 1904-1905, 3e trimestre 1938. Disponible in ligne, public domain edition on the Web
13. Durkheim DÉ (1968) Éducation ct sociologie. Première éditin 1922. Les Presses universitaires de France: Paris; 1968. Disponibles in ligne, public domain edition on the Web
14. Hoskova-Mayerova S (2014) The effect of language preparation on communication skills and growth of students' self-confidence. Procedia Soc Behav Sci 114:644–648. https://doi.org/10.1016/j.sbspro.2013.12.761
15. Hoskova-Mayerova S, Rosicka Z (2012) Programmed learning. Procedia Soc Behav Sci 31:782–787. https://doi.org/10.1016/j.sbspro.2015.04.702
16. Iori V (ed) (2018) Educatori e pedagogisti - Senso dell'agire educativo e riconoscimento professionale, Edizioni Centro studi Erickson, Trento 9788859016038
17. James W (1890) The principles of psychology. 2 vols, Henry Holt and Co., New York. Public domain in the web
18. Le Play PGF (1941) Les Cahiers de l'unité française. Publiés sous la direction de Jacques et René Wittmann, Paris, Éditions d'histoire et d'art, Librairie Plon, Paris
19. Massa R (ed) (1990) Istituzioni di pedagogia e scienze dell'educazione, Laterza, Roma-Bari. ISBN 9788842035695 Sozialpädagogik – Theorie Der Willenserziehung Auf Der Grundlage Der Gemeinschaft. Public domain in the web
20. Orefice P, Corbi E (ed) (2017) Le professioni di Educatore, pedagogista e pedagogista ricercatore nel quadro europeo - Indagine nazionale sulla messa a sistema della filiera dell'educazione informale. Edizioni ETS, Pisa. 9788846746542
21. Peirce CS (1931–1958) Collected papers, 8 vol, Harvard University Press, Cambridge, Massachusetts. Public domain on the Web
22. Popper KR (1935) Logik der Forschung - Zur Erkenntnistheorie der modernen Natur-wissenschaft. Springer, Wien impressum 1935, tatsächlich 1934, edition on the Web, 6. Falsifizierbarkeit als Abgrenzungskriterium. 9783050063782
23. Pólya G (1945) Schule des Denkens. Vom Lösen mathematischer Probleme ("How to solve it"). 3-7720-0608-6
24. Pólya G (1962) Vom Lösen mathematischer Aufgaben. Birkhäuser, Basel. 3-7643-0298-4. Mathematical discovery: on understanding, learning and teaching problem solving, 2 volumes, Wiley. 3-540-04874-X
25. Rosickí, Z, Svoboda, V, Jespersen B, Cheyne C (eds) (2004) Pavel Tichý's collected papers in logic and philosophy. Prague and Otago University Press, Dunedin. 9781877276989
26. Tichý P (1974) On Popper's de finitions of verisimilitude. Br J Philos Sci 25:155–160
27. Visalberghi A (1978) Pedagogia e scienze dell'educazione. Arnoldo Mondadori Editore, Milano. 9788804483656

Architecture Students' Smartphone Use in Design Studio I

Ayten Özsavaş Akçay, Fatma Baysen, and Nermin Çakmak

Abstract The smartphone use is increasing at an unprecedented pace all around the world. Researchers try to reveal its influence on people, particularly the students. The present qualitative research study aims to enlighten smartphone influence on architecture students in Design Studio I (DS I) in an educational perspective. One hundred and thirty-three architecture students coming from different nations attended the present study. The students were in their second semester of architecture department. They answered two open-ended questions. Data is analyzed through content analysis. Except for few applications variation in nationality did not manifest to a significant difference. Students' smartphone use reasons emerged in eight categories, while their self-critics emerged in three categories, and finally recommendations for efficient use of the smartphone in six. Students use the smartphone both for academic and non-academic reasons. Students' success of assignments in the studio I course depends on both academic and nonacademic smartphone uses. Students use the smartphones frequently and would like to use more often. Students would like to use their smartphones to access on-point information, easily, and rapidly without putting in much effort. Students urged actions from their teachers and department managers accordingly, but criticize their use as well. The students may not be aware of the critical targets of DS I. Students need guidance and motivation while using smartphones for DS I to fulfill targets. Recommendations for smartphone use were presented.

Keywords Architecture · Design Studio · Social motivation · Smartphone · Students · Teaching

A. Ö. Akçay
Department of Architecture, Faculty of Architecture, Near East University, Nicosia, North Cyprus

F. Baysen (✉)
Atatürk Education Faculty, Near East University, Nicosia, North Cyprus
e-mail: fatma.baysen@neu.edu.tr

N. Çakmak
Faculty of Letters, Department of Information and Document Management, Ataturk University, Erzurum, Turkey
e-mail: nermin.cakmak@atauni.edu.tr

© The Author(s), under exclusive license to Springer Nature Switzerland AG 2021
D. Soitu et al. (eds.), *Decisions and Trends in Social Systems*,
Lecture Notes in Networks and Systems 189,
https://doi.org/10.1007/978-3-030-69094-6_25

1 Introduction

In 1997, Ericsson presented the term smartphone to the world literature, asserting a new perspective about mobile phones [4, 6]. Since then smartphone technologies improve very fast by adding new features to the smartphones. The additions included Wi-Fi connectivity, full-color screen, touch screen, USB, weight, speed dialing, PC synchronization, number of ringtones, colors, alarm clock, calculator, texting function, embedded camera, calendar applications [4, 6, 31, 35]. Because they possess multi-purposive features and powerful functions, smartphones became part of people lives, and they inevitably changed people behaviors and lives [4, 35]. Smartphones give their users the ability easy access to many services including communication,access information; search information; games; hotel information; reservations of a bus, train, and airplane; health; banking; vacation plans. Smartphones are also used for educational reasons [2, 5, 20, 29, 35]. Research showed that students use mobile technologies more frequently than ever. They want to have the advantage to communicate continuously with their friends and their families, a reflection of social motivation [36]. They want to check messages frequently in emails, Facebook, Twitter and other social networks [8, 35]. Students see a need to use the mobile technologies in the class as well [20, 29, 35]. Working on university students' behaviors [3] found that 95% of the students bring phones to the class, 91% of the students set the phones to vibrate mode, and only 9% turn off their phones. Additionally, 92% of the students stated to send or receive text messages at least once or twice in a class hour. Research also showed that higher education students use their mobiles or smartphones at least 5 h per day [2, 10]. The research found that students are addicted to using smartphone/mobile phones which affect their school performance negatively [1, 2, 10, 13, 17]. Jackson [20] found that students use mobiles mainly for non-academic purposes such as Facebook (39.1%) which is the consequence of social motivation and only 3.1% of the students use for academic reasons. Note-taking, internet searches, and access class materials were the services they use [10, 20]. On the other hand, using smartphones in class is a controversial issue [20, 29]. Some research showed that using smartphones in class distract attention and prevents learning [20, 22, 25]. Additionally, smartphones cause's management problems [29], influence students' academic performance [1–3] and leads to cheating in the exams [20, 29]. On the contrary, researchers stated that students and instructors could use smartphones as a learning tool if used and managed wisely by the educators, students, and school managers. They also stated that the smartphones should be a learner-centric [5, 21, 25, 29, 35].

Research dealt with Design Studio (DS) and the architectural education with a different perspective utilizing mobile technologies in general. Blended learning (combining traditional education with digital and online media) [16, 18, 27, 30], augmented reality-AR [12, 28, 34],m-learning [3, 7, 15] are researched well. Most of these studies stated that mobile technologies support the DS and architecture education. They stated that mobile tools ease the rapid access to information and the materials. Additionally, WhatsApp and Facebook gives students the opportunity to share information with peers, to strength student–teacher relations, to get rapid feedback from the instructor, and forming architectural portfolios [3, 7, 16, 18, 28, 34]. On the contrary, fewer studies showed that students prefer face-to-face teaching because of students' non-academic and unethical behaviors including cheating [3]. Moreover, students declared that they do not prefer to use smartphones, but use a laptop instead, because the screens of smartphones are little [16].

1.1   Design Studio

The DS is the main course and forms the backbone of the Architecture Education. Instructors mostly carry out learning by doing in DS where students acquire design skills and knowledge, under the guidance of the instructor [11, 14]. Students work on projects and try to solve problems created by their teachers. Students continually get critics and feedback from their classmates and tutors [11, 26].

The architecture undergraduate program in the Near East University in Nicosia–Cyprus, Faculty of Architecture, Department of Architecture takes four years, eight semesters including 7 DS. One out of three credits are reserved for DS. DS has a coordinator and students are divided into groups of 15–20 students, and each group has one tutor. Design Studio I (DS I) is a first-year second-semester course in the program and the first design course. Some courses in the curriculum are organized to provide contribution and give support to the DS I. Particular courses supporting DS I include, Architectural Communication, Construction, Building Science, Descriptive Geometry, Materials, History of Architecture, Introduction to Architecture, and Basic Design. Students are expected to reflect the knowledge and skills they acquired in other courses to their designs. Students expected to prepare; two-dimensional drawings, three-dimensional drawings, and models to present their designs [32]. Students submit a final Project at the end of the semester. The evaluation includes all the process during the semester and the final project. Thus through the DS I the students are expected to improve many skills including design-related skills, researching, critiquing, problem-solving, imaginations, synthesizing acquired knowledge, and creative thinking [19].

DS instructors sometimes allow students to work freely in the class. They observed that students use their smartphones during studio. When they ask students to put them away or turn them off, students resist and state that they need the smartphones. Thus depending on student persistence on smartphone use and the innovations in technology, particularly smartphones and growing literature on the usage of mobile phones and smartphones in education in different disciplines, a critical question arises: Should the DS I instruction be changed? Considering students of different nationality in the same teaching environment, the present study tried to enlighten students' understanding to using smartphones in DS I. On the other hand, there is a lack of research dealing with a particular architecture course regarding smartphone in-class use. The present study fulfills a need to consider a course which would welcome smartphone use. The present study focused on smartphone use in DS I from an educational perspective.

2 Method

To explore in-depth ideas about students' smartphone use qualitative research approach was adopted for the study. Data collection, data analysis, and reporting procedures followed the approaches and steps of qualitative research [9].

2.1 Participants

Architecture students (n = 133) attended the study. Of these 84 (53 male and 32 female) are Turkish. Turkish students' age range was 17–24. The other 49 students come from different countries of the Middle East and Africa. Of these 35 are male, and 14 are female. The age range was 18–25. Students were in their second semester. They were taking the Studio I course during the study. Students may find information related to Studio I course on the internet.

2.2 Data Collection

The students answered two open-ended questions. The first question tried to reveal how the students use the smartphone. The question was: "How do you use a smartphone in DS?" Please include the context, the reasoning, how, and timing in your answer. The second question asks for a recommendation to increase the efficiency of smartphone use. The question was, "What do you recommend to ease smartphone use in DS?" Please include the facilities, your tutors, and parents in your answer.

The questions were in two different A4 blank papers giving students big space to answer. Students answered the questions in 15–30 min during their class time. The first researcher was in the class to solve any misunderstanding. She also tried to encourage students to write.

2.3 Data Analysis

Researchers activated content analysis for revealing the themes and the related categories. Researchers worked together reading the raw data first to catch the themes. Three themes appeared in the study. Reasons for using smartphones in the course, students self-critics, and recommendations. After an agreement on the themes and the categories, the researchers finalized the categorization. They then apply the categorization to the remaining data.

3 Results and Discussions

Student responses raised in three themes. Smartphone usage reasons, self-criticism, and recommendations for use. Additionally, students stated duration of use. Smartphone usage reasons emerged in eight, self-criticism in three, and recommendations in six categories. Following section is designed depending on the themes and related categories emerged followed by literature comparisons.

3.1 Theme 1. Smartphone Use Reasons

Students stated smartphone use reasons. The academic and non-academic reasons emerged in eight categories in total. The first two are academic while the rest are non-academic.

Category 1.1. Access academic/architectural information. Students are happy to reach academic/architectural information easily: "I search for project examples", "I search architectural drawing techniques", "I search for models, I tried to be inspired by those models", "I try to solve problems I encounter", "I try to follow novel information about architecture", "I try to get help from Pinterest, which is a good application" "I use Arch digest, "I use Archdaily", "I check my knowledge", "I check the time", "I use helpful files present in the smartphone, stored before." Consistent results can be found in the literature [16, 30].

Category 1.2. Store academic/architectural information. Students use smartphones for storing academic/architectural information about the project process: "I store those my tutors wrote on the whiteboard during the project process."

Category 1.3. Communication. Students use smartphones for communicating which can clearly be explained by the social motivation. "I use my smartphone for emergency situations," "It is important to access others or to be accessible," "I communicate to my family." These results are consistent with previous research [10, 16, 23, 30, 33].

Category 1.4. Rest-relax. Students use smartphones as a resting, relaxing agent. "I relax when I play games during the lesson," "It is a kind of pause for me," "It removes boredom," "I listen to music, it is a kind of refreshment for me."

Category 1.5. Concentration. Students use smartphones as an agent to concentrate on their work, stop the influence of the distractors. "I listen to music which helps me concentrate the process," "I use it to clear my mind."

Category 1.6. Do Operations. Students use smartphones to do operations. "I use a calculator to make some calculations during the process." Other than using calculators international students stated to use dictionaries to translate from English to their own language. "I use a translator to understand meanings of architectural terms written in English."

Category 1.7. Social Media. Students use social media during project critiquing. "There much time, four hours, while our tutor checks our assignments. I enter social media while the instructor controls my classmate's project." Consistent results can be found in the literature [2, 10, 13, 16, 23, 33].

Category 1. 8. Addiction. Students confessed that their smartphone use became an addiction. "I pick my smartphone every hour and unconsciously deal with it," "Smartphone is like a cornerstone for me," "To be honest, I check my smartphone each time the teacher looks somewhere else," "I tried to stay away but never successful." These results are consistent with previous research [1, 2, 13, 17, 23].

3.2 Theme 2. Self-criticism

Students talk about the disadvantages of smartphone use in class. Three hierarchical categories emerged related to self-criticism depending on the intensity of the harm and manageability. Low harm sub-category includes those manageable. Moderate harm category constitutes damages still not vital but more intense and not easy to manage. High harm category includes real damages and uncontrollable.

Category 2. 1. Low Harm. "The smartphone is useful, but I should not use it often."

Category 2. 2. Moderate Harm. "Smartphone distracts me.", "It causes waste of time." These results are consistent with previous research [16, 20, 24].

Category 2. 3. High Harm. "It causes addiction.", "Smartphones are useless." Consistent results can be found in the literature [10, 16, 24].

3.3 Theme 3. Recommendations

Students recommended about smartphone use in class during studio. The recommendations emerged in six categories. The recommendations appeared as Duration of use, Type of Use, Information Share, Support for Research, Infrastructure Development, and Project Group.

Category 3.1. The Duration of Use. Students recommended different durations of smartphone use in Studio. The durations emerged in a hierarchical order: No use at all, limited use, and unlimited use. A student stated: "There is no need to use a smartphone in class, we can use it at breaks." Consistent results can be found in the literature [24]. One other student said: "We may use our smartphones in emergency situations, but limited." One other sated the importance not to distract others attention: "We should use our smartphones with care not to distract our colleagues' attention." Finally, a student for limitless use stated: "Our lessons continue for four hours, teachers should give us permission to use our smartphone." These results are consistent with previous research [10, 23, 33].

Category 3.2. Type of Use. Students suggested two different types of use without a need for permission. The types are Quiet mode and Photographing. Students recommended the quiet mode. They agree to use the smartphone but not disturb others. One student stated, "Teachers should punish those not using quiet mode." Students think that photographing in the class is important. One student said: "They should let us take photos, but not annoying the others." Consistent results can be found in the literature [23, 33].

Category 3.3. Information Share. Students suggested that the tutors construct a site where students and teachers can share information. The information includes examples recommended by the instructors. A student stated: "There should be a site constructed by our instructors, including different kinds of examples." One other student recommended a site specially designed for the course: "The site should be about the Studio, we would ask questions there." A student stated: "Our teachers should share information and help students, they may recommend us sites where we can reach the required information, like Pinterest." One other student stated the role of the department: "The faculty should open an account for the students to help them." Even a student stated the importance of sharing homework: "We can share our homework at the site." Consistent results can be found in the literature [16, 30].

Category 3.4. Support for Research. Students recommended the need for support regarding research. One student stated, "I need information about doing research." One other even go further and stated that: "I want my tutors to be with us while researching and to tell us how to research architectural sites." These results are consistent with previous research [10, 16, 30].

Category 3.5. Infrastructure Development. Students suggested better facilities than what is held: "There is a need for WiFi in the faculty and the classes."

Category 3.6. Project Groups. The students told about the need to form project groups. "We can chat and take support from our tutors, project instructors through Whattsap and Facebook groups." One other student stated the need for project groups: "The project group would make us project focused students." Consistent results can be found in the literature [30].

Moreover, students stated that they use smartphones ranging from every 10 min to continuous use. We can interpret from all above data that architecture students are competent in-class smartphone users. Turkish and the other nationality students shared the same reasoning to use smartphone use in-class. Only one difference appeared between the two cohorts which were about using a smartphone for the dictionary. Students other than Turkish have to use a translator to understand better whats going on in the class. The two groups of students recommendations, critics, and intensity of the smartphone emerged the same.

It is worth to note here that students forwarded access to design related information, to improve design-based skills, and improve design-based research skills. But, they did not mention to improve or to use critical and creative thinking through DS I. Critics is the teacher-centered issue. Additionally, they did not state improving imaginations and synthesizing acquired knowledge.

4 Conclusion

Independent of nationality smartphones became a part of the architecture students' DS I activities. Students find smartphones beneficiary. They use applications special to DS I. Except for few applications nationality does not create a significant difference in student understanding and behaviors. They use smartphones for nonacademic and academic activities in DS I. Students relate nonacademic activities with academic ones. They use nonacademic activities as a refresher and facilitator for success. Academic reasons focused on doing the assignments correct, fast, and smooth. Students would like to use examples, instructor help, and information sharing systems. They focused on accessing not only right information with less effort but to improve design-based skills and design-based research skills. The students may not be aware of the critical targets of DS I including problem solution, critical thinking, creativity, imagination, and synthesizing learned information. They criticize smartphones' wrong use in class, disturbing learning. They have ideas to improve smartphone use in DS I.

5 Recommendations

The DS I students need guidance and support while using smartphones to fulfill targets of the course. They need to put more importance on academic targets, learning by doing; critical thinking; imagination; synthesize learned information to produce original, unique projects; creativity while using smartphones. In that case, the students would use smartphones for academic and real actions, and they may not criticize smartphone use.

References

1. Barks A, Searight HR, Ratwik S (2011) Effects of text messaging on academic performance. J Pedagogy Psychol 4(1):4–9
2. Boumoslesh J, Jaalouk D (2018) Smartphone addiction among university students and its relationship with academic performance. Global J Health Sci 10(1):48–59
3. Brazley MD (2014) How do students learn with mobile technology. US-China Education Review a 4(6):357–371
4. Campbell-Kelly M, Garcia-Swartz D, Lam R, Yang Y (2015) Economic and business perspectives on smartphones as multi-sides platforms. Telecommunication Policiy 39:717–734
5. Casañ-Pitarch R (2018) Smartphone serious games as a motivating resourse in the LSP calssroom. J Foreign Lang Educ Technol 3(2):52–72
6. Cecere G, Corrocher N, Battaglia RD (2015) Innovation and competition in the smartphone industry: Is there a dominant design? Telecommun Policiy 39:162–175

7. Cochrane T, Rhodes D (2011) iArchi[tech]ture: heutagogical approaches to education facilitated by mlearning integration. Education and technology: innovation and research. Proceedings of ICICTE. Retrieved from https://www.researchgate.net/publication/235633454_iArchi techture_Heutagogical_Approaches_to_Education_Facilitated_by_Mlearning_Integration
8. Cole A, Napier T, Marcum B (2015) Generation Z: Fact and fictions. In: Napier T (ed) Not just where to click: teaching students how to think about information. American Library Association, Chicago, IL, pp 107–137
9. Creswell JW (2003) Research design: qualitative, quantitative, and mixed methods approaches, 2nd edn. Sage Publications, Thousand Oaks
10. Derounian JG (2017) Mobiles in class? Active Learn High Educ 1–12. https://doi.org/10.1177/1469787417745214
11. Düzenli T, Alpak EM, Çiğdem A, Tarakçı Eren E (2018) The effect of studios on learning in design education. J Hist Cult Art Res 7(2):191–204. https://doi.org/10.7596/taksad.v7i2.1392
12. Fadzidah A, Mohd Hisyamuddin Bin K, Aliyah Nur Zafirah S (2017) Go virtual: exploring augmented reality application in representation of steel architectural construction for the enhancement of architecture education. Adv Sci Lett 23(2):804–808
13. Felisoni DD, Godoi AS (2018) Cell phone usage and academic performance: an experiment. Comput Educ 117:175–187
14. Goldschmidt G, Hochman H, Dafni I (2010) The design studio "crit": teacher student communication. Artif Intell Eng Des Anal Manuf 24(3):285–302
15. Güler K (2015) Social media-based learning in the design studio: a comparative study. Comput Educ 87:192–203
16. Harpur P (2017) Mobile lecturers, mobile students: an exploratory study in a blended architectural technology context. J Enterprise Inf Manag 30(5):748–778
17. Hawi NZ, Samaha M (2016) To excel or not Excel: strong evidence on the adverse effect of smartphone addiction on academic performance. Comput Educ 98:81–89
18. Hill GA (2017) The "tutorless" design studio: a radical experiment in blended learning. J Probl Based Learn High Educ 5(1):111–125
19. Hoskova S, Mokra T (2010) Alexithymia among students of different disciplines. Procedia-Soc Behav Sci 9:33–37. https://doi.org/10.1016/j.sbspro.2010.12.111
20. Jackson LD (2013) Is mobile technology in the classroom a helpful tool or distraction? A report of university students' attitudes, usage practices, and suggestions for policies. Int J Technol Knowl Soc 8:130–140
21. Jain D, Chakraborty P, Chakraverty S (2018) Smartphone apps for teaching engineering courses: experience and scope. J Educ Technol Syst 1–13
22. Kuznekoff JH, Titsworth S (2013) The impact of mobile phone usage on student learning. Commun Educ 62(3):233–252
23. Mohammad Abu Taleb BR, Coughlin C, Romanowski MH, Semmar Y, Hosny K (2017) Students, mobile devices and classrooms: a comparison of US and Arab undergraduate students in a Middle Eastern University. High Educ Stud 7(3):181–195
24. Mohd Suki N, Mohd Suki N (2011) Using mobile device for learning: from students' perspective. US-China Educ Rev A 1:44–53
25. Odom D (2016) Teaching with smartphones in the higher education classroom. J Youth Ministry 14(2):6–15
26. Oh Y, Ishizaki S, Gross MD, Yi-Luen Do E (2013) A theoretical framework of design critiquing in architecture studios. Des Stud 34(3):302–325
27. Pektaş ŞT (2015) The virtual design studio on the cloud: a blended and distributed approach for technology-mediated design education. Archit Sci Rev 58(3):255–265
28. Redondo E, Fonseca D, Sánchez A, Navarro I (2012) Augmented reality in architecture degree: new approaches in scene illumination and user evaluation. J Inf Technol Appl Educ 1(1):19–27
29. Savery CA (2015) Student mobile devices in class: distruptive or manageable? Int J Humanit Soc Sci Rev 1(1):1–5
30. Schnabel MA, Ham JJ (2012) Virtual design studio within a blended social network. J Inf Technol Constr 17(Special Issue):397–415

31. Shaheen A, Cohen A, Martin E (2017) Smarthone app evolution and early understanding from a multimodel app user survey. In: Meyer G, Shaheen S (eds) Disrupting mobility, lecture notes in mobility. Cham: Springer, pp. 149–164. https://doi.org/10.1007/978-3-319-51602-8_10

32. Svatonova H, Hoskova-Mayerova S (2017) Social aspects of teaching: subjective preconditions and objective evaluation of interpretation of image data. In: Hoskova-Mayerova S, Maturo F, Kacprzyk J (eds) Mathematical-statistical models and qualitative theories for economic and social sciences. Studies in systems, decision and control, vol 104. Springer. https://doi.org/10 1007/978-3-319-54819-7_13

33. Tindell DR, Bohlander RW (2012) The use and abuse of cell phones and text messaging in the classroom: a survey of college students. Coll Teach 60(1):1–9

34. Withell A, Cochrane T, Reay S, Gaziulusoy I, Inder S (2012) Augmenting the design thinking studio. In: Brown M (ed) Ascilite 2012: future challenges, sustainable futures. Wellington, New Zealand: Ascilite, pp 1071–1081. Retrieved from https://aut.researchgateway.ac.nz/han dle/10292/5018

35. Yu F (A.). (n. d.) Mobile/smart phone use in higher education, pp 831–839. https://swdsi.org/swdsi2012/proceedings_2012/papers/Papers/PA144.pdf

36. Zhang KZK, Chen C, Lee MKO (2014) Understanding the role of motives in smartphone addiction. PACIS 2014 Proceedings. Retrieved from https://pdfs.semanticscholar.org/31b0/d1154aa8e11bbcd98fd25fc2f296fdda831e.pdf?_ga=2.39746208.2072551977.1566140252-698949350.1566140252

Gender Equality Between Romanian Difficult Path in Women Political Representation and EU Perspective

Elena Simona Vrânceanu

Abstract Social patterns, cultural specificity or political constraints determined two different perspective regarding gender issues: an occidental one—with a pro-active engagement of women—and an Eastern one—with a declarative reform of women status. In this chapter we propose a theoretical approach of the actual gender political representation in Romania and European Union. The historical path of the Romanian case may help us to understand under-representation phenomenon of women in politics that we are facing in the last three decades. Our chapter looks for an coherent presentation about the terminology and some actions entitled to represent the position of the European Union regarding the gender equality. We state for a European Common Policy when it is about gender equality in legislative and government political presentation which may help states like Romania to assume a pro-active engagement in fulfilling the gender equality objective.

Keywords Gender equality · Romanian path · EU · Women representation · Politics

1 Introduction

The Aftermath of World War II led to e reconfiguration of the gender issues besides numerous life areas. Gender Agenda became intensively important for the political and public decision-makers due to the fact that women were more active in framing the human rights in the international documents and organisations, constituted to represent all individuals and to combat all kinds of discrimination. Our chapter is based on the idea that gender issues is profoundly marked by the two models, Western and Eastern ones, mentioned above and we are facing a huge cleavage between the states inside European Union (EU) because of this differences too, especially in Romanian case. We will start by saying that "gender equality" is a very complex concept which may be presented from different analytical perspectives, starting with

E. S. Vrânceanu (✉)
Faculty of Philosophy and Social-Political Sciences, Alexandru Ioan Cuza University of Iasi, 11 Carol I Boulevard, 700506 Iasi, Romania

© The Author(s), under exclusive license to Springer Nature Switzerland AG 2021
D. Soitu et al. (eds.), *Decisions and Trends in Social Systems*,
Lecture Notes in Networks and Systems 189,
https://doi.org/10.1007/978-3-030-69094-6_26

the ancestral path of the fight against human discrimination, continuing with the fundamental rights and, nevertheless, the issues of political rightness included in gender equality agenda. Gender issue is also a "fashionable" topic that has experienced an extremely interesting evolution in the last century embodied in a real social revolution, from rights that led to changes in the status of women in relation to men, to increasingly varied agendas regarding gender discrimination issues from the ideological, political or social point of view. Gender representation is present in each and every part of our life, starting with anatomical differences and continuing with emotional backgrounds of men and women. We are taught from early childhood that girls and boys have different tastes, different activities, different jobs, different responsibilities in families, in community, different society roles after all. It is all about differences.

2 Gender Topic Through the Lens of Some Evolution Process Challenges

Gender topic has a particularly pronounced resonance in Western societies that share the values of democracy. Gender factuality is no less a matter of politics and institutionally perspective. We assume the idea that gender term is not a synonym for women [1] as we tend to consider when it is about explaining this concept. It is much more a social construct based on behaviour and exceptions provided by different roles of feminity and masculinity: 'Our gender identities, loyalties, interests, and opportunities are intersected and crosscut by countless dimensions of "difference," especially those associated with ethnicity/race, class, national, and sexual identities' [2, pp. 2–3]. Runyan and Peterson have a strong theses regarding the importance of gender topic in the entire spectrum of power distribution by sustaining that worldwide crisis "of representation, insecurity, and sustainability" are determined by important inequalities between race, sex categories, nationality, and class [2, pp. 99–181]. Our cultural heritage, our mentalities, our behaviour or our school education is full of gender differences, some of them being more likely provided in the sense of keeping the men superiority: "societies are configured so that they are in different degrees patriarchal societies (the notable exception to this rule is the Scandinavian countries)" emphasis Mihaela Miroiu, one of the most eminent figure of feminism topic în Romanian actual studies [3, p. 19]. Occidental world developed especially starting with the second part of the 20th Century a real struggle to balance the equilibrium between the main two sex categories: men and women. This kind of struggle has an impressive history of inequalities, of abuses and of ignorance, after all. But, this is not only about men and women, it is about all kind of discrimination towards specific categories, like, race, nationality, class roots, and nevertheless sex appartenance. The feminism of the second wave introduced the gender-gender distinction, the gender taking on the meaning of social and cultural construction. In the same framework of analysis according to "Feminist Lexicon", a Romanian glossary edited by Otilia Dragomir and Mihaela Miroiu, the concept of *gender* is confined to the socio-cultural distinction [4, p. 156]. Gender inequalities are a reality for the evolution of the human being in terms of life opportunities: education, income, rights, etc. Even if new realities are rather to complicated to include gender issues on the main decision-making process agenda, events like migration crisis, Brexit or pandemic crisis being nowadays the most relevant ones, we have to accept that gender topic is correlated with such phenomena at least from the perspective of the vulnerable groups that it includes. In this case we may say that women belong to the category of 'minorities within minorities' [5]. Women as a part of 'minorities within minorities' belong as individuals or groups to a minority culture being vulnerable in front of other competitors in terms of social and political status [6, p. 581]. Until now it is easily to observe that gender equality agenda "followed the first route to equality described in the 'Wollstonecraft dilemma', that of extending to women the same opportunities already enjoyed by men"[7]. Strong voices for gender agenda will miss in European Union (EU) political arena due to the Brexit process: "A circular process is at work: the EU has been a driver for policy and regulation, enhancing the rights of women and minorities of all kinds; the UK, as a long-standing EU Member States (MS), has contributed to that process and in some cases been at the forefront of the drive for change: 'Without the UK as a cog in the larger wheel, change will be determined by a different set of relationships, not always likely to have a positive out-come for women and sexual and gender minorities living around the EU...' [8, p. 468]. UK was an important partner in configuring the EU gender topic and it could have had a significant impact on the new Gender Equality Strategy 2020–2025 document.

3 Reflections Over Romanian Historical Path to Women Political Representation

Women political representation in Romania is correlated with a long and difficult process of social-political reconstruction rules. It started with an increasing incentive of feminist movement, especially after the building of the Romanian nation-state in 1918 and continued intensively during the interwar period and a much greater effort was needed to persuade the politicians to adopt a normative framework that would offer civil and political rights to women. A lot of Associations, Unions, Councils, like Uniunea Femeilor Române (Romanian Women Union), Consiliul Naţional al Femeilor Române (National Council of Romanian Women), Asociaţia pentru Emanciparea Civilă și Politică a Femeilor Române (Association for Civil and Political Emancipation of the Romanian Women) were having a reformatory role in this process of emancipation and an important place in the feminist international movement [9, pp. 11–17]. Calypso Botez-Corneliu is one of the most proeminent Romanian women's rights activist and writer. She founded together with Ella Negruzzi, Elena Meissner and Maria Băiulescu the Association for the Civil and Political Emancipation of Romanian Women (1917). In her paper *Women's Rights in the future*

Constitution [10], Calypso Botez-Corneliu states that the articles from Napoleon's Code present in our national law were totally inappropriate regarding the position of Romanian women, especially since many western states had renounced on them [10, pp. 75–87]. However, only in 1929, the Romanian women won the right to vote and to be voted in the County and Municipal Councils due to the Law of the Administrative Organisation. But, it was just a limited victory because the Electoral Law (1929) contained difficult constraints: 'to have the education of the lower secondary cycle, normal or professional; to be a civil servant of the state, county or communal level; to be widows of war; to be decorated during the war; to have been a part of the Law promulgation stages from the management of a company with legal personality, which was aimed with social claims, cultural propaganda or social assistance' [9, pp. 17–18]. Nevertheless, year 1929 is an important starting point in female political representation process. During this struggle period, newspapers also had a consistent contribution in sustaining the entire emancipation spectrum of women. Publications like Gazeta femeilor (Women Newspaper), Cuvântul femeilor (The Women Word), Graiul femeii (Women Tell), etc. were the written messages of an entire intellectual movement determined to change the Romanian women status in the society. Ten years later, during the dictatorship of Carol II, women over 30 with education involved in industrial, agriculture or intellectual field could participate to parliamentary election and from the age of 40 they could be elected in the Senate Room as it is stipulated in the Law for the Electoral Reform based on the Constitution from 1938 [11]. After another ten years, The Constitution of the Romanian People's Republic since 1948, confirms the universal vote: 'In the Romanian People's Republic the entire state power emanates from the people and belongs to the people. The people exercise their power through representative bodies, elected by universal, equal, direct and secret vote' (Art. 3 of the 1948 Constitution). Unfortunately it was the communist regime that stated declaratively the legislative support, but the reality of the totalitarian era disrupted the continuity of the women social and political representation. Only in December 1989 after the fall of the Nicolae Ceaușescu totalitarian regime, Romania has returned to democracy path. Still, this is not the beginning of a new stage of women political representation, we continue to have 'a masculine definition of the post-communist Romanian political aren' [12, p. 95]. Political under-representation of women in the parliamentary level in the last three decades is correlated with a large area of interpretation, starting with a patriarchal society model, the lack of trust in women political competences and continuing with the educational system which does not encourage political activism. The percentage of parliamentary representation starting with the '90, as they are presented bellow, are showing us a considerable lack in female political representation: 1990–1992—4.9%; 1992–1996—3.7%; 1996–2000—4.7%; 2000–2004—10.8%; 2004–2008—10.2%; 2008–2012—9.8% [12, p. 91], continuing with next legislative 2012–2016—11,40% [13] until now 2016–2020—18,92 [14]. Starting with 2018 the president of the Commission on Equal Opportunities for Women and Men in the Deputies Chamber is Cristina Ionela Iurișniți. She sustain in her official meetings and representations on the national and international level that we should include the representation quotas as affirmative measures in support of equal opportunities and the she also underline the importance of implementing educational policies in this regard, as recommended the European Union and the United Nations [15]: 'We need to eliminate the discriminating stereotypes, which perpetuate from the early ages the reluctance among girls to enter politics or access any other profession traditionally reserved for me'. In 2017 Romanian president explained during The Impact Parity Report meeting in New York that over 1.100 people were trained as experts in equal opportunities which are working with local and central public administration and in 2020, 70% of Romania's national and local public institutions will have experts and technicians in gender equality [16]. We are not talking about a fulfilled objective but it is an important evolution in sustaining the gender equality process.

4 Political Representation of Women in the EU and Future Objectives

It is a huge and impressive historical leap in transforming the social mentalities and politics when it is about gender segregation. More than never this is a time of reflection, debate and decisions for EU which is confronting with a crisis of legitimacy determined by the vulnerable unity that we are having in this unique European construction, marked by cultural diversity, skepticism, conflicts or big breaks, like Brexit. Gender equality is a core value of European Union laid out in the Lisbon Treaty and the EU Charter of Fundamental Rights:

> The Union is founded on the values of respect for human dignity, freedom, democracy, equality, the rule of law and respect for human rights, including the rights of persons belonging to minorities. These values are common to the Member States in a society in which pluralism, non-discrimination, tolerance, justice, solidarity and equality between women and men prevail. (Lisbon Treaty, Article 2, Title I) [17]

As mentioned before, the EU Charter of Fundamental Rights contains two articles regarding gender issues, as follows:

> Any discrimination based on any ground such as sex, race, colour, ethnic or social origin, genetic features, language, religion or belief, political or any other opinion, membership of a national minority, property, birth, disability, age or sexual orientation shall be prohibited. (EU Charter of Fundamental Rights, Article 21—Non-discrimination) and

> Equality between women and men must be ensured in all areas, including employment, work and pay. The principle of equality shall not prevent the maintenance or adoption of measures providing for specific advantages in favour of the under-represented sex. (EU Charter of Fundamental Rights, Article 23—Equality between women and men) [18]

Although in Romania over 51% of citizens are women, their representation at the legislative level is one of the lowest in the European Union. In the same situation are states like Hungary (12%), Malta (15%), Cyprus (18%) or Croatia under 20%. Comparing to Romania, situation in 2019 in EU, women held 32% of seats in national parliaments with an 11% increase from the year of 2003, as Eurostat data are presenting. Only two EU Member States, Finland (58%) and Sweden (52%), have over half of government members female. Spain and Austria (both 50%) as well as France (49%) are also having a high rate of female members in government. In the opposite side of female government representation are the following countries: Malta (9%), Greece (10%), Estonia (13%), Hungary (14%), Poland (15%), Cyprus (17%), Romania (18%) and Croatia (19%) [19]. On March 2020 European Commission adopted *Gender Equality Strategy 2020–2025* as a result of the need to set out a clear perspective regarding the equity between men and women: 'It aims at achieving a gender equal Europe where gender-based violence, sex discrimination and structural inequality between women and men are a thing of the past. A Europe where women and men, girls and boys, in all their diversity, are equal. Where they are free to pursue their chosen path in life, where they have equal opportunities to thrive, and where they can equally participate in and lead our European society' ([20] 152 final). The Gender Equality Strategy 2020–2025 states 6 points:

1. Ending gender-based violence considering that 33% of women in the EU have experienced physical and/or sexual violence;
2. Closing gender gaps in the labour market due to the fact that the difference between women and men's employment rate in the EU is 11.6;
3. Achieving gender balance in decision-making and politics since women represent 7.5% of board chairs and 7.7 of CEOs in the EU's largest listed company and only 32.2% of members of national parliaments in the EU;
4. Integrating a gender perspective in all major Commission initiatives during the current mandate, including the green and digital transitions and demographic change;
5. Funding actions to make progress in gender equality in the EU;
6. Addressing gender equality and women's empowerment across the world (COM[20] 152 final, pp. 3–19).

It is a promising document that focus on a larger area of gender issues and in this effort of putting in practice the Strategy Member States will have a major role in correlating the legislation with the normative framework of the EU. On the other hand it is hard to believe that each state will have the necessary instruments to put in practice such a Strategy which claims professional human resources, institutional design, financial investments, and—why not—mentalities changes. Changing the mentalities seems to be the most difficult process, it takes a lot of time and a lot of determination. In 2019 Gender Equality Index focused on Work-Life Balance scoreboard (WLB scoreboard) evaluated by across three broad areas: paid work, unpaid work (care) and education and training with 15 indicators in six specific areas: parental leave policies; caring for children and childcare services; informal care for older persons and persons with disabilities and long-term care services; transport and infrastructure; flexible working arrangements; and lifelong learning [21]. As far as national perspective with 54.5 out of 100 points, Romania ranks 25th in the EU in terms of gender equality index. The score is 12.9 points lower than the EU average and between 2005 and 2017, Romania increased by 4.6 points [22]

In the last period of time a set of measures have been adopted by the EU in the area of equality between men and women, like The Multi-annual Financial Framework

(MFF 2014–2020) and the program "Rights, equality and citizenship". For 2019 (promoting non-discrimination and equality area) has allocated EUR 37 262 000 in commitment loans. In December 2006 the European Parliament and the Council decided to set up the European Institute for Gender Equality (IEEG). Located in Vilnius, Lithuania, it aims to promote gender equality, including by integrating the gender dimension in all EU policies and in national ones. In 2018, the IEEG chaired the Network of EU Justice and Home Affairs agencies. The network was set up in 2006. It has nine EU agencies dealing with justice and security issues such as migration and border management, combating drug trafficking, organised crime and human trafficking, human rights and fundamental rights, as well as gender equality. An important document was Women's Charter and Strategic Commitment to Gender Equality 2016–2019 which stated: increasing the participation of women in the labor market and equality in what regards economic independence; reducing the difference in pay, earnings and pension differences between men and women; promoting equality between women and men in the decision-making process; combating gender-based violence and protecting and supporting victims; promoting gender equality and women's rights worldwide. Another relevant document is The Gender Equality Action Plan 2016–2020 that stresses the need for women and girls to fully and equally benefit from all human rights and fundamental freedoms. The General Assembly of the United Nations adopted the resolution on the post-2015 Development Agenda, entitled "Transforming our World: The 2030 Agenda for Sustainable Development". Sustainable Development Objectives number five pursues: "Gender equality and empowering all women and girls" [23, pp. 4–6]. On 11 of May 2011 Council of Europe opened for signature, in Istanbul, Turkey *Council of Europe Convention on preventing and combating violence against women and domestic violence* which is already signed by 46 states, including European Union [24]. All this documents and initiatives are showing the EU efforts in sustaining gender equality agenda.

5 Discussions

This chapter emphasise some important perspective regarding the complexity of the gender equality topic in general, and Romanian case in particular. Romanian women struggle for political representation explained here, especially after the Second World War, could be an interesting and useful analyse or maybe at least a premise for next research in this field. For example a future paper might explore in a quantitative perspective a comparisation between Romania and the Eastern block states to whom we share the same political experience during the communist regime and, thus, a lot of similarities determined somehow the same path on the democratisation road, including the field of gender equality. This chapter proposed a qualitative discourse, maybe some more critical perspectives would have been helpful to explain my thesis in this part of the book: it is not enough to have an intellectual assimilation of a terminology like gender political representation, because this debate is correlated with an entire arsenal of practices. In Romania we have a long experience in discussing about a possible solution to impose a quota regime that parties should follow for a much proper political representation, but still, we are in the same stage with a very low participation rate and political representation of women. As mentioned before, EU is a dynamic actor in gender issues representations, thus we are confident in it's power to influence all MS in acting together. Documents like the Gender Equality Strategy 2020–2025 could have a huge impact on the women status if the Member States will implement all the recommendations it has and finally we may achieve the objective of a Common EU Gender Policy.

References

1. Carver T (1996) Gender is not a synonym for women. Lynne Rienner, Boulder, CO. ISBN 13:978-1555873202
2. Runyan AS, Peterson VS (2014) Global gender issues in the new millennium. Westview Press. ISBN 978-0-8133-4917-6 (e-book)
3. Miroiu, M. (2004) Drumul către autonomie. Teorii politice feministe, Polirom. ISBN 973-681-646-X
4. Dragomir O, Miroiu M (2002) Lexicon feminist, Polirom. ISBN 973-683-981-8
5. Eisenberg A, Spinner-Halev J (2005) Minorities within minorities: equality, rights and diversity. Cambridge University Press, Cambridge. ISBN 9780511490224
6. Pinto M (2015) The Absence of the right to culture of minorities within minorities in Israel: a Tale of a cultural dissent case. Laws 4. ISSN 2075-471X, 579–601
7. Lombardo E (2003) EU gender policy: trapped in the 'Wollstonecraft Dilemma'? Eur J Women's Stud 10(2):159–180. Available online at https://journals.sagepub.com/doi/10.1177/1350506803010002003
8. Dustin M, Ferreira N, Millns S (2019) Conclusion: Brexit, gender justice and the overton window in gender and queer perspectives on Brexit. Palgrave Macmillan. (eBook) ISBN 978-3-030-03122-0
9. Mihăilescu Ş (2004) Emanciparea femeii române. Studii și antologie de texte II (1919–1948). ISBN 973-99782-2-3
10. Botez-Corneliu C (1923) Drepturile femeii în Constituția viitoare (Women's Rights in the future Constitution), volume: Noua Constituție a României. 23 de prelegeri publice organizate de Institutul Social Român, Tipografia Cultura Națională, București, (The New Constitution of Romania. 23 Public Lectures organized by the Romanian Social Institute, National Culture Typography, Bucharest,), 75–87, Available at Mihăilescu, Ş. (2004) Emanciparea femeii române. Studii și antologie de texte, (Emancipation of the Romanian woman. Studies and Texts Anthologies), vol II (1919–1948). ISBN 973-99782-2-3, 143-154
11. The Romanian Constitution from 1938. Available online at http://www.cdep.ro/pls/legis/legis_pck.htp_act_text?idt=9206
12. Băluță I (2012) Femeile în spațiul politic din România postcomunistă: De la "jocul" politic la construcția socială (Women within the Political Space of post-Communist Romania: From the Political "game" to Social Construction) Annals of the University of Bucharest/Political science series, 14(2):87–95. ISSN1582-2486. Available online at https://www.ssoar.info/ssoar/handle/document/39002

13. Ministerul muncii, familiei, protecţiei sociale şi persoanelor vârstnice (Ministry of Labor, Family, Social Protection and the Elderly) (2012) Reprezentarea femeilor şi a bărbaţilor în alegerile parlamentare din 2012. Situaţia candidaturilor şi a mandatelor atribuite în urma validării alegerilor (Representation of Women and Men in the 2012 Parliamentary Elections. The Situation of the Applications and Mandates Assigned after the Validation of the Elections) (2013î) Available online at http://www.mmuncii.ro/j33/images/Documente/Familie/ESFB-StudiiAnalizaRapoarte-2012/Reprezentarea_femeilor_si_a_barbatilor_in__alegcrile_parlam entare_din_decembrie_2012.pdf

14. Agenţia Naţională pentru Egalitatea de Şanse (National Agency for Gender Equality) (2016) Reprezentarea femeilor şi a bărbaţilor în alegerile parlamentare din 2016 (Representation of Women and Men in the 2016 Parliamentary Elections). Available online at http://www.mmu ncii.ro/j33/images/Documente/MMPS/Rapoarte_si_studii_institutii_din_subordine_coordo nare_autoritate/ANESFB/2017ANALIZA_ALEGERI_PARLAMENTARE.pdf

15. Comisia pentru egalitatea de şanse pentru femei şi bărbaţi, Camera Deputatilor, (Commission on Equal Opportunities for Women and Men, Deputies Chamber) (2020), Press releases and statements. Available online at http://www.cdep.ro/pls/parlam/structura2015.co?idc=20& cam=2&leg=2016&pag=cprs

16. Intervenţia naţională a Preşedintelui României, domnul Klaus Iohannis, susţinută în cadrul lansării Raportului HeForShe 10x10x10 IMPACT Champions (National intervention of the President of Romania, Mr. Klaus Iohannis, supported during the launch of the HeForShe Report 10x10x10 IMPACT Champions) (2017). Available online at https://www.presidency.ro/ ro/presedinte/agenda-presedintelui/interventia-nationala-a-presedintelui-romaniei-domnul-klaus-iohannis-sustinuta-in-cadrul-lansarii-raportului-heforshe-10x10x10-impact-champions

17. Lisbon Treaty, Article 2, Title I, C 202/1 (2016) Available online at https://eur-lex.europa.eu/ legal-content/EN/TXT/HTML/?uri=CELEX:12016ME/TXT&from=EN

18. EU Charter of Fundamental Rights, C 202/389 (2016) Available online at https://eur-lex.eur opa.eu/legal-content/EN/TXT/HTML/?uri=CELEX:12016P/TXT&from=EN

19. Eurostat (2020) Seats held by women in national parliaments and governments. Available online at https://ec.europa.eu/eurostat/databrowser/view/sdg_05_50/default/table?lang=en

20. European Commission, Communication from the Commission to the European Parliament, the Council, the European Economic and Social Committee and the Committee of the regions. A Union of Equality: Gender Equality Strategy 2020–2025, COM(2020) 152 final. Available online at https://eur-lex.europa.eu/legal-content/EN/TXT/PDF/?uri=CELEX:52020DC0152& from=EN

21. Work-life balance (2019) Available online at https://eige.europa.eu/gender-equality-index/the matic-focus/work-life-balance

22. Gender Equality Index in EU States (2019) The European Institute for Gender Equality. Available online at https://eige.europa.eu/gender-equality-index/2019/compare-countries/index/ map

23. EU technical sheets (2020) Equality between men and women. Available online at www.eur oparl.europa.eu/factsheets/ro

24. Council of Europe (2020) Chart of signatures and ratifications of Treaty 210, Council of Europe Convention on preventing and combating violence against women and domestic violence. Status as of 13/04/2020. Available online at https://www.coe.int/en/web/conventions/full-list/-/ conventions/treaty/210/signatures

25. The Romanian Constitution from 1948. Available online at http://www.cdep.ro/pls/legis/legis_ pck.htp_act_text?idt=1574

Senior Citizen Centres and Sexual Affective Diversity: Homophobia and Residents

Javier Mesas Fernández, Evaristo Barrera Algarín⊙, and Ana Vallejo Andrada⊙

Abstract This chapter talk about the management of sexual diversity in senior citizen centres. It is centred on elderly's opinion about homosexuality and its goal to see if homophobia exists in senior citizen's centres between the residents. It is a part of an extender research about homophobia. We have decided to use the quantitative technique of the questionnaire, it has been created with Homophobia Scale (HS) [1], developed at the University of Georgia and some demographic variables, used in a sample of 23 nursey home residents as our research method. We have make the data correlations using the SPSS program, concluding the existence of homophobia in nursing homes.

Keywords Senior citizen centres · Nursing homes · Elderly · Sexual diversity · Homophobia · Homosexuality and residents

1 Introduction

This study intentionality stems from the evidence of seeing how society does not change, what changes is the problem, previously considered domestic or individual that turn into a social topic. Becoming a necessary social area of study. Society changes, but institutions do not always get adapted at the same time. Because of that, they tend to reproduce erroneous behaviours which could have consequences. In this

J. Mesas Fernández · A. Vallejo Andrada (✉)
Department of Social Work and Social Services, Pablo de Olavide University, Ultra Road, 1, 41013 Seville, Spain
e-mail: avaland@upo.es

J. Mesas Fernández
e-mail: jvrmss@gmail.com

E. Barrera Algarín
Pablo de Olavide University Utrera Road, Ultra Road, 1, 41013 Seville, Spain
e-mail: ebaralg@upo.es

© The Author(s), under exclusive license to Springer Nature Switzerland AG 2021

D. Soitu et al. (eds.), *Decisions and Trends in Social Systems*,
Lecture Notes in Networks and Systems 189,
https://doi.org/10.1007/978-3-030-69094-6_27

case, this situation could be reproduced in the Gerontological Centres. This will be the framework in which we will work and, specifically, in the study of possible cases of homophobia.

Tamam and Diler [8] in his article "Homosexuality and suicide" illustrate how the current violence which is experienced in the Nursing Homes is increasing and how nowadays there are situations of extremely gravity, unfortunately increasing to cases of suicides. In addition, people suffer a setback and are forced to "go back into the closet".

The aim of this chapter is to obtain a sample which will show if there is a possible situation of homophobia and the management degree of sexual diversity in gerontological centres. As our society has experienced a process of growing acceptance of sexual minorities in recent years [3, p. 304].

Even though LGBT associations and foundations have made great advancements to claim rights (acceptance, legal equality, marriage …), it seems that the field of old age has always been left aside. We can say that some of the reasons for this are:

- Prejudices about old age
- The low percentage of people in residences "who have come out of the closet". This fact is going to change and increase with the passing of the years.
- "The Eternal youth" and worry about social, physical, personal care, etc.
- The false belief that older adults are unable to feel desire or pleasure (…) due to the aging of the body it loses its appeal so seduction and sexual activity at this stage of development would have no meaning [11, pp. 122–123].

2   Homosexuality and Transsexuality: Evolution and Consequences

Nowadays, the discrimination suffered by people for reasons of sexual diversity is one of the most serious, crimes based on sexual orientation and sexual identity discrimination have the highest incidence. We can see discrimination in universities, companies, public institutions, NGOs, foundations, health, etc.

Homophobia and transphobia happen when a single person or a group of them, institutions or entities exercise towards other people, due to their sexual orientation or gender identity, actions or attitudes that promote the segregation of the LGBT collective, using prejudices and/or stereotypes. To understand a bit more these behaviours, we are going to analyse the origins and different social aspects and institutions that, with their respective actions, have fed back this discrimination [10].

The origins do not have a source or an exact date. The behaviours of homophobia and transphobia are produced because of different historical events and centuries of social relations, probably due to the fact that they were happening and there were no policies to be able to face them in a holistic way to the historical context and time.

There are following Pichardo's researches [6], different types of homophobia:

- *Cognitive homophobia*: based on ideas and concepts that people have about homosexuality.
- *Affective homophobia*: related to feelings. It is the most unconscious of the three types of homophobia, because rejection develops when a homophobic person imagining another person being related with homosexual people.
- *Behavioural homophobia*: It is the most conscious of the three types of homophobia and the one that it is related to the behaviours towards the LGTB collective. It can manifest from micro homophobia (comments) to physical aggression.

This type of discrimination is based on the ignorance of people about other sexual affective forms and the rejection of everything that comes out of heteronormativity. It is interesting to analyse these behaviours through the two schemes proposed by Rubin [7]. These schemes talk about how moral hierarchies are constituted around sex:

- The first scheme is called "*The magic circle*", It is a circle which shows us how the power is governed by the heteronormativity. The circle is divided in two parts, the smaller one is located at the centre. It is what we call natural and sacred sexuality. This must be heterosexual, monogamous, in marriage, procreative, non-commercial, in a couple, in a relationship, between members of the same generation, in private, only with bodies and without pornography. Outside this small circle we have a bigger one which represents the rest of the sphere, which is considered "the bad sex": homosexual, out of wedlock, promiscuous, not procreative, commercial, solitary or in groups, sporadic, intergenerational, in public, with objects or animals, with pornography and sadomasochistic.
  It shows us how sexual diversity, in the end, focuses on ethnocentrism through a Eurocentric and normative classification. The construction of these dichotomies indispensable to be able to analyse how society sees these behaviours and create identities. The construction of identities, following these practices are based on prejudices and, therefore, people loses their social status, because of the great sexual stigma.
- In the second scheme of Rubin [7], "*the wall scheme*" the author shows us the social borders of sexuality and how rejection displaces people. Again, we see what is good and bad sexually and how the discursive walls of double morality mean that in all social areas the stigmas keep repeating themselves.
  This time, the author instead of dividing the society in spheres, she has divided it into three different walls; the first wall describes the heterosexual, marriage, monogamous, procreative sex, and all this at home. Drawing a line between what is good and what is not. Through this line another wall divides the bad society in two parts, the first one which is more accepted than the other one which is compiled by unmarried heterosexual couples, promiscuous heterosexuals, masturbation, and homosexual stable partners. The last wall is made up by the most despised groups of society: Transvestites, transsexuals, fetishists and sadomasochists.

Moving these walls/frontiers in favour of sexual affective freedom is of vital importance in order to fight against discrimination. We see how these discriminations negatively mark people and the only thing they promote is social rejection. Some of societies examples of impediment on moving or destroying these walls or frontiers are:

- *Religions*: There are different applications according to religions. Some of them are clearly against homosexuality, but others are not. It should be noted that there are different groups of religious homosexuals, such as, for example, "The Catholic Association for Lesbian and Gay Ministry (CALGM) in Berkeley, California".
- *Science*: science considered for many years' homosexuality and transsexuality as a disease treated by different professionals in the health field. Today this is no longer the case, but ignorance and what remains of these ideas in the minds of many people still makes them think that it remains a pathology that must be treated.
- *States*: Countries have for centuries had a great parity between their actions and the opinions of religions and science. An example of this are State laws that typify homosexuality as a crime. Today the death penalty still exists in several countries.

History also shows us how the generation that is currently in nursing homes corresponds to those who have had to suffer to express their orientation and/or gender identity. They have lived past historical moments of the greatest difficulty and how they are now doomed to "hide" again.

Maddux made a documentary called "Gen Silent" in which he recounts six different experiences. It is a documentary about gay, lesbian, bisexual and transgender elderly people. It tells us how they face discrimination in care due to their sexuality and their condition as elderly. You can see how his generation has had to go through the largest historical struggles LGTB in view of the recognition of their rights and their collective identity [5].

The idea that the identities, both individual, collective or public, are rebuilt permanently through social interaction is commonplace both from a theoretical perspective of social construction of protesting as much as from symbolic interaction [9].

3   Methodology

This research is part of an extended research about the possible homophobia existing in the professionals and in the users of Nursing Homes. On this occasion, we have decided to focus on the results gotten from the residents. The survey has been chosen as an instrument to measure, with the intention, to determine if discrimination exists or not being this the main objective, whereas this study has other objectives.

General Objective:

To determine how sexual diversity is perceived and managed in Gerontological Centres, in particular the acceptance of homosexuality by users.

Specific Objectives:

- To determine the social acceptance among users of sexual diversity.

- To identify the possible existence of rejection in the centre of homosexual persons.
- To analyse possible solutions to the existing problems.
- To distinguish manifestations of discrimination and their impact degree.
- To determine the existence, or not, of homophobia both directly and indirectly.
- To identify the different forms of discrimination in the context of gerontological centres.

Having the following hypothesis:

1. Homophobia exists in gerontological centres.
2. There are more manifestations of homophobia in men than women.
3. There is a lack of management of sexual diversity by gerontological centres.

In relation with the survey with the aim to obtain a greater validity, we wanted to use the Homophobia Scale (HS) [1], developed at the University of Georgia. The questionnaire consists of 25 questions which the respondents must answer on a Likert scale, where 5 is completely agree and 1 is total disagreement. The current scales towards homosexual persons affect the attitudes within the heteronormativity, therefore, it is considered the evaluation is needed in a more incisive way.

The questionnaire used differs from others in including elements of social avoidance and aggressive actions towards homosexual people, these attitudinal elements are within homophobia. There lies the importance of this study, to analyse thoughts, feelings and behaviours linked to homosexuality and homophobia.

In Italy, [4, pp. 213–218] in 2015, a validation of said questionnaire was carried out. For the validity of the scale and its optimal results, the scale was used in the study. The internal consistency analysis showed a Cronbach's α coefficient of 0.92 (Table 1).

Our research variables are:

As we can see in Table 2 our sample is composed by men and women of two different Nursing Homes:

In relation with the age of our responders 52% of the people who have done the questionnaire are men and 48% women. In Table 3, we can see the great difference between people between 80 and 89 years old and people aged 60 and 69 years, the former being notably superior in terms of percentage.

Table 1 Research variables

Research variables		Items
VCU001	Total score HS	1–25 (scale no1)
VCU002	Behavior/negative affect	1, 2, 4, 5, 6, 7, 9, 10, 11, 22 (scale no2)
VCU003	Affect/aggressive behavior	12, 13, 14, 15, 17, 19, 21, 23, 24, 25 (scale no3)
VCU004	Negative cognition	3, 8, 16, 18, 20 (scale no4)
VCU005	Sex	1–25
VCU006	Age	1–25

Source Author compilation

Table 2 Sample (age and nursing homes)

	Residents			
Centro de día Virgen de los Reyes	Men	Men	4	
	Women	Women	6	10
SARquavitae Santa Justa	Men	Men	8	13
	Women	Women	5	
Total: 23				

Source Author compilation

Table 3 Sample (age)

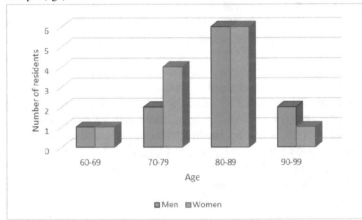

Source: Author Compilation

4   Results

In this section, we are going to focus on the results of our research and the possible relationships between the factors which could be interesting for our conclusion.

In Table 4 we can see the four main variables and the α of Cronbach obtained in each one of them.

We can appreciate that, the index of higher answers with negative connotations refers to the male sex. Therefore, it is estimated that the degree of homophobia is higher among males than females.

The items in which the negative response rate has been higher in women have been the following:

- Item 12: Homosexuality is immoral.
- Item 23: When I see a gay person, I think "what a waste".

The items in which a positive answer rate has been higher in men, have been:

Table 4 General results

VCU001		VCU002		VCU003		VCU004	
HS behavior/negative affect		Behavior/negative affect		Affected/behavior		Negative cognition	
K	25	K	10	K	10	K	5
Vi	49.57	Vi	19.95	Vi	20.25	Vi	9.36
Vt	183.31	Vt	49.90	Vt	132.81	Vt	15.78
Sec. 1	1.042	Sec. 1	1.11	Sec. 1	1.11	Sec. 1	1.25
Sec. 2	0.73	Sec. 2	0.6	Sec. 2	0.85	Sec. 2	0.41
Absolute S2	0.7	Absolute S2	0.6	Absolute S2	0.85	Absolute S2	0.41
α the cronbach	0.76	α	0.67	α	0.94	α	0.51

Source Author compilation

- Item 8: Marriage between homosexual people is acceptable.
- Item 22: It does not bother me to see a gay couple together in public.

We can appreciate that, in all the items, except those above mentioned, homophobia is higher on the part of men than women. Of all the questions 16 have a negative tendency, negative connotations close to homophobia. Versus a total of 9 questions, which result, provides positive tendency towards homosexuality.

With the obtained data, we can see in Table 5; in 87.5% male sex answers, compared to 12.5% of female sex, have closer responses to homophobia. On the other hand, in the total of questions with tendencies closer to homosexuality, we obtain a percentage of 22.2% in the case of the male sex and 77.8% in the case of the female sex.

Table 5 Sex results

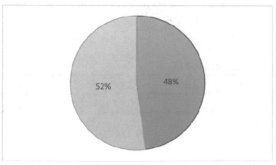

Source: Author Compilation

Table 6 Degree of acceptance of homosexuality

	Total	1	2	3	4	5
Residents	3.4	1	5	7	4	6

Source Author compilation

1.   Most notable answers where the need for action is evident.

Remembering that number 1 corresponds on the Likert scale to totally disagree and number 5 to totally agree.

- Item 3. I accept homosexuality

In Table 6 we can see that most of the residents interviewed, give 3/5 in relation of the degree of acceptance they have of homosexuality, being 5/5 and 2/5 the other two popular answers.

- Item 4. If I discover that a friend of mine is homosexual, I would end my friendship with him/her.

As it is represented in Table 7 the higher percentage of interviewed consider they would not end a friendship with someone because of his/her sexual orientation, however the next popular answer is 5/5, which mean they are totally agreeing to end a relationship with someone because of his/her sexual orientation.

- Item 9. I make negative comments such as "fagot" and "Queer"

Analysing Table 8, we can say that, 11/23 of the elderly people who have answered our survey do not make negative comments referring to homosexual people, follow by 5/23 who has given a "2" to this statement, 1/23 with a "1", 2/23 with a "4" and 4/23 with a "5".

- Item 13. I despise gay people and I make fun of them.

Table 7 Respect of a homosexual friend

	Total	1	2	3	4	5
Residents	2.7	8	5	2	1	7

Source Author compilation

Table 8 Negative comments

	Total	1	2	3	4	5
Residents	2.3	11	5	1	2	4

Source Author compilation

Table 9 Degree of verbal aggressions

	Total	1	2	3	4	5
Residents	1.7	19	1	1	0	3

Source Author compilation

Table 10 Degree of physical aggressions

	Total	1	2	3	4	5
Residents	2.8	6	6	2	5	4

Source Author compilation

In relation if the elderly people despise gay people and make fun of them, as it is show in Table 9, 9/24 completely disagree with this statement, 1/24 disagree and do not know for sure and 0/24 agree and 3/24 completely agree with this affirmation.

- Item 19. I would hit a homosexual if he/she tried to flirt with me.

In Table 10 we can see that 6/23 of people surveyed give a 1/5 to this affirmation, also 6/3 pointed with 2/5 this statement, 3/23 give a 3/5, 5/23 a 4/% and 4/23 a 5/5.

5 Discussion

After a deep results and theory analysis, we can remark some similarities between the previous mentioned theory and our results.

Related with Pichardo's theory about the three types of homophobia, we have: cognitive homophobia, affective homophobia and behavioural homophobia. Despite some of our surveyed answers could be modify depending what they consider the surveyor would consider correct, being the affective homophobia the most unconscious one of the three [6], our survey would not be able to control their answers. As we can see in the answers in the Table 7, a high number of those surveyed would stop a relationship with someone if they discover this person is homosexual. This question is a remarkable link with the affective homophobia.

Tables 9, 10 and 11 are more related with the behavioural homophobia, being this type of homophobia the most conscious ones, the surveyed can control their answers and their behaviours with the homosexual collective, being this the reason of what the answers of Table 7 are so different to the answers of Tables 9 and 10.

All the results are extremely link with Rubbin's theories about the different sexual frontiers and Maddux work Silence Gender [5] As we can see in our results the homophobia still exists in our society, especially in nursing homes, making the homosexual couples who have fight during the last decades for their rights, fight again in this new social sphere.

6   Conclusions

Doing this study, we have tried to determine how sexual diversity is perceived and managed in the Gerontological Centres, focusing the study on the acceptance of homosexuality by the users.

Regarding the objectives of the study, the main objective: how it is perceived and managed diversity in gerontological centres, could be determined that perception and management is very scarce. However, The Centres are starting to perceive it as a necessity. The specific objectives have been covered by the instruments and techniques presented throughout the study. In relation with the degree of management of sexual diversity, it is very low.

If sexuality in the elderly is a taboo element [2], which is often ignored or even stigmatized, in the case of sexual diversity this reality is accentuated.

Homophobia exists, among residents and workers, to a greater or lesser extent, in some aspects the graphics provided after completing the questionnaires show where more emphasis should be placed. To conclude, and as the last specific objective, we see that homophobia, above all, occurs verbally.

In conclusion; As possible solutions to the existing problem, we perceive the need for inclusive measures and regulations, training and awareness days, as well as awareness campaigns for users and professionals, training in sexual diversity for workers of gerontological centres and creation of protocols for action and prevention.

Another measure could be creating more research in this area, about the necessity of create protocols and needs that should be included in gerontological centres, advantages and disadvantages of creating exclusive spaces for LGBT + people, needs that may exist and indicators of acceptance to other types of sexual diversities, family and social support of older people with sexual diversities, feelings of grief and loneliness in LGBT + seniors.

Promoting spaces for reflection on sexual diversity in gerontological centres is of vital importance. In this way, differences will not make us separate, on the contrary, that they unite and complement us.

References

1. Adams HE, Wright LW Jr, Lohr BA (1996) Is homophobia associated with homosexual arousal? J Abnorm Psychol 105(3):440
2. Berlant L, Warner M (1998) Sex in public. Crit Inq 24(2): 547–566
3. Calvo K (2003). Actitudes sociales y homosexualidad en España, in Guasch & Viñuales (2003). Sexualidades. Diversidad y control social. ISBN 978-84-7290
4. Ciocca G, Capuano N, Tuziak B, Mollaioli D, Limoncin E, Valsecchi D, Jannini, E. A. (2015). Italian validation of homophobia scale (HS). Sexual medicine. [online 22-03-2017]. Available online at http://onlinelibrary.wiley.com/doi/10.1002/sm2.68/full
5. Maddux (2010) Gen Silent. [online 27-01-2019]. Available https://www.imdb.com/title/tt1605 721/
6. Pichardo JI (2009) Entender la Diversidad Familiar: Relaciones Homosexuales y Nuevos Modelos de Familia. ISBN 978-84-729-0459-0

7. Rubin G (1989) Reflexionando sobre el sexo: notas para una teoría radical de la sexualidad. Placer y peligro: explorando la sexualidad femenina. [online 27-01-2019]. Available online at http://www.museoetnografico.com/pdf/puntodefuga/150121gaylerubin.pdf
8. Tamam L, Diler RS (2001) Homossexuality and suicide: a case report. Am J Pub Health. [online 08-04-2017]. Available online at http://www.scielo.br/scielo.php?script=sci_arttext&pid=S0104-12902012000300011
9. Tomàs JMM (2003) La construcción de la protesta en el movimiento gay español: la Ley de Peligrosidad Social (1970) como factor precipitante de la acción colectiva. Reis 178. [online 01-01-2019]. Available online at http://www.jstor.org/stable/40184541?seq=1#page_scan_tab_contents
10. Vallejo A, Hoskova-Mayerova S, Krahulec J, Sarasola JL (2017) Risks associated with reality: how society views the current wave of migration: one common problem—two different solutions. Studies in Systems, Decision and Control 104, 283–305. Springer International Publishing. https://doi.org/10.1007/978-3-319-54819-7_19
11. Verdejo C (2009) Sexualidad y envejecimiento. ISBN 978-84-8446-120-3

Promoting Social Inclusion in Romanian Schools

Cristina Ispas and Alina Vişan

Abstract During the time, the history was full of countless cruel facts committed against the disabled people, starting from their physical elimination during antiquity up to their abandonment in the streets, forced to live on the passers-by mercy or to be imprisoned or closed down in social institutions, thus, they should live very far from society's eyes. The end of the twentieth century brings a radical improvement of the way we perceive people with special educational needs, by recognizing the human value and dignity, insisting on their unalienable right to education next to the other members of society. In this context, the schools with inclusive orientation represent the most efficient ways to combat discrimination and the creation of a community attentive to the needs of all its members. The inclusive schools offer the students with special educational needs opportunities to participate in and receive support in all aspects of school life next to their colleagues. The integration of the children with special educational needs in the public schools constitutes a relatively recent practice, specific for the last three decades. In the present paper, we identify and analyse the perception of the didactic staff in Romania referring to the opportunity of the Romanian school to be an inclusive school, a school for all, highlighting the way the Romanian school is perceived as being or not an inclusive school. Therewith the paper also identifies the opinion of the didactic staff from Romania regarding the characteristics of an inclusive school, the degree of promoting the inclusive school in the community by the Romanian educational system, as well as the involvement degree of the family in the school situation improvement of the students with difficulties in learning.

1 Introduction

According to the World Health Organisation the number of the disabled people at world level is of approximately 1 billion people meaning 15% of all the population.

C. Ispas (✉) · A. Vişan
Babes- Bolyai University of Cluj-Napoca, University Center Reşiţa, Reșiţa, Romania
e-mail: c.ispas@uem.ro

© The Author(s), under exclusive license to Springer Nature Switzerland AG 2021
D. Soitu et al. (eds.), *Decisions and Trends in Social Systems*,
Lecture Notes in Networks and Systems 189,
https://doi.org/10.1007/978-3-030-69094-6_28

"Today, around 33 million children with disabilities are not in school—more than 50% of the total number of out-of-school children".[1] In these conditions, we should imagine a careful rethinking of all the policies destined to these people, including the educational ones.

"During the human history, the difference has often had a negative connotation, including in the educational space, until the half of 20th century".[2] In the paper *Le abilità diverse. Percorsi didattici di attività motorie per sogetti diversamente abili,* the teacher Maurizio Sibilio[3] presents a history of the way in which the disabled children were treated during the time. During Antiquity, the malformed children were eliminated; with the coming of Christianity, these were abandoned in the street, being let to live exclusively on the mercy of the passers-by. During the middle ages, the disabled people were marginalised and during the 16th–17th centuries they were inprisoned; they consider that letting them free would provoke a horror, because they were perceived as dangerous for society. With the enlightenment there is a change in the perception of the people with disabilities. The disabled children were hosted (interned) in different institutions, hidden from the public eye, like a bad thing that should be kept away from the others. However, for the first time in the history of society, the abnormality was seen as a human condition which did not affect the human dignity. The Declaration of the rights of man and of the citizen elaborated during the French Revolution (1789) promotes the equality right between all the people regardless of their social, gender, race, religion, physical and psychic conditions etc. For a long period of time the formal education of deficient people was developed in the special schools "destined to receive pupils who by their particular conditions were considered as inappropriate for attending normal schools."[4]

National education law No.1/2011[5] states that "the State ensures to the citizens of Romania equal rights to have access to all the levels and forms of pre-university and higher education and to the lifelong learning, without any form of discrimination", also the "State guarantees the right to education of all the people with special educational requirements."[6]

As a consequence of the Salamanca Declaration (UNESCO 1994), they started to discuss about inclusive education and inclusion in the international context. Thus, many countries promoted the educational policies which award a special attention to the inclusive education and implicitly to the inclusive school with reference especially to the education of the pupils with special educational needs and their integration in the mass school system.

---

[1] The Education Commission. The learning Generation. Investing in Education for a changing world, 2017. https://report.educationcommission.org/.

[2] Horga and Jigău [12].

[3] Sibilio, M. *Le abilità diverse. Percorsi didattici di attività motorie per sogetti diversamente abili,* Gruppo Editoriale Ellissilibri—Simone.

[4] Dovigo [3].

[5] *Legea Educaţiei Naţionale,* Art. 2 (4), disponibil la https://oldsite.edu.ro/index.php/articles/14847, accesat la 04.10.2016.

[6] Idem, Art. 12 (6).

"The rising gaps between children with and without disabilities in developing countries call for stronger policies and interventions to achieve the target of inclusive education adopted under the Sustainable Development Goals."[7]

The definition of the concepts of *inclusion* and *inclusive school* includes a series of difficulties which come from the difference of perspectives and interpretations given to these concepts in the specialised literature. Trying to identify these perspectives which define the concept of *inclusion* Ruth Cigman (2007)[8] highlights three schools/thinking positions at European level: radical school, moderate school and school represented by the UNESCO conception. The radical school/position highlights the need to intervene in the school environment in order to eliminate all the structural, cultural, social and economic barriers which hinder the active participation of all the students to the learning process achieved in school. Likewise, this radical position supports that only the closing of all the special schools can determine the achievement of all the necessary transfers to the common schools, changes meant to ensure a quality education for all the students with or without disabilities. The moderate school/position considers that the educational system must still keep the special schools for the children with severe deficiencies and continuously achieve the integration of those with mild deficiencies in the common educational system. The UNESCO position analyses the concept of inclusion connected to the concept of education for all and "puts the accent on the need to combat discrimination and exclusion from the learning processes and to help the countries to develop schools which should be able to answer the different exigencies of all the students including the disabled students especially those countries who are forced to daily confront with wars, poverty, epidemics, shortages."[9]

Inclusive education is a "permanent process of improving the school institution, aiming at the exploitation of existing resources, especially the human resources, in order to support the participation in the learning process of all the people within a community." Inclusive education offers learning opportunities for all children, youth people and adults regardless of race, ethnic group, language, religion, gender, sexual orientation, economic status, social class, segregation to promote the principles of education for all [4].

Ensuring the equal access to education for all citizens represents an act of high responsibility and involvement for those who activate within the educational system having in view the elaboration and implementation of individualised and personalised teaching–learning-evaluation programs in the spirit of inclusive didactics, centred on the student [6].

Inclusive school/education expresses a large framework which also contains the integrated education. "It is important to clarify a strong difference between the integration and inclusion practices. The integration is addressed to disabled students,

---

[7]Male and Wodon [16].
[8]Medeghini and Fornasa [13].
[9]Medeghini and Fornasa [13].

meaning to a part of those with special educational needs, while the inclusion refers to different practices of individualised answer achieved for all the educational needs of all the students with special educational needs. Thus, inclusion is more ample than integration."[10]

The paradigm of school inclusion puts accent on the formation needs of all students, without exception, respecting the principle of equality of chances and active participation of everyone in their own formation, to contribute to the wellness of society. Consequently, the concept of "inclusion" applied in the educational field sphere must not be reduced (as it sometimes happens) only to the education of the students with special educational needs, but inclusion refers to the education of *all* children/young people/adults with or without special educational needs. Such an approach of education, through the filter of values which support inclusion, stresses the improvement of the educational environment, on the adequate formation of didactic staff in the spirit of inclusive values, on the use of inclusive didactic strategies meant to help each student to develop his/her own potential to his/her adequate capacity, on the encouragement of the active participation of each subject to his/her own formation.

Delivering inclusive education and addressing the global learning crisis require clear solutions and actions.[11]

The paper *L'Index per l'inclusione. Promuovere l'apprendimento e la partecipazione nella scuola,* coordinated by Tony Booth and Mel Ainscow, highlights the main characteristics of inclusion in education[12]:

- the equal valorization of all students and the didactic staff;
- the increase of the students' participation—and reduction of their exclusion—according to culture, curriculum, community;
- reformation of cultures, educational policies and school practices, thus as to correspond to the students' diversity;
- the reduction of the obstacles regarding learning and all the students' participation, not only the disabled students or those with special educational needs;
- the overpassing of obstacles of access and participation felt by certain students, by achieving some changes which bring benefits to all students;
- the approach of differences between students as resources for supporting learning rather than problems in order to overpass them;
- the recognition of the students' rights to be educated in their own community;
- the improvement of school both regarding the teachers and the students;
- the increase of the school's role in the construction of community and the promotion of values beyond the improvement of the educational results;
- the promotion of the mutual support in school and community;

---

[10]Ianes [15].

[11]Global Disability Summit—Inclusion in education, disponibil online https://www.ukaiddirect.org/wp-content/uploads/2018/07/Inclusive-education_GlobalDisabilitySummit2018.pdf. Accesat la 11.02.2019.

[12]Booth and Ainscow [14].

- the recognition that the inclusion in school is a more general aspect of inclusion in society.

In Romania, the children with special educational needs can attend the courses of mass education or of a special school. The children with mild and medium deficiencies are integrated in the mass schools where they can benefit of the specialised help by means of the supporting teachers. The children with serious, severe, profound, associated deficiencies are schooled usually in a form of special education. The responsibility of the identification of the deficiency type and its degree comes to the Commission of Child's Protection subordinated to the County Councils "the children in the special education system can follow the curriculum of the mass school, the adapted mass school curriculum or the curriculum of the special school."[13]

The integration of the children with special educational needs in the mass schools constitutes a relatively recent educational practice, specific to the last three decades, although situations of spontaneous integration of children with special educational needs in the mass education system were registered a long time ago, especially in the rural areas where there were no special schools.

The inclusion was approached during the time in three models: (a) the medical model, (b) the social model and (c) the psycho-social model[14];

(a) *the medical model* (or *the individual model*)—is focused on deficiencies or incapacities of the person which are considered as generating dependency. The model is taken from medicine where the doctor who observes a health issue prescribes a treatment for the healing of the ill person. The person with physical and/or psychic deficiencies was named "patient" and the interventions on him/her have as main objective bringing/adjustment of the person to what is called normality, namely coming back to the common normal condition. The criticisms brought to this model highlighted the fact that it does not take into account the human behavior, the inter-personal and communication abilities, of social and relational environment the person is inserted in. In the relation doctor-patient, the doctor is seen as symbol of authority, he detains knowledge and as consequence he is the only one entitled to make decisions regarding the intervention, while the patient is regarded as a passive receiver of these decisions: the doctor states the health problem, establishes the diagnoses, gives the treatment that must improve the stated problem, the patient's role being that of following strictly the doctor's instructions. The main objectives of the medical model consist in:

- establishing a diagnosis of the illness/deficiency;
- prescribing a specific treatment.

---

[13]*Politici în educație pentru elevii în situație de risc și pentru cei cu dizabilități din Europa de sud-est*, disponibil la https://www.oecd.org/edu/school/38614298.pdf. Accesat la 14.02.2016.

[14]*Inclusione scolastica degli alunni con disabilità*, Universita' degli Studi di Salerno, Facoltà di Scienze della Formazione, Anno accademico 2012/2103, disponibil on-line la adresa https://www.professionistiscuola.it/attachments/article/983/Inclusione%20scolastica.pdf. Accesat la data 18.05.2016, pp. 15–16.

The limits of this model consist especially in ignoring the psychological and social factors and their influence on the general health state of a person.

(b)   *The social model*—appeared in the 70's in the last century as opposed to the medical model. The social model does not deny the role and importance of individual interventions (on the medical, educational or professional level) but especially completes the empty spaces left by the medical model. In other words, the social model draws the attention upon on the limits outlined by the medical model and moves attention on the functional limits of the disabled people towards the problems raised by the hostile environment to inclusion, the cultural barriers which at their turn generate disability forms. The social model is a holistic approach which explains the specific problems experienced by the disabled people referring to the totality of environment and cultural factors which favourises the appearance of disability. Among the identified factors as being generators of disability, according to the social model we can specify: the non-inclusive education, discriminatory communication and information systems, inadequate economic help, discriminatory sanitary services and solidarity, public transport, public institutions, hotels, work environment with architectonic barriers, negative image transmitted by the media.[15]

(c)   *the psycho-social model*—represents a synthesis between the two previous models: the medical model and the social model. According to this model the person's condition which presents an alteration of his/her own body at functional or structural level was not a statical rigid one, on the contrary it is dynamic, active and it can be resumed in two situations:

- the loss or limit of their own level of activity and participation to the life of community due to some obstacles which come from the hostile or indifferent environment (conditions of disability);
- to obtain a good level of performance regarding his/her activity and participation to the community life due to a friendly, facilitating environment (the absence of conditions of disability).

The concept of inclusion is explored by means of three basic dimensions interconnected through which we pursue the improvement of school: (a) cultures, (b) policies, (c) practices[16]:

- *dimension A*: creating the inclusive cultures:

  - building the community;
  - stating the inclusive values;

  - *dimension B*: creating inclusive policies:

[15]*Inclusione scolastica degli alunni con disabilità*, Universita' degli Studi di Salerno, Facoltà di Scienze della Formazione, Anno accademico 2012/2103, disponibil on-line la adresa https://www.professionistiscuola.it/attachments/article/983/Inclusione%20scolastica.pdf. Accesat la data 18.05.2016, p. 16.

[16]Booth and Ainscow [14].

- development of school for all;
- the organization and support of diversity;
  dimension C: development of inclusive practices:

- achieving curricula for all;
- Coordination of the educational system.

Complexity and diversity of the students' educational needs require from the part of the school adequate answers meant to facilitate the access of all students to education preventing and eliminating the risk of exclusion and/or social marginalization. Inclusive school operates changes and modifications at curriculum level, organizing didactic activities, adopted educational strategies, the assumed responsibilities for the recognition of the right to education for all.

The teachers are, in our opinion, the main vectors of change, promoters of inclusive values in education.

2 The Purpose of Research

The purpose of this research was he identification of teachers' opinion about the inclusive school reported to the within the Romanian reality. The purpose was reached by the follow in objectives:

1. The identification of the didactic staff's perception in the counties Caraş-Severin and Mehedinţi regarding the opportunity of the Romanian school manifestation as an inclusive school, a school for all;
2. The identification of the didactic staff's perception in the counties Caraş-Severin and Mehedinţi about the way in which the Romanian school is perceived as being or not an inclusive school;
3. The identification of the didactic staff's perception in the counties Caraş-Severin and Mehedinţi about the characteristics of an inclusive school;
4. The identification of the didactic staff's perception in the counties Caraş-Severin and Mehedinţi about the degree of promotion by the Romanian educational system of the inclusive school in the community;
5. The identification of the didactic staff's perception in the counties Caraş-Severin and Mehedinţi about the involvement degree of the families of the students' with learning difficulties for the improvement of their school situation.

The period of the research development—the research was achieved in the period September 2016–July 2017.

The investigated population—the structure of the sample—in the research there were 190 teachers from the counties Caraş-Severin and Mehedinţi, thus:

- 95 people (50% of the total sample) from the Caraş-Severin county and 95 people (50% of the total sample) from Mehedinţi county;

- regarding *the structure on gender* of the sample in the research there were 130 women (respectively 68.42% of the people investigated out of the total sample) and 60 men (respectively 31,58% of the investigated people out of the total sample), thus:

  - out of the 95 teachers from the Caraş-Severin county 63 were women (respectively 66.32% of the investigated people in Caraş-Severin), and 32 were men (respectively 33,68% of the investigated people in Caraş-Severin);
  - out of the 95 teachers from the Mehedinţi county 67 were women (respectively 70.53% of the investigated people in Mehedinţi), and 28 men (respectively 29,47% of the investigated people in Mehedinţi);

  regarding *the structure according to the area in which the teachers achieve their activity*, in the research there were 72 people from the rural area (respectively 37.89% out of the investigated people from the total sample) and 118 people from the urban area (respectively 62.11% out of the investigated people of the total sample), thus:

  - out of the 95 teachers from the Caraş-Severin county, 37 people were from the rural area (respectively 38.95% of the investigated people in Caraş-Severin), and 58 people were from the urban area (respectively 61.05% of the investigated people in Caraş-Severin).
  - out of the 95 teachers from the Mehedinţi county, 35 people were from the rural area (respectively 36.84% of the investigated people in Mehedinţi), and 60 people were from the urban area (respectively 63.16% of the investigated people in Mehedinţi).

3 Research Methodology

The research was based on the data collected with the help of the *sociologic inquiry* based on a questionnaire which contained 5 items corresponding to the set objectives. For the accuracy of the data provided, the teachers involved in the research have anonymously completed the questionnaires.

3.1 Presentation and Interpretation of Results

The presentation and interpretation of data contained in this research are achieved based on the rendering of objectives and items in the questionnaire.

Objective 1: The identificationof the didactic staff's perception in the counties Caras-Severin and Mehedinti counties on th characteristics regarding the opportunity of the Romanian school manifestation as an inclusive school, a school for all.

Table 1 Results of item 1

Variants of answer—Item 1	Number of people	Percentage (out of the total sample) [%]
yes	176	92.63%
no	4	2.11%
I don't know/I don't answer	10	5.26%

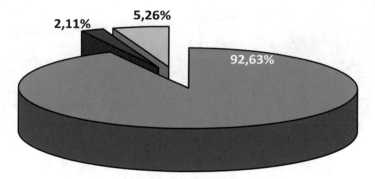

Fig. 1 Results of item 1

- *Item 1. Do you consider as pertinent that the Romanian school should manifest as an inclusive school, a school for all?*

As the answers offered by the teachers participating in the research reveal over 92% of them consider that the Romanian school should manifest as an inclusive school, while only 2.11% offer a negative answer (Table 1, Fig. 1).

Each child is an individuality, a person different from the others. The inclusive school is a space in which individuality, diversity and differences find numerous convergence points in the educational plan, the students and teachers live together some common experiences which bring a plus of value and knowledge to all those involved. "An inclusive environment tends to move the obstacles which block the person from the full participation to the social, didactic, educational life of school. Inclusion means to have the same opportunities to participate, providing their own and personal contribution. The inclusive school values, creates space, builds resources".[17]

Objective 2: The identification of the didactic staff's perception in the counties Caraş-Severin and Mehedinţi about the way in which the Romanian school is perceived as being or not an inclusive school;

Item 2. In your opinion the Romanian school is an inclusive school ?

---

[17]*La scuola incluziva*, disponibil on-line la https://www.icscastelfocognano.gov.it/joomla/attachments/article/94/La%20scuola%20Incluziva.pdf la data 14.11.2014.

Table 2 Results of item 2

Variants of answer—Item 2	Number of people	Percentage (out of the total sample) [%]
Yes	143	75.26%
No	42	22.11%
I don't know/I don't answer	5	2.63%

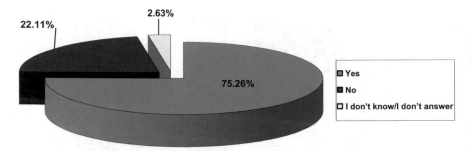

Fig. 2 Results of item 2

As compared to the previous item, the number of those who answer positively and increase the number of those who choose a negative answer. Thus, 75.26% of the total of teachers included in the research consider that the Romanian school can be characterized as an *inclusive school*, while 22.11% do not agree with this statement. We consider that during the last decades some important steps were taken regarding the inclusive education, respectively the integration with the children with deficiencies in the mass schools. However the concept *inclusive school* aims at a certain attitude towards all the students implicitly those with special educational needs, respectively the students who are faced with learning difficulties (Table 2, Fig. 2).

The inclusive school is attentive to its students in order to help them develop the specific competences meant to facilitate their integration in society, as people worthy of respect and appreciation from the other members of society. It is notorious that in a common class we can integrate 1–2 children certified with deficiencies, but the number of the students who require an increased attention during the demarche of learning, respectively the students with special educational needs (with learning difficulties) is much higher and it seems to be in a continuous increase.

Objective 3: The identification of the didactic staff's perception in the counties Caraş-Severin and Mehedinţi about the characteristics of an inclusive school;

Item 3. We kindly ask you to state in what measure you agree with the following statements. Inclusive school ….

According to the opinion of the majority of the teachers participating in the research the *inclusive school develops specific programs meant to help the students* with behavioural disorders or who have remained behind in learning (Totally agree—93.68%;, Chiefly agree—6.32%), values the potential of every child (Totally agree—84.21%; Chiefly agree—14.21%), uses an individualized curriculum for the students with special educational needs (Totally agree—81.58%; Chiefly agree— 16.32%), uses an individualized curriculum for the students with special educational needs (Totally agree—81.58%; Chiefly agree—16.32%), uses preponderantly active-participative didactic strategies (Totally agree—70%; Chiefly agree— 28.95%), gives a special attention to deficient students (Totally agree—67.37%; Chiefly agree—32.63%). Very few answers used the variant undecided while the variants which are at the opposite pole, respectively Totally disagree, Chiefly disagree recorded no answer (Table 3).

In our opinion, these characteristics are essential for a school that wants to be an inclusive school or a school for all, to which a positive attitude of the didactic staff is added, doubled by a permanent preoccupation for their own development on the professional and personal plan.

Objective 4: the identification of the didactic staff's perception in the counties Caraş-Severin and Mehedinți about the degree of promotion by the Romanian educational system of the inclusive school in the community.

- Item 4. In what extent do you consider that the Romanian educational system promotes the inclusive school (school for all) at the community level?

Table 3 Results of item 3

Inclusive school	Totally disagree	Chiefly disagree	Undecided	Chiefly agree	Totally agree	Total
Values the potential of every child	0	0	3	27	160	190
	0.00%	0.00%	1.58%	14.21%	84.21%	100.00%
Uses preponderantly active-participative didactic strategies	0	0	2	55	133	190
	0.00%	0.00%	1.05%	28.95%	70.00%	100.00%
Gives a special attention to deficient students	0	0	0	62	128	190
	0.00%	0.00%	0.00%	32.63%	67.37%	100.00%
Uses an individualized curriculum for the students with special educational needs	0	0	4	31	155	190
	0.00%	0.00%	2.11%	16.32%	81.58%	100.00%
Develops specific programs meant to help the students with behavioural disturbances or who remain behind in learning	0	0	0	12	178	190
	0.00%	0.00%	0.00%	6.32%	93.68%	100.00%

The success of the activities developed by the inclusive school depends in a great extent on the support that it has in the community. In our opinion neglecting the activities to promote the principles of the inclusive school in the community reflects negatively on the process of integration of the children with special educational needs. A significant number of the teachers participating in the present research consider that the Romanian educational system promotes the inclusive school *much* (44.21%) and *very much* (28.42%). Very few options are given to the *medium* (12.11%), *little* (10.53%), *at all* (4.74%).

Obiectiv 5: The identification of didactic staff's perception in the counties Caraş-Severin and Mehedinţi about the degree of involvement of the families of the students with learning difficulties for the improvement of their school situation (Table 4, Fig. 3).

Item 5 In your opinion the families of the students with learning difficulties are actively involved in the improvement of their situation in learning?

One of the basic activities of every school and implicitly of the inclusive school is constituted by the specific activity of improving the situation of the students' learning (of those who record remaining behind according to the level of the school curriculum). The family involvement in the support of this process plays an important part for the success of the activity. Most teachers participating in the research choose the variants *Totally agree* (46.84%) and *Chiefly agree* (22.63%) regarding

Table 4 Results of item 4

Very much	Much	Medium	Little	At all	Total
54	84	23	20	9	190
28.42%	44.21%	12.11%	10.53%	4.74%	100.00%

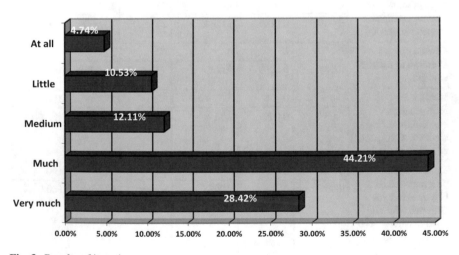

Fig. 3 Results of item 4

Table 5 Results of item 5

Totally disagree	Chiefly disagree	Undecided	Chiefly agree	Totally agree	Total
15	12	31	89	43	190
7.89%	6.32%	16.32%	46.84%	22.63%	100.00%

the involvement of the family in the support of their own child who faces difficulties in learning, while 13.32% declare themselves *undecided*, 6.32% choose the variant *chiefly disagree* and 7.89% choose *totally disagree*. The cases responsible for the non-involvement of the family in the support of the students' school activity are numerous. Only a careful attention of these ones for each case in particular can lead to school taking some concrete measures by which to attract the family on the side of the school. Then, when the school contents itself by only stating that the family is not receptive to the problems on the educational plan of the student without coming to meet half way in an adequate and professional way (for instance without reproaches and accusations) the situation can degenerate especially that in some cases the family, based on a deficient communication with school, is not even acquainted to the student's problems (Table 5).

4 Conclusions

"An inclusive school must always promote the right to be considered equal to the others and in the same time different from the others."[18]

The diversity in the light of inclusion is understood as a resource which allows each one to contribute to his/her development and of the others. The process of educational inclusion has as basis the right to education of every child, without being limited by its ethnic or cultural origin, by gender or religion, by the abilities/capacities they dispose of. The inclusive school recognises the inestimable value of the human being; it is based on the full participation to the school life of every student implicitly of those with special educational needs/requirements.

School for all or inclusive school is attentive to all its students, to their aptitudes, interests and aspirations, to the efficient didactic ways which should valorise the potential of every child.

The students with special educational needs require some individualised interventions, adapted to their differences and needs. The concept of *students with special educational needs/requirements* very well illustrated in the Anglo-Saxon literature and not only is very close as significance to another concept, respectively *students with learning difficulties* that they face either for a limited period of time or during their whole educational journey. The term of learning difficulties refers to every

---

[18]*La scuola incluziva*, disponibil on-line la adresa https://www.icscastelfocognano.gov.it/joomla/attachments/article/94/La%20scuola%20Incluziva.pdf. Accesat la data 14.11.2014.

difficulty met by a student during his/her school career".[19] The students with special educational needs "live a particular situation, which blocks their learning and development: this negative situation can be at organic, biologic or familiar level, social, environment, contextual or combinations of these ones. A student with special educational needs may be on the organic, biologic, family, social, environmental or contextual level or combination of the above. A student with special educational needs can may have a very serious cerebral lesion, Down syndrome, a certain cerebral or perceptive dysfunctionality, serious family conflicts or a different or deprived social and cultural background and emotional reactions and/or disturbing behaviours etc.[20]"

References

1. Booth T, Ainscow M (2008) L'Index per l'inclusione. Promuovere l'apprendimento e la partecipazione nella scuola, Italian edition curated by Dovigo, F. și Ianes, D., Editura Erikson, Gardolo
2. Cornoldi C (2016) Le difficolta' di apprendimento (L'apprendimento difficile) https://www.aid aiassociazione.com/documents/Cornoldi-Le_difficolta_di_apprendimento.pdf. Accessed on 25 Sept 2016
3. Dovigo F (2008) L'Index per l'inclusione: una proposta per lo sviluppo incluzivo della scuola. în: Booth T, Ainscow M (eds) L'Index per l'inclusione. Promuovere l'apprendimento e la partecipazione nella scuola, Italian edition curated by Dovigo, F. și Ianes, D., Editura Erikson, Gardolo
4. Hošková Š (2009) Experience with blended (distance) learning study materials. In: Conference distance learning, simulation and communication—DLSC. CATE, Brno. pp 70–77. ISBN 978-80-7231-638-0
5. Horga I, Jigău M (2011) Situația copiilor cu cerințe educative speciale incluși în învățământul de masă, Editura Vanemonde, București. https://www.unicef.ro/wp-content/uploads/Situatia-cop iilor-cu-cerinte-educative-speciale-inclusi-in-invatamantul-de-masa.pdf. Accessed on 12 Nov 2016
6. Hoskova-Mayerova S (2011) Operational programm "Education for Competitive Advantage", preparation of Study Materials for Teaching in English, Procedia-Social and Behavioral Sciences, vol 15, pp 3800–3804. https://doi.org/10.1016/j.sbspro.2011.04.376.
7. Ianes D (2016) I Bisogni Educativi Speciali. https://www.darioianes.it/site/articoli/i-bisogni-educativi-speciali. Accessed on 14 May 2016
8. Male C, Wodon Q (2017) The price of exclusion: disability and education. Disability gaps in educational attainment and literacy. links.org/sites/default/files/media/file/Disability_gaps_in_educational_attainment_and_literacy.pdf. Accessed on 14 Aug 2019
9. Medeghini R, Fornasa W (2011) L'educazione incluziva. Culture e pratiche nei contesti educativi e scolastici: una prospettiva psicopedagogica, Franco Angeli, Milano
10. The Education Commission. The learning Generation. Investing in Education for a changing world, 2017. https://report.educationcommission.org/. Accessed on 14 May 2019
11. Inclusione: DSA—BES, https://ictito.gov.it/index.php?option=com_content&view=article&id=587&Itemid=1522. Accessed on 25 Sept 2016
12. Horga I, Jigău M (2011) Situația copiilor cu cerințe educative speciale incluși în învățământul de masă. Editura Vanemonde, București, disponibil la https://www.unicef.ro/wp-content/uploads/Situatia-copiilor-cu-cerinte-educative-speciale-inclusi-in-invatamantul-de-masa.pdf. Accesat la data 12.11.2016

---

[19]Cornoldi (2016).

[20]Ianes (2016).

13. Medeghini R, Fornasa W (2011) L'educazione incluziva. Culture e pratiche nei contesti educativi e scolastici: una prospettiva psicopedagogica. Franco Angeli, Milano, p 72
14. Booth T, Ainscow M (2008) L'Index per l'inclusione. Promuovere l'apprendimento e la partecipazione nella scuola, Italian edition curated by Dovigo, F. și Ianes, D., Editura Erikson, Gardolo, pp 115–116
15. Ianes D (2016) I Bisogni Educativi Speciali, disponibil on-line la adresa https://www.darioi anes.it/site/articoli/i-bisogni-educativi-speciali. Accesat la data 14.05.2016
16. Male C, Wodon Q (2017) The price of exclusion: disability and Education. Disability gaps in educational attainment and literacy. links.org/sites/default/files/media/file/Disability_gaps_in_educational_attainment_and _literacy.pdf
17. Cornoldi C (2016) Le difficolta' di apprendimento (L'apprendimento difficile), p.8 disponibil on-line la adresa https://www.aidaiassociazione.com/documents/Cornoldi-Le_difficolta_di_a pprendimento.pdf. Accesat la data 25.09.2016

Societal Perception and Germane Factors Promoting Female Genital Mutilation in Oyo State, Nigeria

Bolanle Oyundoyin⬡, Fatai Adebayo⬡, and Olatunbosun Soetan⬡

Abstract This chapter examined societal perception and germane factors promoting female genital mutilation, in Oyo State, Nigeria. Four research questions and one null hypothesis were raised to guide the study. The perception on FGM by the respondents were FGM: guarantee better marriage prospects for women ($\dot{x} = 3.33$) and it is a pre-requisite for marriage ($\dot{x} = 3.61$). Factors promoting FGM were gender discrimination ($\dot{x} = 2.29$) and religious belief ($\dot{x} = 1.82$). The hypothesis tested showed a significant association between age ($X^2 = 14.506$); sex ($X^2 = 10.591$); religion ($X^2 = 11.647$); marital status ($X^2 = 11.200$); educational level ($X^2 = 15.335$) and perception of the parents on FGM with ($p < 0.05$). The study concluded that FGM has an adverse effect on girls' and women's health. Hence, the study recommends sensitive campaign at the grass root and inculcating consequences of FGM into primary and secondary schools curriculum.

Keywords Societal · Perception · Female genital mutilation · Germane factors

1 Introduction

Female genital mutilation (FGM) refers to a variety of operations on the private parts of young girls and women that involves partial or total removal of the external genitalia. The practice causes injury to female genital organs for cultural or non-therapeutic reasons (World Health Organization (WHO, 2010) [5, 18, 19]). Whatever the purpose, FGM is a dangerous and potentially life threatening procedure that causes unspeakable pain and suffering to the victim. According to Black (2000), the

B. Oyundoyin (✉) · F. Adebayo
Department of Home Science and Management, Federal University of Agriculture, Abeokuta, Abeokuta, Nigeria
e-mail: oyundoyinbm@funaab.edu.ng

O. Soetan
Department of Agricultural Extension and Rural Development, Federal University of Agriculture, Abeokuta, Abeokuta, Nigeria
e-mail: soetanoj@funaab.edu.ng

© The Author(s), under exclusive license to Springer Nature Switzerland AG 2021   361
D. Soitu et al. (eds.), *Decisions and Trends in Social Systems*,
Lecture Notes in Networks and Systems 189,
https://doi.org/10.1007/978-3-030-69094-6_29

practice is no more rampant in the western world but still exists in many African countries including Nigeria. It continues to be one of the most persistent, pervasive and solemnly endured human rights violations in the developing world [11]. Also FGM has been widely condemned by international organisations and feminist groups all over the world due to the licalth outcomes it has for women, and because it is an abuse of the women's Fundamental Human Rights [3].

Internationally female genital mutilation is recognized as a violation of the human rights of girls who are forced to undergo this procedure (WHO 2010). It has been compared to foot-binding because it is a harmful traditional practice, ethically indefensible due to its permanent physical and psychological damage (Mackie 1996). With all the medical complications prevalent among the circumcised female, the obnoxious practice is still in existence especially in some rural areas in most developing countries like Nigeria. The World Health Organization (WHO) has directed its advocacy and research efforts towards the termination of this practice, in conjunction with local governments that have attempted a variety of policy interventions (WHO 2010).

It is estimated that 130–140 million females in the world today have been the victim of female mutilation in any form; while in Africa an estimated 92 million girls and women have undergone female circumcision (WHO 2010); and at least 200 million girls and women living in 31 countries have undergone FGM, UNICEF data [14]. At the current rates of population increase and with the slow decline in these procedures, it is estimated that each year at most 2 million girls are at risk from the practice, and the women and girls affected live in 28 African countries and a few in the Middle East and Asia (World Health Organization (WHO 2002)).

More so, in some cultures it is carried out in adolescence, just before marriage, or at childbirth. The 'operator' is usually an older woman in the community, either a relative or a traditional birth attendant. This practice is being carried out by followers of many different religions, including Muslims, Christians and Judaist, it is not a practice of religion because, neither the Bible nor the Quran prescribed female genital mutilation (FGM). The practice certainly preceded the founding of Christianity and Islam, and there is no basis for the belief about FGM [7].

It was confirmed that the form of genital mutilation varies in different cultures [8], Habtemarian (2003) and the consequences of FGM is at variance because of the social factors, social structures and religious belief, coupled with its wide publicity without evidence, [15]. The practitioners (mostly traditional birth attendant (TBA) and the person that experienced it do not know the consequences, except the health professionals and health workers involved in maternal care in the areas where it is practiced (Teare 2000).

In Nigeria, the prevalence of female genital mutilation differs from one state to another state (Onuh et al. 2006). According to the Nigerian Democratic Health Survey (NDHS 2003), prevalence of female genital mutilation is highest in the South-Western and South-Eastern regions of the country and lowest in the North-Western part of the country.

1.1 Statement of the Problem

Despite the numerous lectures, seminars and conferences about medical complications of FGM, as well as other policies and health campaigns against this unacceptable practice, it is still practiced by some people. FGM negatively affects the psychological, social health and well-being of women. The influence of some cultures about the need for FGM has today carry serious health and welfare risks, with regard to Human Immunodeficiency virus (HIV) and sexually transmitted infections (STIs). The United Nations has labelled FGM as one of the harmful cultural practices that need to be eliminated in society not only in Africa but also in the African immigrant communities in Europe. FGM has immediate and long term effect on the women.

The immediate problems faced by women are death, the threat of death from haemorrhaging, and shock from the pain and inhuman of violence associated with the procedure. The long term effects were frequent urinary tract infections, urinary retention and abscess formation, septicaemia, increased pain during menstruation, with related buildup of menstrual fluid as time passes, and the development of scar tissue are some of the documented long-term side effects of the procedure [13].

A recent study involving 28,509 women by the WHO found that women with FGM are higher than those without FGM to have difficult obstetric outcomes. Extensive FGM takes more risk [17]. Women with FGM Types I, II and III were more likely to experience the following: caesarean section, post-partum haemorrhage, extended maternal hospital stay, infant resuscitation, stillbirth or neonatal death, and low birth weight.

Post-traumatic stress disorder (PTSD) in both women and children who have undergone FGM has been widely documented [1, 9, 16]. Behrendt and Moritz [4] in their pilot study on 23 circumcised Senegalese women in Dakar found that the circumcised women showed a significantly higher prevalence of PTSD (30.4%) and other psychiatric syndromes (47.9%) than the uncircumcised women. For instance, during pregnancy, cases have been noted whereby what are seen as 'routine' procedures, for example internal examinations, can cause a traumatic flashback of the event.

Further studies by El-Defrawi et al. [6] found that circumcised women were statistically more likely to experience psychosexual difficulties than non-circumcised women. Statistically notable difficulties were found in circumcised women which include a lack of sexual desire, slow or inability initialize sexual activity with husbands and unsatisfied sex or inability to reach orgasm or peak of sexual pleasure [2, 12]). Research has also indicated that the destruction of clitoris by FGM women compensate by 'adjusting' which is the most sensitive part of their bodies and instead identify their breasts as the most sensitive area [10]. This study seeks to investigate societal perception and germane factors promoting female genital mutilation, in Oyo State, Nigeria. Following the aforementioned, this study will proffer answers to the following research questions:

i.   What are socio-demographic characteristics of the parents in the study area?
ii. What are the effects of female genital mutilation on female child health?

iii.   What are the factors that promote FGM in the study area?
iv.   What is the perception of the respondents on FGM?

1.2   Hypothesis of the Study

H_{01}: there is no significant association between socio-demographic characteristics of the respondents and their perception on FGM.

2   Research Methodology

2.1   Description of the Study Area

This study was carried out in Ibadan North Government Area. Ibadan North Local Government is one of the twenty six Local Governments in Oyo State. The Headquarter is at Gate, Ibadan. It has an area of 27 km^2 and a population of 306,795 at the 2006 census. It also has bustling academic and economy activities with the presence of the First Premier University in Nigeria, the University of Ibadan, founded in 1948, and The Polytechnic, Ibadan in 1970 creates an aura of lively place to live in.

2.2   Population of the Study

The population of this study were male and female parents in Ibadan North Local Government Area, Oyo state, Nigeria.

2.3   Research Design

This study employs a cross sectional research design to investigate the perception of parents on female genital mutilation in Ibadan North Local Government Area, Oyo state, Nigeria.

2.4   Method of Data Collection

Data was collected with the aid of structured questionnaire. The structured questionnaire was divided into three (3) categories in which the first contain the perception on FGN; the second contain the effect of FGN on female child health while the third contain factors promoting FGN.

2.5   Sample Size and Sample Procedure

Purposive was used to select respondents for this study. Purposive sampling was used because FGM is common in the study area. Five (5) wards out of the twenty six (26) wards that constitute the Local Government Area were purposively selected due to the aforementioned. The five wards selected randomly include Gate, Basorun, Ikolaba, Bodija, and Total garden. Thirty respondents that have practiced or know anybody that have done the practice were purposively selected in each of the selected wards to give a total of 150 respondents.

2.6   Research Instruments

A well-structured questionnaire is designed to collect data from respondents. However, the questionnaire contained three sections marked A, B and C. Section A is based on socio-demographic of respondents and B based on effect of female genital mutilation while section C based on factors promoting female genital mutilation.

2.7   Validity and Reliability of the Research Instrument

Face validity use to determine the validity of the research instrument. This was assured by submitting a drafted questionnaire to expert in the field of health and Home Science and Management, any bias and ambiguity was removed and reconstructed. Reliability of the research instrument was determined using Cronbach alpha test the reliability. The overall reliability coefficient of the research instrument was 0.78 which was adjudged reliable according to international standard.

2.8   Method of Data Analysis

Data was analysed using descriptive statistics such as frequency counts, percentages and inferential statistics such as Chi square and student t-test analysis.

2.9   Measurement of Variables

Perception on FGM and effects of FGM were measure on 4 point Likert type scale of strongly agree (4), agree (3), disagree (2) and strongly disagree (1). Factors promoting FGM was measure on 3 points rating scale of major factor (3), minor factor (2) and not a factors (1).

3   Results and Discussion

3.1   Research Questions

RQ1 What is the Socio-economic Characteristic of the Respondents?

The result in Fig. 1 shows the distribution of respondents by age. 46.7% were between age of 20 and 30 years, 14.0% were between 31 and 40 years while 28.0% were between age of 41 and 50 years and 11.3% were between 51 and 60 years. It implies that majority of the parents were young and were still in their reproductive age brackets. One of the factors that affect the practice of FGM is age. FGM is common among people within the age bracket of 20–30 years.

Figure 2 above show the distribution of respondents by sex. The results revealed that 87.3% were females while 12.7% were males. Research has shown that FGM is common among girls and women. This is in line with WHO (2002) report that revealed that at most 2 million girls and women are at risk from the practice.

Figure 3 above show the distribution of respondents by religion. The results revealed that 0.7% were Africa Traditional worshippers, 22.6% were Muslims while 76.7% were Christians. It implies that majority of the parents were Christians.

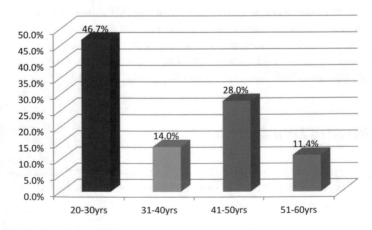

Fig. 1 Distribution of the respondents by age

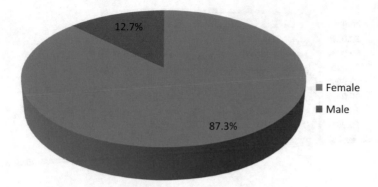

Fig. 2 Distribution of the respondents by sex

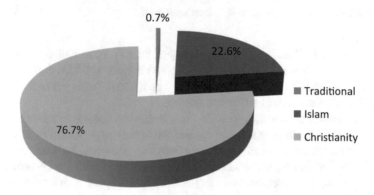

Fig. 3 Distribution of the respondents by religion

Research as shown that religion is one of the major factors that influences the practices of FGM. This practice is being carried out by followers of many different religions. Karim and Ammar [7] posited that the practice is common among different religion groups.

Figure 4 above show the distribution of respondents by level of education. The results revealed 2.7% of the respondents had Ph.D, 20.0% had M.Sc. education, 70.0% had Tertiary education and 7.3% had secondary education. It implies that majority of the parents were educated. One of the factors that promote FGM is illiteracy.

RQ2 What is the Perception of the Respondents on Female Genital Mutilation (FGM)? (Table 1).

The perception of the respondents on FGM was ranked according to mean as follows: Female circumcision is a pre-requisite for marriage (mean = 3.61); female circumcision guarantee income when the daughter is married and dowry is paid (mean = 3.46); Female circumcision is pre-requisite to a trustworthy marriage (3.43) were ranked first, second and third respectively. This implies that some culture promote the

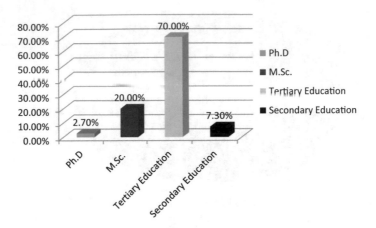

Fig. 4 Distribution of the respondents by educational level

practice of FGM. Culture is said to be the way of life of people. Culture and/or belief are one of the key factors in health utilization and behaviours. This corroborates the finding of McCloud [8], Habtemarian (2003) who independently stated that the form of genital mutilation varies in different culture.

Female circumcision makes women more feminine and more attractive to men (mean = 3.41); Female circumcision reduce the beauty of a woman (mean = 3.39) and Female circumcision guarantee better marriage prospects for women (mean = 3.33) were ranked fourth, fifth and sixth respectively. According to Uwasomba [15] social factors, social structures and religious belief, coupled with it wide publicity without evidence promote the practice of FGM.

Female circumcision prevent pre-marital sex (mean = 3.28); Circumcise woman fulfil religion obligation (mean = 3.24) and Female circumcision enhance fertility on and promote child survival (mean = 3.20) were ranked 7th, 8th and 9th respectively. This practice is done without considering the negative effect on the victims. WHO (2010) estimated that 130–140 million females in the world today have being the victim of female mutilation in any form.

RQ3 What are the Factors Promoting FGM? (Table 2).

Factors promoting FGM were uncircumcised woman can easily be recognized (mean = 2.68); Gender discrimination contribute to continuation of female circumcision (mean = 2.29) and practice of female circumcision serve as source of income for the practices (mean = 2.04) were ranked 1st, 2nd and 3rd respectively. This implies that gender discrimination is one of the key factors that promote FGM.

Religious belief female circumcision is right for a women (mean = 1.97); parent belief circumcision is good for girls (mean = 1.82) and tradition assist in promoting female circumcision (mean = 1.41) were ranked 4th, 5th and 6th respectively. This shows that the role of culture/belief and religion cannot be overemphasized in the practice of FGM. The influence of some cultures about the need for FGM has today carry serious health and welfare risks on girls and women. It has been noted that the

Table 1 Perception of the respondents on female genital mutilation (FGM)

S/N	Items	SA F (%)	A F (%)	D F (%)	SD F (%)	Mean	Rank
1	Female circumcision enhance fertility on and promote child survival	75(50.0)	40(26.7)	25(16.7)	10(6.7)	3.20	9th
2	Female circumcision is done in order to initiate girls into womanhood	65(43.3)	43(28.7)	30(20.0)	12(8.0)	3.07	11th
3	Female circumcision guarantee income when the daughter is married and dowry is paid	93(62.0)	40(26.7)	11(7.3)	6(4.0)	3.46	2nd
4	Female circumcision preserve family honour and prevent immorality	70(46.7)	49(32.7)	21(14.0)	10(6.7)	3.19	10th
5	Female circumcision guarantee better marriage prospects for women	75(50.0)	51(34.0)	13(8.7)	11(7.3)	3.33	6th
6	Female circumcision is a pre-requisite for marriage	90(60.0)	47(31.3)	9(6.0)	4(2.7)	3.61	1st
7	Female circumcision makes women more feminine and more attractive to men	83(55.3)	49(32.7)	14(9.3)	4(2.7)	3.41	4th
8	Female circumcision reduce the beauty of a woman	85(56.7)	48(32.0)	7(4.7)	10(6.7)	3.39	5th
9	Circumcise woman fulfil religion obligation	64(42.7)	65(43.3)	14(9.3)	7(4.7)	3.24	8th
10	Female circumcision is pre-requisite to a trustworthy marriage	84(56.0)	51(34.0)	11(7.3)	4(2.7)	3.43	3rd
11	Female circumcision prevent pre-marital sex	80(53.3)	44(29.3)	14(9.3)	12(8.0)	3.28	7th

form of genital mutilation varies in different culture [8], Habtemarian (2003) and the consequences of FGM is at variance because of the social factors, social structures and religious belief, coupled with it wide publicity without evidence, [15].

RQ4 What are the Effects of FGM on Girls' and Women's Health? (Table 3).

Effects of FGM on girls and women health were women who are circumcised may have problem during childbirth (mean = 2.74); female circumcision can delay conception (mean = 2.67) and women who undergo female circumcision may experience pain during sexual intercourse (mean = 2.51) were ranked 1st, 2nd and 3rd respectively. This means that FGM negatively affects the psychological, social health and well-being of girls and women. This agree with Thierfelder [13] who posited that immediate problems faced by girls and women that were circumcised are death,

Table 2 Factors promoting FGM

S/N	Items	MAF F (%)	MIF F (%)	NAF F (%)	Mean	Rank
1	Gender discrimination contribute to continuation of female circumcision	73(48.7)	47(31.3)	30(20.0)	2.29	2nd
2	Tradition assist in promoting female circumcision	15(10.0)	32(21.3)	103(68.7)	1.41	6th
3	Parent belief circumcision is good for girls	34(22.7)	55(36.7)	61(40.6)	1.82	5th
4	Lack of education contribute to continuous practice of female circumcision	13(8.7)	28(18.7)	109(72.6)	1.36	7th
5	Lack of awareness is attach to female circumcision	8(5.3)	32(21.3)	110(73.3)	1.32	8th
6	Religious belief female circumcision is right for a women	47(31.3)	52(34.7)	51(34.0)	1.97	4th
7	Uncircumcised woman can easily be recognized	113(75.3)	26(17.4)	11(7.3)	2.68	1st
8	The practice of female circumcision serve as source of income for the practices	55(36.7)	46(30.6)	49(32.7)	2.04	3rd

the threat of death from haemorrhaging, and shock from the pain and inhuman act of violence associated with the procedure.

Female circumcision can cause increased pain during menstruation and often experience difficulty in passing urine (mean $= 2.20$); female circumcision can cause injury to the genital thereby causing failure to heal (mean $= 2.16$) and women who undergo female circumcision experience haemorrhage (bleeding) during the procedure (mean $= 1.80$) were ranked 4th, 5th and 6th respectively. This support El-Defrawi et al. [6] who found that circumcised women were statistically more likely to experience psychosexual difficulties than non-circumcised women.

3.2 Test of Hypothesis

Ho$_1$: There is no significant association between socio-economic characteristics and their perception on FGM (Table 4).

The result revealed that there is significant association ($p < 0.05$) between age (Chi square $= 14.506$); sex (Chi square $= 10.591$); religion (Chi square $= 11.647$), marital status (Chi square $= 11.200$); education (Chi square $= 15.335$) and perception towards FGM. This implies that age, sex, religion, marital status and education associated with the practice of FGM.

Table 3 Effect of FGM on girls' and women's health

S/N	Items	SA F (%)	A F (%)	D F (%)	SD F (%)	Mean	Rank
1	Women who undergo female circumcision experience haemorrhage (bleeding) during the procedure	11(7.3)	16(10.7)	55(36.7)	68(45.3)	1.80	6th
2	Female circumcision cab cause increased pain during menstruation and often experience difficulty in passing urine	28(18.7)	59(39.3)	38(25.3)	25(16.7)	2.20	4th
3	Women who undergo female circumcision may experience pain during sexual intercourse	26(17.3)	52(34.7)	44(29.0)	28(18.7)	2.51	3rd
4	Women who are circumcised may have problem during childbirth	34(22.7)	65(43.3)	30(20.0)	21(14.0)	2.74	1st
5	Female circumcision can cause injury to the genital thereby causing failure to heal	19(12.7)	27(18.0)	63(42.0)	41(27.3)	2.16	5th
6	Female circumcision can delay conception	24(16.0)	73(48.7)	32(21.3)	21(14.0)	2.67	2nd
7	Female circumcision can increase the susceptibility to HIV when non sterile instrument is used to circumcise several people	9(6.0)	12(8.0)	39(26.0)	90(60.0)	1.60	7th
8	The practice of female circumcision has more health hazards and therefore be abolished	13(8.7)	11(7.3)	32(21.3)	94(62.7)	1.42	8th

Table 4 Result of test association between socio-economic characteristics and their perception on FGM

S/N	Variables	Df	Chi square value	P-value	Decision
1	Age	3	14.506	0.021	Significant
2	Sex	1	10.591	0.044	Significant
3	Religion	2	11.647	0.044	Significant
4	Ethnicity	2	0.232	0.891	Not significant
5	Marital status	1	11.200	0.027	Significant
6	Occupation	3	1.126	0.771	Not significant
7	Level of education	3	15.335	0.015	Significant

Decision criteria: reject null hypothesis if $p < 0.05$, accept null hypothesis if $p > 0.05$

4   Conclusion and Recommendations

The study concluded that FGM has negative effects on girls' and women's health. The negative effects revealed in the study were lack of or slow sexual desire in women, unsatisfied sex or inability to reach orgasm or peak of sexual pleasure, problem during child birth, delay in conception, increased pain during menstruation, difficulty in passing urine and injury to the genital thereby causing failure to heal. Factors promoting FGM in the study area are age, sex, religion, marital status and level of education. This chapter therefore, recommended that government at all levels and non-governmental organisations and other stakeholders should work conglomerate to eliminate this act, and make policies, promulgate laws, punish offenders of the practice of FGM and their supporters even at the grass roots. Also sensitisation on media houses such as radio, television, social media, internet and newspapers should be made on negative effects of FGM on girls' and women's health.

It is imperative to know that many studies have widely documented the Post-traumatic stress disorder (PTSD) in both women and children who have undergone FGM for example [1, 4, 9, 16] but this very chapter specifically dealt with the societal perception and germane factors promoting female genital mutilation, in Oyo State, Nigeria.

References

1. Almqvist K, Brandell-Forsberg M (1997) Refugee Children in Sweden: Post-Traumatic Stress Disorder in Iranian pre-school children exposed to organised violence. Child Abuse Negl 21:351–366
2. Almroth L et al (2001) Male complications of female genital mutilation. Soc Sci Med 53(11):1455–1460
3. Alo OA, Adetula GA (2005) Myths and realities surrounding female genital mutilation (FGM) in Ekiti State of Nigeria. Int J Violence Relat Stud 2(1):314–333
4. Behrendt A, Moritz S (2005) Post traumatic stress disorder and memory problems after female genital mutilation. Am J Psychiatry 162(5):1000–1002
5. Black JA, Debelle GD (1995) Female genital mutilation in Britain. Br Med J 310:1590–1592
6. El-Defrawi M et al (2001) Female genital mutilation and its psychosexual impact. J Sex Marital Ther 27:465–473
7. Karim M, Ammar R (2003) Female circumcision and sexual desire. Shams University PRESS, Cairo
8. McCloud P (2003) Promoting FGM abandonment in Egypt. CEDPA, Washington
9. Momoh C (ed) (2005) Female genital mutilation. Radcliffe, Abingdon
10. Nwajei SD, Dtiono AI (2003) Female genital mutilation: implications for female
11. Onodu WU (2014) Assessing socio-cultural factors that still preserve female genital mutilation practice among women in selected rural communities of Enugu State
12. Thabet SM, Thabet AS (2003) Defective sexuality and female circumcision: the cause and the possible management. J Obstet Gynaecol Res
13. Thierfelder C (2005) Female genital mutilation in the context of migration: experiences of African women in the Swedish healthcare system. Eur J Public Health 15(1):86–90
14. UNICEF Data (2020) Female genital mutilation

15. Uwasomba NC (2003) Knowledge, attitude, practice and effect of female genital mutilation on women of childbearing age in Enugu Urban (Unpublished)
16. Whitehorn J et al (2002) Female genital mutilation: cultural and psychological implications. Sex Relat Ther 17(2):161–170
17. World Health Organization (2006) Female genital mutilation and obstetric outcome: a collaborative prospective study in six African countries. Lancet 367:1835–1841
18. World Health Organization (2000) Female genital mutilation, fact sheet No. 241. World Health Organization, Geneva
19. World Health Organisation (WHO) (2012) Female genital mutilation: understanding and addressing violence against women

The Etymology of Romanian Place Names. Case Study: The Vrancea Region

Theodora Flaut

Abstract Toponomastics or toponymy is a branch of linguistics which focuses on the study of toponyms. The study of the places names is the result of the researches in various domains as for example: linguistic, dialectology, sociology, geography, history, economics, etc. Vrancea is a region situated at the boundary between the three historical Romanian provinces, Moldavia, Wallachia and Transylvania. This area has been populated since time immemorial. In this chapter, we talk about some geographical and historical aspects of the Vrancea region, we explore the origin of several place names from this area, highlighting their evolution in time, as well as influences exerted on them throughout the course of history, with special emphasis on river names (hydronyms). Moreover, we discuss several family names that are common in this area and emphasize the way in which these surnames have turned into toponyms (anthroponyms).

Keywords Historical Romanian provinces · Toponyms · The places names · Dialectology

1 Introduction

Toponomastics or toponymy is a branch of linguistics which focuses on the study of toponyms. The toponymy studies the origin of the meaning of the places names in a given language. The study of the places names is the result of the researches in various domains as for example: linguistic, dialectology, sociology, geography, history, economics, etc. Therefore, toponymy can be considered as a border discipline. There is an important relationship between toponymy and the above mentioned social science branches since the good knowledge of the origin of the places names give us information regarding the evolution and the development of the society in that area.

T. Flaut (✉)
Faculty of Letters, Ovidius University, Constanţa, România
e-mail: theodora_flaut@yahoo.com

© The Author(s), under exclusive license to Springer Nature Switzerland AG 2021
D. Soitu et al. (eds.), *Decisions and Trends in Social Systems*,
Lecture Notes in Networks and Systems 189,
https://doi.org/10.1007/978-3-030-69094-6_30

In the following, as a study case, we choose Vrancea as the region we are going to talk about. Vrancea is a full of history region situated at the boundary between the three historical Romanian provinces, Moldavia, Wallachia and Transylvania, and has been populated since time immemorial.

This chapter focuses on a number of geographical and historical aspects of the Vrancea region. It moreover explores the origin of several place names from this area, highlighting their evolution in time, as well as influences exerted on them throughout the course of history.

In the following section we will analyse the toponymy of Vrancea County. Firstly, we will elaborate on the county's name, then highlight the origin of place names in this region, with special emphasis on river names (hydronyms). We will moreover discuss several family names that are common in this area and emphasize the way in which these surnames have turned into toponyms (anthroponyms). Such names can reveal the depth of the relationship between people and places, the level of concern, affection, attachment and gratitude felt towards one's environment, as well as the friendly coexistence of the inhabitants, irrespective of their ethnic origin; they thus provide a unique insight into the lives and history of dwellers of this region.

2 Vrancea—Geographical Aspects

I. Geographic location

Vrancea is a Romanian county situated at the boundary between the historical regions of Moldavia and Muntenia (Greater Wallachia), two territories divided by the Milcov River. Its seat is the town of Focșani. The geographical coordinates circumscribing this county are 45° and 46° 11′ northern latitude and 26° 23′ and 27° 32′ eastern longitude. Situated in the south-east of the country, in the Eastern Carpathian curve, Vrancea can be seen as a transition area connecting the three historical provinces, Moldavia, Wallachia and Transylvania. Its immediate neighbours are Bacău County to the north, Vaslui County to the north-east, Galați County to the east, Brăila County to the south-east, Buzău County to the south and Covasna County to the west (Fig. 1).

II. Relief

The landscape consists of steps descending from west to east. The western level is the highest and consists of mountain peaks and massifs. The altitudes of Vrancea Mountains range from 960 to 1783 m. There is a depression area stretching from the north to the west of Soveja, divided into hilly peaks by transversal river valleys. The altitudes range from 50 to 850 m. Forms of relief such as the Vrancea Mountains, the Subcarpathian Hills and the Siret Plain are distributed in decreasing order of height and age from west to east.

The territory is crossed by several waterways: Siret, Putna, Milcov, etc. The Siret is the most important river from an economic point of view and serves as the county's natural eastern boundary. The Putna River flows from west to east and is 144 km in

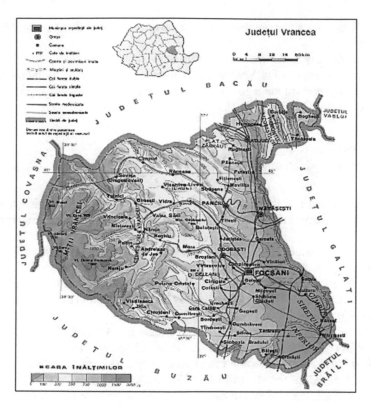

Fig. 1 Vrancea county [29]

length. Its tributaries include Zăbala (64 km in length), Milcov (68 km) and Râmna (56 km), as well as 15 other rivers and streams with lengths ranging from 5 to 25 km. The southern part of the county is partly crossed by the Râmnic River, whereas the north is dominated by the 68-km-long Sușița River [27].

III. Climate

As far as the climate is concerned, the flat areas of Vrancea County have an average yearly temperature higher than 9 °C. The area of the Subcarpathian Hills has average yearly temperatures ranging from 6 to 9 °C, and the average yearly temperatures in the mountains range from 2 to 6 °C. The average annual rainfall of this region is above 400 l/m^2, but the varied landscape ensures that precipitation is unevenly distributed across the county. Thus, the average annual rainfall is below 600 l/m^2 in the Siret Plain, the lowest relief area, no higher than 800 l/m^2 in the Subcarpathian Hills and as high as 1200 l/m^2 in the mountains.

The rainiest period is between May and June and the driest between December and February, occasionally lasting until March. Heavy rainfalls (above 30 l/m^2 in 24 h) are very common throughout the county [27].

The climate is continental, influenced by air masses from Eastern, Northern and Southern Europe.

Vrancea County is the most seismically active Romanian region; its epicentre is located in the small town of Vrâncioaia. Statistics show that two earthquakes occur here every day; most of them go unnoticed by the population, according to the local seismic station [42].

3 Vrancea—History Overview

Vrancea is situated at the boundary between the three historical Romanian provinces, Moldavia, Wallachia and Transylvania, and has been populated since time immemorial.

Between the 3rd and 11th centuries, together with the emergence of the Romanian language and people, the first clustered settlements appeared; these village communities lasted from the Middle Ages until the late nineteenth century. Vrancea thus became an administrative division of Moldavia and went on to evolve into the Land of Putna [43].

The Land of Putna made its first appearance in recorded history in 1431. It increased in size by merging with the Land of Adjud in 1591, but kept its former name. It is interesting to note that the Land of Adjud had first appeared in recorded history in 1460. Until the 1859 Union of Moldavia and Wallachia, this region had a local ruler [33]. Parts of this territory answered to Wallachia's rulers until 1482, when it became part of Moldavia, after Stephen the Great set the Milcov River as the boundary between the two provinces [28].

In 1864, the first territorial-administrative legislation passed by ruler Alexandru Ioan Cuza divided the Romania of those times into 33 counties. Putna County was one of them and took its name from the Putna River. The newly established Putna was a frontier county, whose immediate neighbours at the time were the counties of Bacău, Tecuci, Râmnicu-Sărat and Buzău [33].

The territorial-administrative legislation of the Socialist Republic of Romania passed in 1968 (and still applicable nowadays) led to the emergence of the current Vrancea County, comprising most of the former Land of Putna and parts of the former Râmnicu-Sărat County [28].

4 Vrancea's Toponomastics

In linguistics, toponyms comprise the names of places, settlements and forms of relief. Anthroponyms are words representing the names of people or toponyms originating in such names, and oikonyms are names of human settlements. As we said above, toponomastics or toponymy is a branch of linguistics which focuses on the study of toponyms.

In the following section, we will analyse the toponymy of Vrancea County. Firstly, we will elaborate on the county's name and its origin, then highlight the origin of place names in this region, with special emphasis on river names (hydronyms). We will moreover introduce several family names that are common in this area and emphasize the way in which these names have turned into toponyms (anthroponyms).

5 Vrancea or Vrancea Country

In this context, the notion of "country" refers to an area defining the territorial identity of a Romanian region, being thus considered an early form of organization of the Romanian landscape in the pre-state period [14], p. 5–6.

Studies on the toponymy of Vrancea region prompted a variety of explanations for the origin and significance of its name. According to one of the local legends, the region's name is derived from that of an old woman, Tudora Vrâncioaia, who allegedly welcomed Stephen the Great, ruler of Moldavia, into her home. Her dwelling is supposed to have been located on the current territory of Bârseşti, on Dumbrava Hill. In the Battle of Războieni of 1476, the Moldavian army was defeated by the much more numerous Ottoman army, led by Sultan Mehmed II (Mehmed the Conqueror). Legend has it that the ruler went into the mountains in search of aid. He thus reached Tudora's threshold and asked to be put up for the night. Although he did not reveal his identity, the old woman recognized the great ruler and decided to help him. As such, she sent her seven sons (Bârsan, Bodea, Pavel, Negrilă, Spulber, Nistor and Spânea) to call out to all the young men of Vrancea and thus raise a new army [23]. The new legion fought bravely to help the ruler defeat his enemies. Stephen the Great eventually managed to come victorious, forcing Mehmed to withdraw from Moldavia without conquering a single citadel [25]. At the end of the battle, in payment for their help, the ruler gifted the seven lads with the seven mountains of Vrancea, to fully own and pass down from father to son. The inhabitants of Vrancea argue that the great ruler certified this gift through a charter, leaf written with gold on vellum, signed and confirmed with the royal seal [2], p. 25. The seven young men set up prosperous households, giving rise to actual villages which they named after themselves: Bîrseşti, Bodeşti, Păuleşti, Negrileşti, Spulber, Nistoreşti and Spineşti [41].

The name "Vrancea", spelled "Varancha" (a Thracian-Getic name which might mean "forest") [21], makes its first appearance in writing during the reign of ruler Alexander the Good, within the text of a letter written in Latin by Ladislau Apor, Voivode of Transylvania, to the leader of Braşov and dated 2 July 1431 [5], p. 41. In this document, the name Vrancea is used to refer to what was then known as "The Woodlands" or "The Black Country".

Other approaches to the name suggest that it might have originated in the Sanskrit word "vran", meaning "mountain", or the Slavic word "vran", meaning "raven" [12], p. 110–112. It would therefore follow that the name Vrancea is in fact an anthroponym, made up by adding the suffix *-cea* to the Slavic word *vran*, which would translate as "Corbea" or "Country of Ravens" [9], p. 535.

According to another possible interpretation, the name Vrancea originates in the Old Romanian word "franc" or "frâncu", meaning "Westerner" or "Frank" [22] The hypothesis that the name Vrancea might stem from the word "francea" can also be taken into consideration, given Vrancea's proximity to the area inhabited in the Middle Ages by the Saxons [13], p. 16.

6 Adjud Town

Adjud is situated in the north of Vrancea County, at the confluence of the rivers Siret and Trotuş [8] and gained municipal status in the year 2000 (Law 223 of 28 November 2000). It consists of Adjud and Burcioaia towns and the villages of Adjudu Vechi and Şişcani. It has a population of 16,045 inhabitants according to the 2011 census.

Adjud emerged in close connection with the medieval towns of Wallachia and, according to archaeological findings, was undergoing a change from rural to urban settlement in the fourteenth century [4], p. 19. Its appearance and development were influenced by various factors, such as the end of the Mongolian dominion, the migration from Transylvania of a Hungarian and Saxon population engaged in trading and crafts, its position at the crossroads of two major Eastern-European trade routes, etc. The routes in question were the Moldavian one, connecting Poland to cities along the Danube River, following the Siret Valley, and the one coming from Braşov, following the Trotuş Valley and turning Adjud into a transport hub; one path led to Wallachia, the other to Bârlad-Iaşi-Botoşani [4], p. 20.

The Saxon but especially the Hungarian colonists greatly contributed to toponymic developments, as many of the place names in the area are of Magyar origin [10], p. 2201–2202. There are two main views regarding the name of Adjud. The first belongs to Academy member Iorgu Iordan, who believes the name is related to the term "jude", the old word for "judge", a dignitary with judicial responsibilities; he supports this argument by means of other place names, such as for example "Balta Giudelui" (Judge's Pond) in Iaşi County [4], p. 12. Iorgu Iordan (1888, Tecuci-1986, Bucharest) was a linguist, philologist, full member and Vice President of the Romanian Academy. He wrote studies on Romanian linguistics and his main areas of interest concerned contemporary aspects of Romanian language and onomastics [32].

The second line of thought argues that the town's name is of Magyar origin and related to the name of Saint Giles, whose Latin name was Aegidius [17], p. 1–2. Saint Aegidius also features in the Catholic calendar, and *Egidius* was a very popular and widespread first name in the Middle Ages. The same saint can be recognized under the Magyar name "Egyed Szente" [11].

The town of Adjud was first mentioned in the list of privileges granted by Ilias Voivode, the son of ruler Alexander the Good, to Transylvanian merchants in a document issued on 9 April 1433 which contains a reference to *"Egyd halma" (Egyd's*

Knoll) [11]. It would appear that Adjud is the Romanian version of the Magyar name Egyd—Egidius [4], p. 12. Egyd's Knoll was a mound situated on the left side of the road to Bacău, approximately 4 km out of town, which was visible until the 1990s. Archaeological excavations in the location yielded findings dating back to medieval times [4], p. 13.

Since there are other settlements in the area whose names are of Magyar origin [16], p. 66, most historians and linguists appear to favour the second point of view. One such example is Sascut, a small town in Bacău County, located 17 km north of Adjud on the E85 road from Adjud to Bacău. The name comes from "Fântâna Sasului" (meaning Saxon's Well), that is Szászkút in Magyar [48]. The name of the river Trotuș, situated in the vicinity of Adjud, is of Magyar origin too, Tatáros or Tátros in Magyar or "tătărăşti" in Romanian [48], and means "Tartars' River" [3], p. 55. The name of Siret River is also believed to be of Magyar origin, a derivation of "Szeret", meaning "Dear one" [38].

It is also important to point out that the first Adjud merchant ever mentioned in an official document was also of Magyar origin. His name, Kelemen of Antal, appears in a court sentence of 24 January 1547, passed by a Braşov judge. Kelemen was a livestock trader from Agiud Borough [24]. What is particularly interesting to note is that the older inhabitants of neighbouring villages still use the name "Agiud" in reference to the town of Adjud.

7 Focşani Town

Focşani, Vrancea County's seat, is situated in the vicinity of the Milcov River. It comprises the towns of Focşani (the administrative seat), Mândreşti-Moldova and Mândreşti-Munteni [26]. It was this river that ruler Stephen the Great set as the boundary between Wallachia and Moldavia in 1482, after he occupied Crăciuna Citadel (10 March 1482) [7], p. 145.

After the 1859 Union of Moldavia and Wallachia, the town entered collective memory as "The Town on the Milcov" or "Union Town" [44]. On 6 July 1862, Alexandru Ioan Cuza signed the decree ratifying the union of the town's two parts, the Moldavian Focşani and the Wallachian Focşani, the town being split by the Milcov River between the two Romanian principalities.

As we have seen in other cases, there are several hypotheses regarding the name of this town. It would appear that the name of Focşani might be related to the name of a boyar family, Focşa, living at the time of Stephen the Great [26]. The town's first mention in recorded history took place on 30 January 1575, in a document issued by the then ruler of Wallachia, Alexander Voivode (1568–1577) [30], in which the latter writes that he was "struck" by John III the Terrible (1572–1574) [18], p. 86–107, ruler of Moldavia, "in the vicinity of Focşani" [26]. This supposition seems to be the most plausible theory given the fact that in the Middle Ages settlements were named after the estates on which they were set up [1], p. 2.

Another hypothesis states that the town might date back to the times of the Moldavian ruler Vasile Lupu (1634–1653 and 1653–1653) [40]. The latter is believed to have built the settlement together with Matei Basarab (1632–1654) [34], ruler of Wallachia, as testimony of their decision to end their conflicts; the word "focşani" is used in this context to describe a "peripheral location" [1], p. 2.

8 Mărăşeşti Town

Mărăşeşti, a town of great historical significance, is the place where the greatest battle fought on the Romanian front in the First World War took place in August 1917, culminating in the victory of the Romanian troops. It consists of the towns of Mărăşeşti (the administrative seat), Siretu, Tişiţa and the villages of Călimăneşti, Haret, Modruzeni and Pădureni [31].

The name of Tişiţa is of commercial provenance and means "the selling price of vessels" [9], p. 533.

The name "Mărăşeşti" evolved from the archaic form "Mărişeuţi" and was first mentioned in 1392, in a grant deed [45] signed by the ruler of Moldavia, Roman I (1392–1394) [39]. Through this deed, the ruler endowed Ioanăş Viteazul with a section of this village: "I, Roman Voivode [...] give our servant Ioanăş Viteazul, for faithful service, 3 villages on the Siret [...] their boundary being below *Mărişeuţi*" [15].

9 Odobeşti Town

Odobeşti is situated on the left banks of the Milcov, approximately 10 km away from Focşani, and comprises the town itself and the village of Unirea; the area is well known for its vineyards [36]. The town's name is believed to be based on the surname of a respected former member of the local community, Odoabă [46]. The name of the Milcov River, which runs past the town, is of Bulgarian origin and means "caressing", "graceful" [9], p. 478.

In 1227, a Catholic diocese, *Civitas de Mylco* (Milcov Citadel), was set up in the area. Crăciuna Citadel, which was conquered by Stephen the Great in 1482 and which marked the boundary between Moldavia and Wallachia, was also located in the Odobeşti area [36]. Nevertheless, the town made its first appearance in recorded history [46] in 1626, in a document issued by the Moldavian ruler of the time, Miron Barnovschi (1626–1629 and 1633) [35].

10   Panciu Town

Panciu Town also comprises the settlements of Crucea de Jos (Lower Crucea), Crucea de Sus (Upper Crucea), Dumbrava, Neicu and Satu Nou (New Village). It is situated on the left banks of Şuşiţa River, approximately 30 km north of Focşani, in an area renowned for its vineyards [37]. The name of Şuşiţa River is of Slavic origin, from the word "suh", and means "dry", "dried up". The same word features in Czech, Croatian and Ukrainian [9], p. 128.

The town made its first appearance in recorded history in 1798, but one of its satellite villages, Crucea, appeared in documents as early as 1589 [19]. At the time, the then ruler of Moldavia, Petru Şchiopul (Peter VI the Lame) [6], endowed the governor of the Lowlands with "the vineyards of Crucea, former crown property" [47]. It would appear that the town's name is based on that of a local greengrocer, "Penciu", pronounced "Panciu" [20].

The Vrancea area is characterised by a number of specific family names, some related to sheep farming or woodworking, common occupations around those parts. Here are some of them: Baciu ("shepherd"), Jitaru, Cojocaru ("furrier"), Olaru ("potter"), Ungureanu, Cornea, Nistor, Doldor, Tuvene, Dantiş, Aga, Marcu, Oancea, Vatră ("hearth"), Bahnă, Bâtcă, Găină ("chicken"), Bratie, Predoi, Roman, Fega, etc. [2], p. 61.

The names of Tudora Vrâncioaia's seven sons became the names of the villages they started at the foot of the West Vrancea mountains; the latter constituted their reward from Stephen the Great for their bravery and contribution to the Battle of Războieni [2], p. 25. These Vrancea County settlements are "Bodeşti" (after Bodea), "Spineşti" (after Spânea), "Negrileşti" (after Negrilă), "Bârseşti" (after Bârsan), "Spulber" (after Spulber), "Păuleşti" (after Paul) and "Nistoreşti" (after Nistor) [41].

11   Conclusions

The relationship between toponymy and some social science branches is very well known that's why the good knowledge of the origin of the places names give us information regarding the evolution and the development of the society in that area.

As a case study, we chosen the Vrancea zone. Therefore, this chapter has tackled the toponymy of Vrancea County. To begin with, it detailed some etymological aspects of the county's name, highlighting the origin of some place names and river names (hydronyms) from this region. It moreover elaborated on several family names common in this area, detailing the way in which they became toponyms (anthroponyms).

References

1. Andreiaş A (2012) Monografia oraşului Focşani. Geografie umană generală-Portofoliu semestrial
2. Antimirescu I (2003) Contribuţii la Monografia comunei Nistoreşti
3. Artimon A (2003) Oraşul Medieval Trotuş. Direcţia Judeţeană pentru cultură, culte, şi patrimoniul cultural naţional Bacău, Corgal Press
4. Buculei PC (coord), Virgil Cobileac V, Avram AI (1998) Oraşul Adjud. Monografie istorică – geografică, Adjud
5. Conea I (1993) Vrancea. Editura Academiei Române, Bucureşti
6. David G (1984) Petru Şchiopul: (1574-1577; 1578-1579; 1582-1591). Editura Militară, Bucureşti
7. Drăguţ V (1976) Dicţionar enciclopedic de artă medievală românească. Editura Ştiinţifică şi Enciclopedică, Bucureşti
8. Grumăzescu H, Ştefănescu I (1970) Judeţul Vrancea. Editura Academiei Române, Bucureşti
9. Iordan I (1963) Toponimia românească. Editura Academiei Romane, Bucureşti
10. Iorga N (1937) Istoria poporului românesc, vol 1. Editura Academiei Române, Bucureşti
11. Iliescu I, Apostu AE (2013) Evoluţia Adjudului până la formarea târgului în secolul al XVI-lea in Cronica Vrancei, vol XV. Editura Pallas, Focşani
12. Mocanca TD (2013) Ţara Vrancei, dincolo de legenda Tudorei Vrâncioaia şi a celor şapte feciori. In: Metafore ale devenirii din perspectiva migraţiei contemporane. Naţional şi internaţional în limba şi cultura română, volume taken care of Luminiţa Botoşineanu, Daniela Butnaru, Ofelia Ichim, Cecilia Maticiuc. Elena Tamba. Editura Alfa, Iaşi
13. Mitu S (2000) Imagini europene şi mentalităţi româneşti. Presa Universitară Clujeană, Cluj-Napoca
14. Mureşan A (2012) Tara Vrancei. Studiu de geografie regional. PhD thesis, Facultatea de Geografie, Universitatea Babes-Bolyai, Cluj-Napoca
15. Neculai A, Ion S (2017) Mărăşeşti în fapte şi imagini - O istorie în date. Editura ATEC, Focşani
16. Paragină A (2002) Habitatul medieval la Curbura exterioară a Carpaţilor în secolele X –XV. Editura Istros, Brăila
17. Sfantul Egidius (1936) In "Dumineca", anul XIII, No. 10, 8 Martie
18. Xenopol AD (1925) Istoria românilor din Dacia Traiană, vol. V. Editura Cartea Românească, Bucureşti. Legea 223 din 28 Noiembrie 2000, regarding the declaration of the town of Adjud, Vrancea county, municipality

Sites

19. http://adafinimihailiviu.blogspot.com/p/istoria-orasului-panciu-scrisa-de-mine.html
20. https://adevarul.ro/locale/focsani/cel-mai-mic-oras-judetul-vrancea-panciu-localitatea-vii-ren umita-sampanie-invechita-hrubele-Stefan-mare-1_55223ffd448e03c0fd437c30/index.html
21. https://adevarul.ro/locale/focsani/de-vine-numele-vrancea-trecut-tinutului-i-spunea-Tara-cor bilor-negri-1_54ec3cef448e03c0fdef1b70/index.html
22. https://adevarul.ro/locale/focsani/de-vine-numele-vrancea-trecut-tinutului-i-spunea-Tara-cor bilor-negri-1_54ec3cef448e03c0fdef1b70/comment/771470.html
23. https://adevarul.ro/locale/focsani/legatura-putin-stiuta-Stefan-mare-baba-vrancioaia-reusit-domnitorul-reformeze-dreptul-proprietate-Tara-romaneasca-1_56177597f5eaafab2c77ab05/index.html
24. http://adjud.ro/comertul-adjudean-de-a-lungul-secolelor-partea-ii
25. https://doxologia.ro/viata-bisericii/documentar/amintirea-jertfei-de-la-razboieni-legenda-vra ncioaiei

26. http://enciclopediaromaniei.ro/wiki/Focşani
27. http://enciclopediaromaniei.ro/wiki/Judeţul_Vrancea
28. http://enciclopediaromaniei.ro/wiki/Jude%C5%A3ul_Vrancea#Istoric
29. https://pe-harta.ro/vrancea/
30. https://ro.wikipedia.org/wiki/Alexandru_al_II-lea_Mircea
31. https://ro.wikipedia.org/wiki/Bătălia_de_la_Mărăşeşti
32. https://ro.wikipedia.org/wiki/Iorgu_Iordan
33. https://ro.wikipedia.org/wiki/Jude%C8%9Bul_Putna_(interbelic)
34. https://ro.wikipedia.org/wiki/Matei_Basarab
35. https://ro.wikipedia.org/wiki/Miron_Barnovschi
36. https://ro.wikipedia.org/wiki/Odobeşti
37. https://ro.wikipedia.org/wiki/Panciu
38. https://ro.wikipedia.org/wiki/Râul_Siret
39. https://ro.wikipedia.org/wiki/Roman_I
40. https://ro.wikipedia.org/wiki/Vasile_Lupu
41. http://tara-vrancei.ro/legenda-babei-vrancioaia.html
42. https://www.camping.info/docs/Campsite/27561/zona-turistica-vrancea.pdf
43. http://www.muzeulvrancei.ro/despre-noi/
44. http://www.muzeulvrancei.ro/focsani/
45. http://www.muzeulvrancei.ro/marasesti/
46. http://www.muzeulvrancei.ro/odobesti/
47. http://www.muzeulvrancei.ro/panciu/
48. http://www.wikiwand.com/ro/Toponimia_maghiară_în_Moldova

Performance Management Risk-Based Approach in the European Research Infrastructures (RIs): An Introduction to an Integrated Perspective. Case Study at EMSO ERIC

Maria Incoronata Fredella, Roberto Jannelli, Paola Materia, Maria Grazia Olivieri, Juan José Dañobeitia, and Massimo Squillante

Abstract Combining three essential management elements (Risk Management System, Internal Control Process and Management Accounting) to strengthen the process of achieving strategic objectives in the management of a Research Infrastructure (RI) is the methodology the authors recommend in running Research Infrastructures. The key performance indicators intend to be multi-functional tools in a more complex system, and in integrating them into the RI governance will guarantee effective management. The next phase will focus on how the organisation can leverage risk management throughout the RI, by incorporating it into its governance structure and processes.

M. I. Fredella (✉) · P. Materia · J. J. Dañobeitia
European Multidisciplinary Seafloor and water-column Observatory European Research
Infrastructure Consortium (EMSO ERIC), Via di Vigna Murata, 605, 00143 Rome, Italy
e-mail: maria.fredella@emso-eu.org

Via G. A. Resti, 63, 00143 Rome, Italy

P. Materia
e-mail: paola.materia@emso-eu.org

J. J. Dañobeitia
e-mail: juanjo.danobeitia@emso-eu.org; jjdanobeitia@cmima.csic.es

R. Jannelli · M. G. Olivieri · M. Squillante
Department of Law, Economics, Management and Quantitative Methods, University of Sannio,
Palazzo de Simone, 82100 Benevento, Italy
e-mail: jannelli@unisannio.it

M. G. Olivieri
e-mail: mgolivieri@unisannio.it

M. Squillante
e-mail: squillan@unisannio.it

J. J. Dañobeitia
Spanish National Research Council (CSIC), Paseo Marítimo de La Barceloneta, 37-49, 08003
Barcelona, Spain

© The Author(s), under exclusive license to Springer Nature Switzerland AG 2021   387
D. Soitu et al. (eds.), *Decisions and Trends in Social Systems*,
Lecture Notes in Networks and Systems 189,
https://doi.org/10.1007/978-3-030-69094-6_31

Keywords Performance management · Internal audit · Integrated risk approach · Balance scorecard · Research infrastructures

1 Introduction

Managing Research Infrastructures (RIs) has generated considerable recent research interest due to the strategic meaning of this new type of organisations. According to the Organisation for Economic Co-operation and Development (OECD) definition, they are recognised as long-term enterprises, they represent strategic investments in enabling and developing research and may offer significant socio-economic returns.

EU considers that science and technology are favourable strategic factors for growth to boost the European economy, and Research Infrastructures represent the primary tools to fill the scientific and technological gaps with regards to other international Countries as they play an essential role in achieving scientific progress, technological advancement and knowledge transfer. Indeed, across the EU timeline, the RIs have been a subject of strategic discussions. This condition reflects the fact that even if the importance of creating organisations representing a unique source of knowledge-based competitiveness is undisputed, on the other hand, maintaining them requires a considerable economic effort. The direct budget foreseen and dedicated by the EC to the Research Infrastructures for the next seven years (2021–2027) is around €2.4 billion. Research Infrastructures are, de facto, the synthesis of European and national growth strategies in terms of research and innovation. To act as an incubator for RI projects with pan-European interest, the European Strategic Forum on Research Infrastructures (ESFRI) was set up in 2002; it provides both the strategic policies within a broad framework for the Research Infrastructures and a strategic interface as it is composed of national delegates appointed by the research ministers of the EU countries and includes Commission representatives. The ESFRI roadmap 2018 includes 18 ESFRI Projects in the development phase; the estimated investment is around €2.9 billion in the coming years. Moreover, thirty-seven Landmark already set represent a total capital value of around €14.4 billion.[1]

Since funding agencies are expected to make key decisions about funding to invest, implementing new RIs, the competitiveness between different RIs is increasing. Meanwhile, the importance of strengthen methodologies to making the RIs comparable, assessing the overall impact of RIs, is becoming a central issue of concern for the EU. The evidence is revealed by the fact that the Competitiveness Council of 29 May 2018 adopted conclusions on Accelerating knowledge circulation in the EU, in which "invites Member States and the Commission within the framework of ESFRI to develop a common approach for monitoring their (RIs) performance and invites the Pan-European Research Infrastructures (RIs), on a voluntary basis, to include it in their governance and explore options to support this through the use of Key Performance Indicators (KPIs) …" As a result of this situation, the game is

---

[1]WEB: https://ec.europa.eu/info/research-and-innovation/strategy/european-research-infrastructures/esfri_en.

played in three different perspectives reflecting the different player roles: the first one is the European Commission one, which through a High-Level Expert Group (HLEG) assess the effectiveness of the Research Infrastructures funding instruments under the Union Framework Programme and the Research Infrastructures progress towards implementation and their long-term sustainability[2].

Furthermore, it assesses the Research Infrastructures progress towards implementation and their long-term sustainability. In this respect the HLEG works interacting with ESFRI, the second one, to ensure complementarity with its ongoing assessment and evaluations of the ESFRI ad hoc working group (WG) which is developing a set of KPIs as a basis to monitoring RI performance and to developing a methodology for the periodic update of ESFRI pan-European RIs'. The third actors are the RIs; they are associated with the long-term enterprises, and as consequence they have to identify policies and procedures to be performant and guarantee their financial autonomy.

A broader literature refers to the importance of managing enterprises according to a harmonised view between strategy and risk management framework. Moreover, integrating Internal Audit within Risk Management is equally suitable and required. Nevertheless, organisations often are managed according to the silo mentality, in which each section or department does not interact with the others. Here we are also arguing this silo approach is time-consuming and inefficient if applied to a Distributed RIs.

The Multidisciplinary Seafloor and water-column Observatory European Research Infrastructure Consortium (EMSO ERIC) management team, and the Department of Law, Economics, Management and Quantitative methods of the University of Sannio, are jointly developing the methodology presented here as a tool aimed to optimise processes and procedures to ensure allocating resources most effectively and efficiently. This approach includes the implementation of a new control system with a risk-based approach integrating commonly used control systems, to make more efficient, equitable, and sustainable use of RIs human, financial, and other resources, and to strengthen the process of achieving EMSO ERIC strategic objectives. This approach underpins the definition and the choice of our KPIs in the living process of self-learning to support strategies, also considering the interests and requirements of the stakeholders and beneficiaries of investment in the RI.

---

[2]Register of Commission expert groups and other similar entities—*Horizon 2020 High-Level Expert Group to assess the progress of ESFRI infrastructures towards their implementation and long-term sustainability (E03641.* Policy area: Research and Innovation-Lead DG: RTD-DG Research and Innovation—Last updated: 23 Jul 2019

2 Methodology

In the RI environment we believe that segregating strategy/performance management, risk management and internal control within separate organizational boxes is unfertile. The approach here discussed is designed to merge strategy and, consequently, the related objectives with risk management into a unique framework, thereby creating the "Risk Governance". In this perspective, it is possible to align strategic goals and sustainable exposure to risk. The Risk Governance (RG) framework and methodology push into a holistic approach the Strategy/performance management, the Risk Management and the Internal Control of the RIs. Therefore, RG enables RIs to evolve their previous approaches into an integrated view, rather than having a deconstructed process.

In this chapter, we will trace how a RI organization is looking to incorporate risk management into the Strategy/performance plan [11]. It is a step by step process in which some elements are foreseen to identify and manage risk at the strategic objective level. Top management should focus the RIs on obtaining its strategic objectives, including undertaking risks that may hinder to achieve the goals at all levels. One of the most crucial steps is to identify risks for each strategic objective of the RIs Strategy/Performances plan. Thus, this step provides a view for implementing strategy/performance measurement system and risk management plan. Subsequent phase defines the environmental context at both the internal level, such as strategy, organization, risks, and the external level, such as stakeholders, and applying a SWOT alongside PESTLE analyses system. Then combined strategy/performances with risks using Risk assessment. Reviewing strategy goals and KPIs, it is possible to identify, to analyse, to evaluate and to bear the risks. Internal control allows monitoring phase about the execution of RIs strategic and operating plans and risks status. Subsequent phase defines the risk treatment strategy related to strategy/performance plan. Usually, risk treatment obtained through a mitigating initiative when not focused on reducing the likelihood of changing the impacts.

In order to put the RIs on the basis of complete efficiency and effectiveness, it is necessary to jointly to develop a proper system of programming and control and systematic risk management. First, it is essential to have a valid system to define objectives, to determine resources, to define the measure of the impact and a reliable performance measuring system. Subsequently, it should be incorporated the risk management system into this planning and control framework, creating only one integrated system.

3 The Multidisciplinary Seafloor and Water-Column Observatory European Research Infrastructure Consortium (EMSO ERIC)

EMSO ERIC is a pan-European Distributed Research Infrastructure, hosted the head-quarters at Istituto Nazionale di Geofisica e Vulcanologia, in Rome. It consists of an integrated system of Regional Facilities that now involves 8 European Countries with 11 fixed-point multi-sensor platforms deployed in 'key environmental sites' across the European seas, in the Atlantic, in the Mediterranean to the Black Sea in the eastern side of the region. The RI is part of a broader scientific research landscape, it complies with the national and the European research strategies in the marine domain, producing scientific services and products demanded to improve Europe´s ability to achieve societal benefits, increasing knowledge in pressing environmental issues, such as Global Change, Biodiversity and Marine Ecosystems, Geo-hazards, which already affects the daily life and the economy of the European and Worldwide societies.

Since it is a Distributed RI, the only EMSO ERIC physical facility is the Central Office. The term EMSO Regional Facility (RF) indicates a set of resources in terms of expertise (Human Resources: skills, knowledge…), data, research activities, hardware (that comprises the Observatory infrastructure and other types of equipment). Furthermore, an EMSO RF operates its services, responding to the demands of its own users. Data, as well as other resources, are generated by the EMSO RF and a specific agreement defines specifically those resources each RF dedicate to the RI. EMSO ERIC Service Groups (SGs) provide the service function. These *units are located in one or more countries that are tasked with specific activities of transversal interest* and guarantee access to data, technology, innovation expertise and scientific research.

The EMSO ERIC SGs ensure that excellent research leads and promotes innovation. The integration of activities carried out by the Regional Teams and the Service Groups constitute part of the value proposition and impacts on the sustainability of the RI.

The Chief Executive Officer and the legal representative (Director-General) is responsible for defining the global strategy and coordination of the EMSO RFs [7], assisted in the performance of his functions by the staff of the Central Management Office (CMO) and by the Executive Committee (the Advisory Body to support the DG). The CMO also needs to help in optimising the framework conditions and instruments to enabling high-quality services, products and activities and facilitate the interlinkages between the different SGs.

The main objectives of the EMSO ERIC mission statement are enabling excellent research; driving technological developments; guaranteeing both physical and virtual access to infrastructure information/data and services; promote innovation and transfer of knowledge and engage in partnerships with industry; developing an

EMSO ERIC Multidisciplinary users community and facilitating regional and international interdisciplinarity activities; providing advanced tools for decision processes
at the level of policymakers, business, civil society and other stakeholders.

Moreover, consolidating the establishment of EMSO ERIC management governance rules and procedures, coordinating the EMSO Regional Facilities operations,
extension or new construction of them, maintenance, as well as providing the integrated services and ensuring an added value, attracting new members, are the main
challenges that the RI has to face to ensure its long-term sustainability, together with
the access to high-quality information, which represents the core of the activities and
contributes to the implementation of the European Research Area (ERA).

4   Phase I. the *Strategy-Driven Performance Management* as a Tool to Translate EMSO ERIC Strategy in Actions

The Risk Governance Approach is part of the EMSO ERIC governing tasks setting up
to support general management to strengthen both the efficiency and the effectiveness
of the RI. It has been designed in cross perspective and designed in distinct phases.

A crucial first element, faced in this chapter, is the adoption of two management systems: the Management By Objectives (MBO) and the Balanced Scorecard
(BSC), to align concrete objectives with EMSO ERIC vision. The model of BSC
proposed is conceived not just as a performance management tool but as a means for
creating a strategy-driven performance management RI (Kaplan and Norton 2001).
This approach implies that not only traditional financial and monitoring measures are
foreseen for the RI, but also such qualitative measures as the internal processes, the
employee activities/growth, the corporate mission and the concerns of shareholders,
together with the user's issues. In this framework, the external perspective component will allow us, in the further phase, to move to the RI impact analysis by turning
the performance indicators into impact indicators.

The BSC here has been applied to a Research Infrastructure and describes, through
causal relationships between its main strategic objectives, how EMSO ERIC creates
value. Indeed, it represents a roadmap to transform the Strategy into operational
terms, and also to align the organization to the Strategy by linking different functions of the structure (finance, services, communication, and so on). Also, the tool
is conceived to support employees develop personal objectives by promoting their
adherence to and implementation of the business' strategies. It is also needed to link
Strategy to the budget and to establish a process for learning and adapting RI in its
different life-phases.

The roadmap is formulated according to four perspectives, financial, stakeholders'
outcomes (users), internal processes, and learning and growth (RI capacity to create
add-value in terms of social and human capital). Each perspective, here outlined,
has its own objectives, measurements, targets and initiatives. *Financial perspective*
outlines the objectives to be accomplished to satisfy Members Countries. In this

perspective, the organization plans for growth, increasing revenues, handling risk, and managing revenues are considered. It is possible, then, to evaluate both how well revenues are managed, and the strategy is being executed, as reflected in bottom-line results. *User Perspective* is the strategy for creating value and differentiation from the perspective of the RI users. It represents the RI ability to attain its strategic objectives in terms of users and its attractiveness. This perspective gives the magnitude of the capability of the RI to serve users' needs in order to achieve the main financial objectives. Users view is necessary because the only route to long-term financial success is to deliver the services demanded by users. The ultimate goal is to create a portfolio of services that deliver superior value for the targeted users. *Internal Process Perspective* focuses on internal operations and focuses on how the RI delivers services. It means in which internal processes the organisation must excel to satisfy its users and Members. In this category, physical factors with a direct impact on economics are described, such as schedule efficiency, wage level, behaviours. It also includes services quality, and indicators on accessibility, availability, reliability, safety and transparency are required. Furthermore, a RI needs well-trained, highly skilled employees if it is to excel in the first three strategic perspectives. The *Learning and Growth Perspectives/Innovation* is oriented to the priorities to create a climate that supports organizational change, innovation and growth. Invest in human capital by spending money on employee capabilities, information/knowledge systems and motivation, empowerment and alignment are crucial points, especially for a Research Infrastructure. These factors can be measured by assessing employee satisfaction, skills and productivity, training and skill levels.

EMSO ERIC strategic priorities aim to the long-term sustainability in terms of growth, user services (by increasing differentiation and supply) and awareness. Linking the intangible and tangible assets in value-creating activities and tracking progress to the objectives is the idea underpinning the RI structure. A division of organisational structure has been dedicated to fulfilling its mission in terms of delivery services to the users. The Central Management Office reflects this approach, and it is structured for specific objectives, and decisions are made based on the identified purposes, set priorities (Fig. 1). The process of specification of the relevant objectives

Mission statement	Science Enabling excellent science	Technology Driving technological developments. Physical access to the RI	Data Guarantee the openness and the H-Q of research data. Making data meet standards of findability, accessibility, interoperability, and reusability.	Innovation Promoting innovation and transfer of knowledge and engage in partnerships with industry	Communication Effective communication and dissemination
Officers	SCIENCE OFFICER	TECHNOLOGY OFFICER	IT DIRECTOR	INNOVATION OFFICER	COMMUNICATION OFFICER
Service Groups	SCIENCE SERVICE GROUP	ENGINEERING AND LOGISTIC SERVICE GROUP	DATA SERVICE GROUP	INNOVATION SERVICE GROUP	COMMUNICATION SERVICE GROUP

Fig. 1 Part of EMSO ERIC organizational components and their link with the main objectives related to mission statement

allows the transition from the general objectives to the operational choices, activities and subjects responsible for the goals developed by the executive management.

The objectives in EMSO are formulated referring to a multi-year period and ordered in terms of assigned priorities, time horizon and available resources. The directional objectives (relating to the year) arise from the strategic ones, as the result of a mediation between *strategic* needs and *practical* constraints. They are articulated in sub-objectives, achievable and measurable. Once expected results are outlined, an internal evaluation of the performances for each Office within the CMO and each Service Group is needed, so that each organizational component is tracked to achieving goals in compliance with timing agreed at general managerial level, and in the four-perspective approach.

Since the RIs are the expression of social investments, they have to demonstrate the sustainability and the effectiveness for science and society. Moreover, the request for normalizing behaviours expresses the need to make them comparable and to evaluate the goodness of their actions. On the other hand, we also understand that each RI is unique (even if we compare RIs at the same stage of its life cycle) because it reflects a specific scientific community, with own needs and character. We have faced the dichotomic thought of universality and uniqueness by designing a model consisting of two kinds of objectives to monitor, the so called common and the specific ones.

The facilitate comparison of the activities of EMSO ERIC, we have chosen to apply to the RI a Strategy-Focused model described above. This approach allows the usage of a "double-loop" process, integrating the management of budgets and operations with the management of strategies. Moreover, a monitoring system allows progress to be monitored and corrected. The list of possible indicators to be implemented KPIs has been outlined in line with the EMSO ERIC strategic objectives and reviewed on a comparative basis to work carried out by the Global Science Forum both in the Expert Group meeting on reference framework for assessing the socio-economic impact on RIs (March 2018) and Reference framework for assessing the scientific and socio-economic impact of Research Infrastructures (March 2019).

We have chosen to compare our work to the list of indicators they prepared because it is related to the common strategic objectives reported by many European RIs. The Indicators they propose respect following principles: neutrality (an external subject define them); generality (they are valid for many RIs); specificity (they are valid for homogeneous bodies and allow both temporal and spatial comparison). Furthermore, the OECD has a global echo. Concerning the specific objectives, we are planning setting up another kind of KPI, so-called *Enterprise KPI* (specific for EMSO ERIC). Here we will discuss the *common KPIs*.

Even if the RI have to keep the focus grounded on the question of how to achieve the greatest impact on the mission, their activities have inherent economic value. Indeed, non-profits can generate multiple funding streams, as well as also different program activities have potentials for revenue. An orientation toward sustainability is required to maintain and making growth organisation, but decisions about growth must be supported by both the potential for advancing the mission and the potential for financial sustainability (Fig. 1).

EMSO ERIC is a Distributed RI, so its mission impact, between other factors, also depends on the geographic scale [12]. This character is also reflected into the model implementation, which will be focused on the activities centralised as well as on the distributed ones, and addressed to both the RI Services and activities carried out at the regional level.

Figure 2 shows a conceptual framework comprising a range of observable direct and indirect effects. *Strategic critical factors* and *strategic perspectives* work together to create an impact in different aspects of science, society and economy, fostering the RI financial sustainability. Here, an RI is designed as a holistic system in which research institutions, industry, culture-based public and civil society are interconnected within a natural environmental perspective to support government/policy-making processes. For an analytic study on the methodology to implement strategic perspectives within public institutions [19].

The component objectives, as described through the BSC, are characterized by an aligned scorecard of KPIs and targets, as well as strategic initiatives. The management team identified in the KPIs a simple and reliable means to measure achievement, and changes connected to corrective intervention. The implementation of the qualitative variables as well as quantitative listed below will help management in assessing the performance of the RI development actors. Qualitative KPIs have been selected concerning the objectives related to the EMSO ERIC mission statement, while the quantitative ones refer to the Financial perspective (Fig. 3).

4.1 Implementing Financial KPIs

In this session, we started to look at how the Consortium is supported financially. This preliminary analysis aims to test the model, and the results presented below reflect the organization as a whole to date. The study has been built on some basic financial documents which concern the Central Office activities. Data to generate the following summaries have necessary been collected since 2016, the year in which EMSO started its activities as a European Research Infrastructure Consortium. Since 2017 EMSO is in the pre-operational phase even if its implementation project, EMSO-Link is still running. As the Consortium is in the early stage, some financial data are under development. This does not compromise on providing critical information necessary to create a sound future business model.

EMSO ERIC Income Statement—Performance Measurement—Comparing Total Revenues with Total Operating Costs permits us to understand both the ability of EMSO ERIC to break even and also its capacity to bear the expenses strictly related to the operative functioning of the Consortium (Fig. 4).

Moving from 2016 to 2017, we illustrate a 20% increase in both revenues and costs, moreover in 2017 Tot. Revenue exceeds Tot. Operating Costs of 2524€, which could be then used to cover depreciation and taxation.

EMSO ERIC Balance Sheet—Performance Measurement—Comparing Total Assets with Total Liabilities indicates the level of the entity's financial leverage.

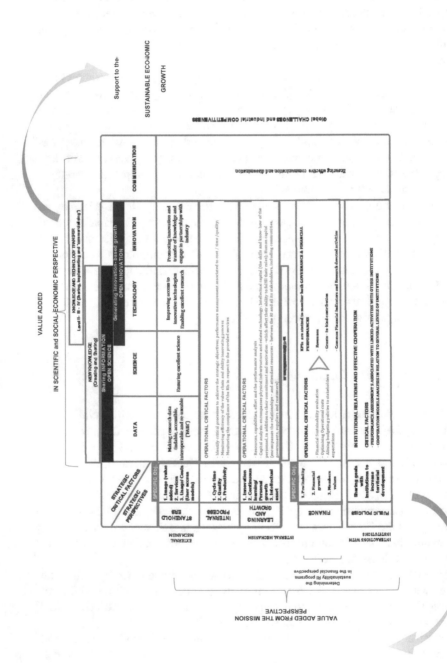

Fig. 2 Matrix representing the EMSO ERIC sustainability and growth process

NON FINANCIAL PEFORMANCES

MEASURES		KPIs	
A. Scientific Excellence			
A.1	Experimental time available or size of resources database	A.1.1	$\dfrac{Number\ of\ charged\ hours\ of\ experiment}{\sum Available\ hours}$
		A 1.2	*Number of Links to other Databases*
		A 1.3	*Number articles in Database*
A.2	Number of proposals/user requests or number of registered users of data, services	A.2.1	*Number of accounts registered / year*
		A.2.2	*Δ Accounts registered / year*
		A.2.3	*Number of logins / year or months*
		A.2.4	*Number of downloads / years*
A.3	Number of granted proposals/accepted users or number of logins/months; number of downloads, number of studies or services	A.3.1	*Total $ grants amount*
		A.3.2	*N grants awarded / Total number of proposals*
A.4	Number of publications	A.4.1	$\dfrac{Number\ of\ publication\ per\ researcher}{\sum of\ experimental\ hours}$
		A.4.2	*Number of publications in Top-ranked Journals (Top 10%)*
A.5	Proportion of publications in top 10% in comparable field	A.5.1	$\dfrac{Number\ of\ publication\ per\ researcher}{Total\ experimental\ hours}$ *(Top 10%)*
		A.5.2	*Number of publications in Top-ranked Journals (Top 10%)*
B. Education and Training			
B.1	Person-hour for staff receiving training	B.1.1	$\dfrac{Total\ annual\ training\ hours}{N\ of\ annual\ trainees}$
B.2	Number MSC and PhD Thesis	B.2.1	*Number of annual PhD Thesis*
		B.2.2	$\dfrac{Number\ of\ PhDs}{Total\ staff\ involved}$
B.3	Number of hours of participants in training events or through online services	B.3.1	*Total Training Hours*
		B.3.2	*Total hours of e-learning/Total training hour*

Fig. 3 Key performance indicators prospect

C. **Facilitating regional and transnational collaboration and activity in Europe**

C.1	Number of members (from other EU countries)	C.1.1	$$\frac{Number\ of\ staff\ from\ other\ EU\ countries}{Total\ staff\ available}$$
		C.1.2	$$\frac{Number\ of\ staff\ from\ other\ EU\ countries}{Total\ FMSO\ members}$$
C.2	Share of users and publication per EU Country	C.2.1	$$\frac{Total\ number\ of\ EU\ publications}{Total\ number\ of\ publications}$$
		C.2.2	$$\frac{Number\ of\ staff\ from\ other\ EU\ countries}{Total\ number\ of\ publications}$$
		C.2.3	$$\frac{Number\ of\ users\ from\ other\ EU\ countries}{Total\ number\ of\ users}$$

D. **Innovation and Knowledge transfer**

D.1	Share of publications/co-publication with industry	D.1.1	$$\frac{Number\ (co)publications}{Total\ number\ of\ publications}$$
		D.1.2	$$\frac{Number\ co/publications\ Top-Ranked\ Journals}{Total\ number\ publications}$$

E. **Outreach to public and policy makers**

E.1	Number of events organized for target groups and number of participants	E.1.1	Total number of events organized
		E.1.2	Total number of participants
		E.1.3	$$\frac{Number\ of\ partecipants}{Total\ number\ events}$$
		E.1.4	$$\Delta\% = \frac{Partecipants\ EVENT\ n+1 - Partecipants\ EV\ n}{Total\ partecipants\ Evento\ n}$$
E.2	Number of times the RIs is mentioned in media, articles, radio or TV broadcasts or web-based media	E.2.1	Number of times the RIs is mentioned in media, articles, radio or TV broadcasts or web-based media
		E.2.2	$\Delta\%$ Number of times the RIs is mentioned in media, articles, radio or TV broadcasts or web-based media / year
E.3	Website popularity and level of social media engagement	E.3.1	$\Delta\%$Number of followers
		E.3.2	$\Delta\%$Mentioning on social media
		E.3.3	$\Delta\%$ Retweets, likes, views and comments
		E.3.4	$\Delta\%$ Engagement rate

Fig. 3 (continued)

F.   Data

	F.1.1	*Number of reports published/year*
F.1 Number of publicly available data sets used externally	F.1.2	*Total Reports downloads or citations/year*
	F.1.3	*Growth rate of data published*

G.   Support to public policies and Standards

G.1 Participation by RIs in policy related events, committees & advisory boards	G.1.1	*Number of policy-related & advisory events lunched/attended*
G.2 Number of times the RIs or its projects are cited in policy related publications	G.2.1	*Yearly references in political Journals/Magazines/Newspapers*

H.   International cooperation

H.1 Share of research projects with one or more partners outside the EU	H.1.1	*Number of Agreements – Partnership/Year*
	H.1.2	$\dfrac{Number\ of\ projects\ with\ non-EU\ Partners}{Total\ Projects\ lunched}$
H.2 Training in an international context (participant-days)	H.2.1	$\dfrac{Total\ annual\ international\ training\ days}{N\ of\ annual\ trainees}$

I.   Governance

I.1 Revenues	I 1.1	*Total revenues / Total approved budget*

FINANCIAL PERFORMANCES

J.1 Research (i.e. RI access costs)	J.1.1	$\dfrac{Research\ expertiments\ costs}{Total\ number\ of\ experiments}$
	J.1.2	$\dfrac{Total\ Research\ expenditures}{Total\ Reasearc\ budget}$
	J.1.3	$\dfrac{Researchers\ costs}{Total\ personnel\ costs}$

Fig. 3 (continued)

		J.1.4	$$\frac{Amount\ of\ research\ investements}{Total\ amount\ on\ investements}$$
		J.1.5	$$\frac{Total\ reserach\ expenditure}{Total\ number\ of\ publications}$$
J.2	Grants	J.2.1	$$\frac{\$\ Total\ grants}{Total\ number\ of\ grants}$$
		J.2.2	$$\frac{\$\ Total\ grants}{Total\ number\ of\ researchers}$$
		J.2.3	$$\frac{\$\ Total\ grants}{Total\ revenues}$$
		J.2.4	$$\frac{Total\ number\ of\ grants}{Total\ number\ of\ researchers}$$
J.3	In-kind contribution	J.3.1	$$\frac{\$\ Total\ in-kind\ contributions}{Total\ revenues}$$
		J.3.2	$$\frac{Contribution\ in-kind\ by\ Country}{Total\ in-kind\ contribution}$$
J.4	Main Financial Indicators	J.4.1	$$\frac{Current\ Asset}{Current\ liabilities}$$
		J.4.2	$$\frac{Cash\ and\ Cash\ Equivalents}{Short\ term\ debt\ and\ other\ liabilities}$$
		J.4.3	$$\frac{EBIT}{Total\ Revenues}$$
		J.4.4	$$\frac{Staff\ Disposable\ by\ Hosting\ country/Others}{Total\ Staff\ Disposable}$$
		J.4.5	$$\frac{Revenues\ from\ rendered\ services\ to\ industry}{Total\ Revenues}$$
		J.4.6	$$\frac{Revenues\ from\ EU\ and\ NON-EU\ Projects}{Total\ Revenues}$$

Fig. 3 (continued)

This analysis provides information on the proportion of Tot. Assets that have been financed by creditors, namely through S-T and L-T liabilities (Fig. 5).

The absence of equity, the amount of the total assets matches exactly the value of the liabilities, both current and non-current. Therefore, the Consortium in 2016 and 2017 has fully relied on debt sources to support the investments.

Subset J.2: GRANTS

J.2.1—This ratio gives information on the average proportion of grants that EMSO ERIC can attract each year.

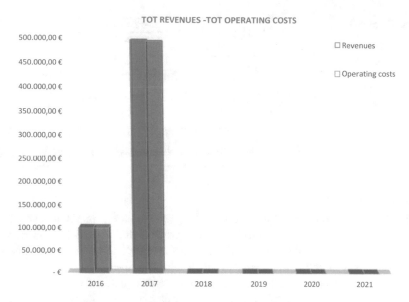

Fig. 4 EMSO ERIC income statement—performance measurement

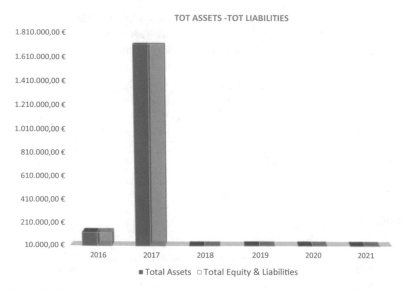

Fig. 5 EMSO ERIC balance sheet—performance measurement

The ratio total grants versus total number of grants showing that EMSO ERIC has a sustainable growth in the first three years reaching its peak in 2018. Practically, the Consortium has been able to obtain an average amount of subsidies that has supported the financing of Program. Some data is in progress for 2019 (Figs. 6

Fig. 6 J.2 grants

J.2.1	$\dfrac{€\,Total\ grants}{Total\ number\ of\ grants}$
J.2.2	$\dfrac{€\,Total\ grants}{Total\ number\ of\ researchers}$
J.2.3	$\dfrac{€\,Total\ grants}{Total\ revenues}$
J.2.4	$\dfrac{Total\ number\ of\ grants}{Total\ number\ of\ researchers}$

and 7).

J.2.2—This ratio gives information on the proportion of grants that each people in the Consortium can use throughout the projects and activities. Higher is the ratio greater will be the amount of resources available to the research function, consequently affecting the ability to reach expected results (Fig. 8).

Total grants versus total number of people ratio equally shows a sustainable increase in the number of grants per researcher reached its peak in 2018 by ~€44.139, the proportion subsequently decreases but this is attributable to the partial information of the year 2019.

J.2.3—This ratio compares the amount of income from awarded grants, with the total revenues of the Consortium. It gives information on the ability of EMSO ERIC to sustain its activity by using grants rather than all the other sources of

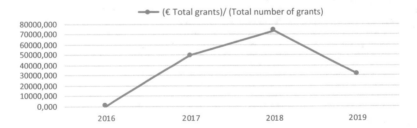

Fig. 7 J.2.1 ratio

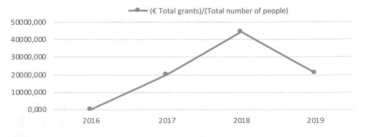

Fig. 8 J.2.2 ratio

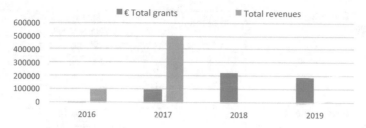

Fig. 9 J.2.3 ratio

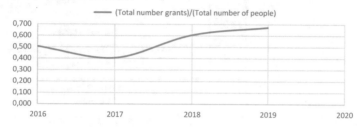

Fig. 10 J.2.4 ratio

income. Meanwhile, the ratio can also be interpreted by considering the degree of dependence that EMSO ERIC has on the Grants it gets each year (Fig. 9).

In 2017 EMSO ERIC had both an increase in grants and a boost in revenues. At that time, the grants awarded amounted to 20% of the total revenues, which means the Consortium had a moderate degree of dependence on grants. Vertical axis numbers are thousands of euros.

J.2.4—An increase in the number of people should be sided by a proportionate rise in the number of grants awarded. This will inevitably demonstrate that activities are well-financed and supported by external funds (Fig. 10).

Ratio from total number of grants versus total number of people illustrates, since 2017, EMSO ERIC has experienced a positive trajectory in the average number of grants per person, which has respectively increased by 50 and 11% in 2018 and 2019.

Subset J.3: In-Kind contribution

J.3.1—It measures the degree of incidence of the total nominal amount of the in-kind contributions in EUR (€) on Total Revenues. In 2017 the ratio amounted 47.9%, which means that in 2017, almost half of the total Revenues derive from the Total in-Kind contribution (Fig. 11).

J.3.2—It measures the extent to which each Member contributes to the Total in-kind contribution. From data for 2017, two Countries gave their in-kind contribution, the host country, namely Italy, with 92% and Spain with 8% (Fig. 12).

Fig. 11 J.3 In-kind
contribution

J.3.1	$\dfrac{\text{€ Total in − kind contributions}}{\text{Total revenues}}$
J.3.2	$\dfrac{\text{Contribution in − kind by Country}}{\text{Total in − kind contribution}}$

Fig. 12 J.4 main financial
indicators

J.4.1	$\dfrac{\text{Current Asset}}{\text{Current liabilities}}$
J.4.2	$\dfrac{\text{Cash and Cash Equivalents}}{\text{Short term debt and other liabilities}}$
J.4.3	$\dfrac{\text{EBIT}}{\text{Total Revenues}}$
J.4.4	$\dfrac{\text{Staff Disposable by Hosting country}}{\text{Total Staff Disposable}}$
J.4.5	$\dfrac{\text{Revenues from rendered services to industry}}{\text{Total Revenues}}$
J.4.6	$\dfrac{\text{Revenues from EU and NON − EU Projects}}{\text{Total Revenues}}$

Subset J.4: Main Financial Indicators

J.4.1—The current ratio is a liquidity ratio that measures the entity's ability to pay short-term obligations or those due within one year. It tells investors and analysts how a company can maximize the current assets on its balance sheet to satisfy its current debt and other payables. According to the data provided in 2016, there has been a strong financial performance. In 2017, the amount of current assets did not match current liabilities. Therefore, EMSO ERIC solvency capacity has slightly lowered of 0.6%

J.4.2—The cash ratio is a measurement of the Consortium's liquidity, specifically the ratio of a company's total cash and cash equivalents to its current liabilities. The metric calculates an organisation's ability to repay its short-term debt with cash or near-cash resources, such as readily marketable securities. This information is useful to creditors when they decide how much money, if any, they would be willing to loan an organisation (Fig. 13).
The S-T liabilities has increased between 2016 and 2017, the cash and cash equivalent have raised more than proportionally, enabling the entity to speed up the repayment of its obligations.
J.4.3—EBIT margin is an assessment of a firm's operating profitability as a percentage of its total revenue. It is equal to earnings before interest and taxes (EBIT) divided by total revenue. EBIT margin can provide a significant indicator of the entity's operating profitability and cash flow. For an even more complete

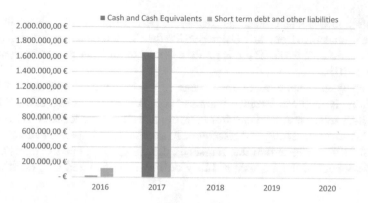

Fig. 13 J.4.2 the cash ratio

view it is possible to calculate the EBITDA Margin. Being a not for profit organization EMSO ERIC EBIT margin will logically tend to 0. However, we observe a positive value in 2017, which means that the consortium has been able to cover interest and taxes expenses fully.

J.4.4—With this ratio, it is possible to understand the extent to which the hosting country (Italy) provides staff with respect to the overall staff disposable.

J.4.5—Both in the EMSO ERIC activity report 2016 and in the EMSO ERIC balance sheet 2017 it is remarked how services rendered to the industry are one of the three sources foreseen of income pivotal for EMSO ERIC financial sustainability. With this ratio we understand the extent to which the services provided generate liquidity the entity. However, so far, no revenues have been generated by the services mentioned above.

J.4.6—With this ratio, it is possible to understand the degree of incidence of revenues arising from EU and NON-EU grants on the Total revenues (Fig. 14).

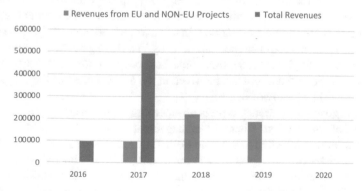

Fig. 14 J.4.6 ratio

Illustration of income-revenues for 4-year period (2016–2019). Note the sharp increase in subsidies from the EU and outside the EU in total revenue that goes from u 5% 2016, to 20% 2017. There is no data available on other types of revenues for 2018 and 2019.

The analysis wants to represent a mean to leverage the effectiveness of linking strategic goals and KPIs. Besides, it aims to complement the EMSO ERIC Financial Plan, which solves issues arising from interrelated strategic decisions, verifying the overall internal consistency of their effects. After tested and implemented the selected financial KPIs (as well as the other foreseen), the assessment of factors likely to affect the achievement of the financial stability will be identified and integrated into the model.

5 Concluding Remarks

We analysed the first elements of a more sophisticated methodology. The work has been focused on the identification and integration of some selected KPIs and in integrating them in the EMSO governance for effective management. In the second step, risk management will be introduced and integrated into the methodology to finalise the Risk Governance Model.

Step to shift towards risk culture and so contribute in defining a shared direction to leverage the effectiveness of the RIs means reviewing their strategic goals and KPIs in term of identifying, analysing, evaluating and bearing the risks that may influence success and failure in achieving the strategic objectives.

Even if already proposed and welcomed by the scientific community, at the moment the methodology we recommend, in which the KPIs intend to be multi-functional tools in a more complex system, is still at an experimental stage, promising to be innovative and fruitful.

We stress that the efficacy of the model grows up if we intersect and integrate into the approach several conceptual tools and if we consider in the same framework several themes of the domain of interest, in this case, RIs and the related governance. Further developments could be then obtained by introducing methodology as Multi-criteria methods (MCDM) [1, 9]; this could be a powerful tool in order to evaluate the weights of the different criteria involved in the Management of the Ris, see also Maturo et al. [22], Hoskova-Mayerova [17].

The "risk governance" perspective is an emerging approach at companies and RI's and all kind of organizations where the management of risks is integrated and coordinated across the organization as a whole [19]. The "risk governance" takes a holistic approach to risk—moving from a piecemeal view, related with risk management, performance management and internal control, to integrated and multidisciplinary approach. The "risk governance" approach spreads risk view to include financial, strategic, operational, environmental, human resources, social and other risks.

Acknowledgements The work presented in this paper has been supported by all EMSO Community colleagues.

References

1. Cavallo B, Canfora G, D'Apuzzo L, Squillante M (2014) Reasoning under uncertainty and multi-criteria decision making in data privacy. Qual Quant 48(4):1957–1972
2. Carfí D, Donato A, Fredella MI, Squillante M (2019) Plastic ocean and climate change: a coopetitive integrated risk management approach. Am Geophys Union Meeting Fall. Res Abst. OS53B-1518
3. Carfí D, Donato A, Fredella MI, Squillante M (2020) Coopetitive games for environmental sustainability: climate change and decision global policies. Socio-Economic Planning Sciences. Elsevier
4. Committee of Sponsoring Organizations of the Treadway Commission COSO (2013) Internal control—integrated framework
5. Committee of Sponsoring Organizations of the Treadway Commission COSO (2018) Enterprise Risk Management. Applying enterprise risk management to environmental, social and governance-related risks. Introduction
6. Committee of Sponsoring Organizations of the Treadway Commission COSO (2017) Enterprise risk management integrating with strategy and performance
7. Dañobeitia JJ, Favali P, Beranzoli L, Best M, Blandin J, Hernandez-Brito J, Cannat M, Gates A, Fredella M, Furlani A, Materia P, Petihakis G, Picard J, Ruhl H, Ugolini F, EMSO ERIC CMO (2018) EMSO ERIC—development of a strategic plan for a pan-European distributed research infrastructure. Geophys Res Abst 20. EGU2018-11974-2, 2018-EGU General Assembly
8. European Commission (2015) Better regulation guideline presented in SWD (2015) 111. Available online at http://ec.europa.eu/smart-regulation/guidelines/docs/br_toolbox_en.pdf
9. Fattoruso G, Marcarelli G, Olivieri MG, Squillante M (2019) Using Electre to analyze the behaviour of economic agents. Soft Comput. https://doi.org/10.1007/s00500-019-04397-
10. Fiorani G, Jannelli R, Meneguzzo M (2012) CSR 2.0 proattiva e sostenibile. Tra mercati globali e gestione della crisi, Egea, Milano
11. Fredella MI, Materia P, Favali P, Furlani A, Ruhl H, Gates AR, Blandin J, Cannat M, Chavrit D, Petihakis G, Dañobeitia J (2018) EMSO ERIC strategy towards integration of seafloor and water column observatories. Geophys Res Abst 20. EGU2018-18906-EGU General Assembly 2018b
12. Fredella MI, Jannelli R, Materia P, Olivieri MG, Dañobeitia JJ, Squillante M (2018a) Risk management systems to underpin enterprise performance. Innovative methodologies in management of European Research Infrastructures (RIs). Case study on the EMSO ERIC. In: Proceedings of DYSES2018, La Sorbonne, Paris
13. Fredella MI, Materia P, Beranzoli L, Blandin J, Cannat M, Magnifico G, Petihakis G, Rodero I, Ruhl H, Danobeitia JJ (2019) Growth strategy to increase the value of existing European ocean observatories optimizing interrelated scientific resources and benefits. Abs OS53B-1524, American Geophysical Union Meeting Fall
14. Giorgino M (2015) Risk management. Egea, MI
15. Hajdinjak M et al (2019) A practical guide: assessment of socio-economic impacts of research infrastructures. ResInfra@DR Project
16. Horlings E et al (2012) The societal footprint of big science. Report of the Rathenau Instituut, The Hague, The Netherlands
17. Hoskova-Mayerova S, Talhofer V, Hofmann A et al (2013) Mathematical model used in decision-making process with respect to the reliability of geodatabase advanced dynamic modeling of economic and social systems. In: Studies in computational intelligence, vol 448, pp 143

18. Jannelli R (2012) L'azienda Regione. In: AA.VV. L'economia dell'azienda: paradigmi e declinazioni. Giuffrè, Milano
19. Jannelli R (2006) Governance e misurazione della performance nell'Azienda Pubblica. Un possibile approccio. Aracne, Roma
20. Lauras M, Marquès G, Gourc D (2017) Towards a multi-dimensional project performance measurement system. Decis Supp Syst 48(2), 342–353. https://doi.org/10.1016/j.dss.2009.09. 002emse00682715
21. Materia P, Bozzoli S, Beranzoli L, Cocco M, Favali P, Freda C, Sangianantoni A (2017) EMSODEV and EPOS-IP: key findings for effective management of EU research infrastructure projects. EGU 2017 abstract—European geosciences union general assembly
22. Maturo F, Hoskova-Mayerova S (2018) Analyzing research impact via functional data analysis: a powerful tool for scholars, insiders, and research organizations. In: Conference: innovation management and education excellence through vision 2020, VOLS I-XI, pp 1832–1842
23. Merchant KA, van der Stede W (2011) Management control system: performance measurament. Evaluation and Incentives Pearson, NY
24. OECD (2010) Global science forum reports: large research infrastructures. Available online at http://www.oecd.org/sti/inno/47057832.pdf
25. OECD (2017) Strengthening the effectiveness and sustainability of international research infrastructures. OECD science, technology and industry policy papers, No. 48. OECD Publishing, Paris. https://doi.org/10.1787/fa11a0e0-en
26. OECD (2018) Global science forum. Expert group meeting on reference framework for assessing the socio-economic impact of research infrastructures
27. OECD (2019) Reference framework for assessing the scientific and socio-economic impact of research infrastructures. OECD science, technology and industry policy papers, No. 65. OECD Publishing, Paris, https://doi.org/10.1787/3ffee43b-en
28. Özdemir Aİ, Janelli R, Simonetti B (2015) Determining critical success factors related to the supply chain integration and competition capabilities on business performance. Qual Quant 49(4):1621–1632
29. Rizzuto C (2012) Benefits of research infrastructures beyond Science, presentation at ERF workshop. The socio-economic relevance of research infrastructures, 31 May-1 June 2012, Hamburg
30. Sarbanes–Oxley Act (2002) Pub.L. 107–204, 116 Stat. 745, Enacted July 30

Security Management in the Emergency Medical Services of the Czech Republic—Pre-case Study

Irena Tušer and Jiří Jánský

Abstract Security management is an area of management that deals with the security of assets (resources) in an organization, i.e. both physical security and digital security. The key areas also include human resources management, occupational safety and health protection. Security management is very closely linked to risk management and aims to create conditions that will help to prevent or minimize identified risks, mainly through internal policies, processes, standards and tools. Risks also include accidents at work of employees. As part of the prevention of accidents, measures that would prevent the occurrence or repetition of the occurrence of an injury with the same cause or source must always be applied. The first phase of risk management must identify the origin and causal link of occupational accidents, which was the intention of the implementation of the submitted pre-case study in the field of the Emergency Medical Services of the Czech Republic. The authors dealt with a quantitative method consisting of statistical processing of data from the accident book and identifying the interdependencies (correlations) between them.

Keywords Emergency medical service · Risk management · Safety · Security management · Work accidents

1 Introduction

The owner, the statutory body, and the top management of the organization (company) have the highest responsibility for internal security. Internal security is generally managed by professional departments and experts, but it is primarily part of the daily work of the manager and the statutory body. The Chief Security Officer (CSO) is usually responsible for security management [5, 15].

I. Tušer (✉)
Department of Security and Law, AMBIS College, Lindnerova 1, 180 00 Prague, Czech Republic
e-mail: irena.tuser@ambis.cz

J. Jánský
Faculty of Military Technology, University of Defence, Brno, Kounicova 65, 662 10 Brno, Czech Republic

© The Author(s), under exclusive license to Springer Nature Switzerland AG 2021
D. Soitu et al. (eds.), *Decisions and Trends in Social Systems*,
Lecture Notes in Networks and Systems 189,
https://doi.org/10.1007/978-3-030-69094-6_32

Promoting security in an organization is a set of approaches, methods, and measures. It consists of several areas, disciplines and their activities. The overall security of an organization includes both actions aimed at the safety of people (employees) and everything connected with it, as well as actions of managerial and organizational aspects of security [10].

Security management can be understood as a system on the basis of which integrated security management, which is part of the security policy, is implemented in an organization. Its workplace subsidy is the implementation of an effective safety management system, minimizing the organization's loss due to the prevention of potential emergency situations, support of working and social well-being in the workplace, minimization of risks leading to damage to employees' health and loss of life, and many others [12, 13, 17, 19]. It can therefore be stated that the individual areas, approaches, methods and measures related to security in an organization are interrelated and complementary [2, 7, 16].

The aim of the pre-case study was to determine whether it is possible to improve current aspects of internal security in a selected emergency medical service in the Czech Republic (in Prague, the capital of the Czech Republic), and from the analysis of the principles of internal security organization of this emergency medical service then suggest possible measures leading to the improvement of the current situation so that it is applicable in all emergency medical services (EMS) in the Czech Republic.

2 Statistical Data Analysis of the Emergency Medical Service in Prague

This chapter will analyze work-related injuries causing incapacity for work that occurred in the organization of the Emergency Medical Rescue Service (EMRS) of the Capital City of Prague in the years 2016–2019. The data were obtained from records in the accident book, which lists the number of accidents at work without incapacity for work, with incapacity for work, and also their causes. Based on the analysis of these data, the most risky activities were evaluated.

In the EMRS of the Capital City of Prague, all accidents at work are carefully recorded. Thanks to this, a unique opportunity arises to prepare a static analysis of occupational accidents of the employees of the EMRS of the Capital City of Prague.

Pursuant to the legislation in effect (Act No. 262/2006, Government Regulation No. 201/2010 Coll.), all accidents in an organization must be recorded, investigated, evaluated, and reported to superior management [1, 4].

2.1 Methodology

The aim of the chapter is to analyze the level of internal safety of the medical unit of pre-hospital emergency care during its activities in the capital city of Prague, or rather to know whether it is possible to improve current aspects of safety [9, 18].

Due to the ongoing state of emergency in the Czech Republic and the high workload of potential respondents (paramedics of the EMRS of the Capital City of Prague), a quantitative method was used consisting of statistical data processing and identifying the dependencies (correlations) between them. In the data analysis, the statistical software MAPLE [6] was used, the types of data probability distributions were determined, and the Spearman correlation coefficient was used to find their internal dependence [3, 8].

2.2 Data Set

The following data were available for each occupational accident from accident book of the EMRS of the Capital city of Prague:

1. **Length of employment of the injured** paramedic (data broken down after 3 years).
2. **Location of the fire station of the EMRS of the Capital City of Prague**. A record of the location of the workplace to which the injured employee belongs.
3. **The time of the incident**. Rescue crews work 12-h shifts in continuous operation, i.e. 24 h 7 days a week, 365 days a year. The data presents the time of the incident according to the hours worked in the given shift.
4. **Type of crew activity**. The data from the accident book distinguish the following activities of the paramedic:

   - Walking—a normal walk that involves carrying a load (crew equipment, such as a defibrillator), but not the patient.
   - Stumbling
   - Lifting—crews lift the patient from the ground onto a stretcher or other equipment.
   - Transport—the operations required to transport a patient to a hospital, such as loading a patient into an ambulance, transporting the patient and then unloading them at the hospital.
   - Assault—an action in which a patient or a third party directly physically attacks the crew.
   - Closing a door—an activity in which a crew member has been injured while closing a door.
   - Others—other activities that are difficult to categorize or have a rare frequency, such as vehicle cleaning or a traffic accident.

5. **Injury severity**—the injury is classified according to severity on a scale of 1 to 3. Level 1 is the smallest injury with low severity (muscle strain, bruising, joint sprain). Level 2 is a moderate injury (open wound, dislocation of the bone, contusion of the head, fracture of the bone). Level 3 is a serious injury with possible permanent consequences up to eventual death (limb amputation, internal bleeding, intracranial bleeding).

6. **Source of injury, sorted by cause**:

   - The paramedic—the accident was caused by the employee, either he did not follow the principles of work safety, or it was an unfortunate accident. It may or may not be in direct connection with the patient.
   - The patient—the paramedic's injury was caused by the patient, either indirectly, for example during lifting and handling of the patient, or directly - he attacked the paramedic.
   - An obese patient—a patient whose body weight exceeds 130 kg.
   - A traffic accident—while driving to the patient, with the patient to the hospital, or on the way back to the fire station.
   - Third party—This can be a person who has nothing to do with the patient, such as a random witness, a family member or an acquaintance of the patient.

7. **Injured body part** (head, abdomen, back, upper and lower limbs).

8. **Place of injury**—here the injury occurred (location):

   - In the patient's home.
   - Outdoors—exterior.
   - Fire station.
   - In the outpatient area of an ambulance.
   - In the hospital—when handing over the patient.

The following dependencies were determined from the above information:

1. the dependence of the number of hours worked until the accident on the number of years worked at the EMRS (Sect. 2.3.1),
2. the dependence of the type of injury on the number of years worked in the service (Sect. 2.3.2),
3. the dependence of the type of injury on the hours worked in the shift (Sect. 2.3.3),
4. the dependence of the number of work accidents on the age of the paramedic (Sect. 2.3.4),
5. the dependence of the number of work accidents on the time that elapses before the injury (Sect. 2.3.5).

2.3 Analysis of Occupational Accidents in the EMRS of the Capital City of Prague

Consistent data were available for the period 2016–2019, therefore the accidents at the EMRS of the Capital City of Prague that occurred in those years were analyzed.

In the monitored period, an average of 480 employees worked at the City of Prague's EMRS. More precisely, the numbers ranged from 475 to 487 employees. In the monitored period, a total of 78 occupational accidents occurred (internal accident book of the City of Prague's EMRS). The aim of the analysis was to determine the dependence of the occurrence of a work accident on other data obtained from the book of accidents.

2.3.1 Analysis of the Dependence of the Number of Hours Worked Until the Accident on the Number of years Worked for the EMRS

Table 1 shows the frequency of accidents depending on the number of hours worked in a shift until the accident and the number of years worked for the EMRS.

First, a preliminary data analysis was performed. For individual age categories, the weighted average number of hours worked until the accident was calculated according to the formula:

$$\overline{x_i} = \frac{\sum_j j \cdot n_{i,j}}{\sum_j n_{i,j}}, i, j \in N$$

where $n_{i,j}$ are the frequencies of injuries. The resulting values are listed in the last column of Table 1 and are also indicated in Graph 1.

The values plotted in Graph 1 are interpolated by the least squares method with the equation $y = 0.11x + 4.42$. A coefficient of 0.11 indicates a weak linear dependence, which indicates that more experienced employees become injured a little later than employees who have been in the organization for a shorter period of time (fewer years worked).

Subsequent analysis of the data was to show whether the trend tendency is indeed statistically conclusive or whether it can only be attributed to random noise. The data do not have a normal distribution, therefore a Spearman's correlation coefficient r was used:

$$r = \frac{\Omega_x + \Omega_y - \delta^2}{2\sqrt{\Omega_x \Omega_y}},$$

where

$$\delta^2 = \sum_{i=1}^{R} \sum_{j=1}^{S} n_{i,j} \left(a_i - b_j \right)^2, \Omega_x = \frac{1}{S} \left(n^3 - \sum_{i=1}^{R} x_i^3 \right),$$

$$\Omega_y = \frac{1}{S} \left(n^3 - \sum_{j=1}^{S} y_j^3 \right),$$

Table 1 Number of hours worked in a shift until the accident depending on the number of years worked

i	years worked	j number of hours worked in a shift until the accident												x_i
		1	2	3	4	5	6	7	8	9	10	11	12	
1	0–3	4	1	1	1	1	0	1	0	1	1	0	0	4
2	3–6	1	2	0	1	1	0	2	1	1	0	0	0	5
3	6–9	1	1	1	0	0	2	0	1	0	1	0	0	5.1
4	9–12	0	0	2	0	3	0	1	0	2	2	1	0	7
5	12–15	1	1	0	0	0	1	2	1	0	0	1	2	7.3
6	15–18	1	3	0	0	0	0	3	2	0	0	1	0	5.5
7	18–21	1	0	2	0	3	0	0	0	1	3	0	0	6.1
8	21–24	1	1	0	0	0	0	0	1	2	0	0	0	5.8
9	24–27	0	0	0	1	0	0	2	0	1	0	0	2	8.5

Source own

while

$$a_1 = \frac{x_1 + 1}{2}, a_i = \sum_{l=1}^{i-1} x_l + \frac{x_i + 1}{2}, \text{ for } 2 \leq i \leq R,$$

$$b_1 = \frac{y_1 + 1}{2}, b_j = \sum_{l=1}^{j-1} y_l + \frac{y_j + 1}{2}, \text{ for } 2 \leq j \leq S$$

and $n = \sum_{i,j} n_{i,j} = 78$ is the total number of injuries. In previous relations, it is denoted by a symbol $i = 1, \ldots, R$ row index, $j = 1, \ldots, S$ column index, the letter R denotes the number of rows and S denotes the number of columns.

In the actual data analysis, which was performed in MAPLE software [6], the following values were obtained: $\delta^2 = 58,776.5$, $\Omega_x = 38,979$, $\Omega_y = 39,129$, and the resulting Spearman coefficient $r = 0.247$.

We use the Spearman's correlation coefficient $r \in [-1, 1]$ most often to measure the strength of the relationship for such quantities when we can not assume the linearity of the expected relationship or the normal distribution of the observed variables X and Y. The dependence of variables can generally be ascending or descending. If $r = 1$, resp. $r = -1$, the correlation pair (x_i, y_i) lies on some ascending, or declining, function.

If its absolute value is close to one, it means that the given quantities have a strong dependence. In our case, its value is close to zero, which indicates a weak dependence.

To find out that this dependence is not caused only by random deviations, the zero hypothesis was tested: $r = 0$ against the alternative hypothesis: $r \neq 0$. The following applies to the value of the test statistic t:

$$t = r\sqrt{\frac{n-2}{1-r^2}} = 0.247\sqrt{\frac{78-2}{1-0,247^2}} = 2.22.$$

See e.g. [3]. This hypothesis was then compared at the significance level $\alpha = 0.05$ with the table value for the student's t distribution with $n - 2$ degrees of freedom. Since $2.22 > 1.99$, we reject hypothesis H_0 about the zeroness of the correlation coefficient.

In other words, with a probability of 95%, the independence of the columns of Table 1 from the rows is not valid, and thus the number of hours worked in a shift until the accident is weakly dependent on the number of years the employee has worked for the EMRS.

The task of future research will be to verify the invalidity of the hypothesis: "the time of the paramedic's injury does not depend on the number of years worked", on a larger sample of data. In case of its rejection, the validity of the alternative hypothesis will be discussed: "the time of the paramedic's injury depends on the number of years worked".

Only the number of years worked is shown in Table 1. However, it is likely that emergency employees are also older. We therefore do not know whether the later number of injuries is affected by the years worked for the EMRS or the age of the employee. Therefore, in the following research, we will analyze the number of hours worked until the accident also based on the age of the employee.

2.3.2 Analysis of the Dependence of the Type of Injury on the Number of years Worked in the Service

In this chapter, the question will be studied whether employees who have been working in the EMRS for longer have the same or different types of injuries compared to those who have been working for a shorter period of time. Only the most common injuries were analyzed.

The frequencies of occupational accidents are presented in Table 2, which has $R = 9$ rows and $S = 5$ columns. The frequencies of occupational accidents are again marked $n_{i,j}$, where index i denotes the row number and corresponds to the number of years the employee has worked in the EMRS, index j is the column number and corresponds to the type of injury. Row and column sums of frequencies $n_{i,j}$ are denoted x_i, y_j.

The variables in the column are nominal variables, so it is not possible to use Spearman's correlation coefficient to determine the relationship between the rows and columns. Pearson's chi-square statistic [14] is most often used to analyze such data. If the relationship between the variables is random, then the frequencies of $n_{i,j}$ given in Table 2 do not differ much from the so-called expected frequencies $e_{i,j}$.

Table 2 Dependence of the type of injury on the number of years worked in the service

Number of years	Strain		Fracture		Sprain		Wound		Dislocation		
	$n_{i,1}$	$e_{i,1}$	$n_{i,2}$	$e_{i,2}$	$n_{i,3}$	$e_{i,3}$	$n_{i,4}$	$e_{i,4}$	$n_{i,5}$	$e_{i,5}$	x_i
0–3	3	5	4	1.1	2	1.9	1	0.9	0	0.9	10
3–6	7	4.5	0	1.0	2	1.7	0	0.8	0	0.8	9
6–9	4	4.5	1	0.8	1	1.3	0	0.6	1	0.6	7
9–12	3	5	1	1.1	3	1.9	3	0.9	0	0.9	10
12–15	3	4	1	0.9	1	1.5	1	0.7	2	0.7	8
15–18	8	5.5	0	1.3	1	2.1	1	1.0	1	1.0	12
18–21	4	5	2	1.1	2	1.9	1	0.9	1	0.9	10
21–24	3	2.5	0	0.5	2	0.9	0	0.4	0	0.4	5
24–27	3	3	0	0.7	1	1.1	2	0.5	0	0.5	6
y_j	38		9		15		9		5		76

Source own

$$e_{i,j} = \frac{x_i y_j}{n}.$$

The Goodman–Kruskal's τ was used for the analysis of these data [3]. After checking the condition that non-zero frequencies occur in more than one column, the following was calculated

$$\tau = \frac{n \sum_{i=1}^{R} \sum_{j=1}^{S} \frac{(n_{ij} - m_{i,j})^2}{x_i}}{n^2 - \sum_{j=1}^{S} y_j^2} = 0.11.$$

The number τ takes values from the interval [0,1], while it takes on the value 0 when the quantities are independent of each other. In our case, therefore, it is a very weak dependency.

This is a number close to zero, so the null hypothesis was tested at the significance level of $\alpha = 0.05$: $\tau = 0$ "The type of injury is independent of the number of years worked." compared to the alternative hypothesis: $\tau \neq 0$ "The type of injury is dependent on the number of years worked." The test statistic is of the form $(n-1)(S-1)\tau$ and take the value of 33.1. This statistic has a chi-square distribution with $(S-1)(R-1)$ degrees of freedom [20]. By comparing the tabulated value $\chi_{1-\alpha}^2((S-1)(R-1)) = \chi_{0.95}^2(28) = 46.1$ with the calculated value, we get the inequality $33.1 < 46.1$ and do not reject the null hypothesis at the level of significance $\alpha = 0.05$. In other words, it was found that with a probability of 95%, the type of injury is independent of the number of years the employee has been employed by the emergency services.

However, this hypothesis will also need to be studied in the future in a case study on a larger sample of data.

2.3.3 Analysis of the Dependence of the Type of Injury on the Hours Worked in a Shift

Another aspect examined was the question of the relationship between the types of injuries at the beginning and end of a shift. The same types of injuries were studied as in the previous chapter, and their analysis was performed analogously. Table 3 shows the frequencies of accidents at work. The rows correspond to the number of hours from the beginning of a work shift, i.e. $R = 12$, the columns indicate the type of injury, $S = 5$.

Then, in a similar manner as in chapter 2.3.2, a Goodman–Kruskal's $\tau = 0.183$ was obtained. This coefficient is again close to zero, which means a very weak dependence.

Finally, the hypothesis that this coefficient is equal to zero was tested, i.e. the hypothesis: $\tau = 0$ compared to the alternative hypothesis H_1: $\tau \neq 0$. The test statistic in the form of $(n-1)(S-1)\tau \cong 54.9$ was calculated and compared with the table value for $\chi_{0.95}^2((S-1)(R-1))$. Since $54.9 < 56.3$ is valid, the hypothesis

Table 3 Dependence of the type of injury on the hours worked in a shift

No. of working hours	Strain	Fracture	Sprain	Wound	Dislocation	x_i
1	8	0	1	1	0	10
2	5	1	2	0	2	10
3	0	3	0	1	0	4
4	1	1	0	0	1	3
5	4	0	2	2	0	8
6	2	0	0	0	1	3
7	5	1	3	0	2	11
8	5	0	1	0	0	6
9	2	0	3	1	1	7
10	2	3	2	0	0	7
11	2	0	0	1	0	3
12	2	0	1	1	0	4
y_j	38	9	15	7	7	76

Source own

that $\tau = 0$ at the level of significance $\alpha = 0.05$ was rejected. In other words, it has been calculated that with a probability of 95%, the type of injury does not depend on the number of hours since the start of the work shift.

2.3.4 Analysis of the Dependence of the Number of Work Accidents on the Age of the Paramedic

The data in Table 4 show the numbers of injured paramedics at a given age and in the years 2017–2019. The age of the employees is divided into $k = 4$ groups of 10 years, from 25 to 65 years, when employees retire. A total of $n = 54$ values are available.

If the probability of an accident for each employee is the same (regardless of their age), then the probability that the accident will occur in a given age group depends only on its percentage in the EMRS. These probabilities $p_i, i = 1, \ldots, k$ are given in the fourth column of Table 4. In addition to these, Table 4 also shows the empirical

Table 4 The dependence of the number of work accidents on the age of the paramedic

Age (years)	% of employees (%)	Number of accidents	p_i	x_i	$x_i - p_i$
<35	30	18	0.3	0.3333	0.033
35–45	35	19	0.35	0.3518	0.0018
45–55	25	13	0.25	0.2407	-0.0093
>55	10	4	0.1	0.0740	-0.026

Source own

Table 5 The dependence of work accidents on the time that elapses until the injury

i	1	2	3	4	5	6	7	8	9	10	11	12
t_i	[0, 1)	[1, 2)	[2, 3)	[3, 4)	[4, 5)	[5, 6)	[6, 7)	[7, 8)	[8, 9)	[9, 10)	[10, 11)	[11, 12)
x_i	10	9	6	3	8	3	11	6	8	7	3	4

values x_i of the probability of an injury for the given age groups, obtained from the accident book of the EMRS of the Capital City of Prague. The last column of Table 4 presents the differences between theoretical and empirical probabilities. There is a slight declining trend in this difference for older employees. In other words, they have fewer injuries than younger ones. It was tested whether this decrease is statistically significant.

The following hypothesis was formulated: "the probability of injury is the same for each age group", i.e. that "the probability of injury is given by the values in column p_i of Table 4". Contrary to the alternative hypothesis: "the probability of injury is not the same for every age group". Since the condition $np_i > 5$ is met for all $i = 1, \ldots, 4$, a χ^2 test was performed and at the significance level $\alpha = 0.05$, which did not reject the null hypothesis (n number of accidnets).

Thus, the probability of injury does not change with the age of the paramedic. This result can be taken as a hypothesis for a future case study, where it will be necessary to test the hypothesis on a larger sample of data. In particular, the number of 4 injuries in people over 55 is insufficient for serious conclusions.

2.3.5 Analysis of the Dependence of Work Accidents on the Time that Elapses Until the Injury

In this chapter we will deal with the role played by the time elapsed since the beginning of the shift in the number of injuries. At the end of the shifts, fatigue, hunger and stress can occur if the crew does not have time to eat or go to the restroom. Table 5 and Graph 1 show the time at which the accident occurred. The shift was divided into $k = 12$ hour intervals in which a total of $n = 78$ work accidents occurred. The first row of Table 5 shows the sequence numbers of the intervals. The second row lists the individual time intervals t_i, and the last row shows the number of injuries x_i that occurred in them.

Unfortunately, the available data do not distinguish between morning and night shifts. It is probable that the number (and type) of accidents differs depending on the time of day—however, this cannot be determined by analyzing this data. It would be better to analyze the time until an injury occurs separately for day shifts and separately for night shifts.

3   Results and Recommendation

Based on the obtained data, a detailed analysis of the types, times and causes of injuries was performed. In Sect. 2.3.1, it was found that the longer an employee works for the EMRS, the later a work accident occurs in his/her shift (see Graph 1). However, this dependence is weak. The detected phenomenon can be interpreted as meaning that they are older people whose experience is overcome by fatigue at the end of their shift. This results in a recommendation to improve communication and the transfer of experience to younger, while at the same time trying to ease the burden at the end of work shifts to older colleagues.

In Sect. 2.3.2, it was found that the type of injury is independent of the number of years the employee has worked for the EMRS. It was statistically significantly proven that a less experienced paramedic and an employee with many years of experience are equally prone to suffering any of the monitored occupational accidents. It is interesting that the fatigue of the body of paramedics did not manifest itself statistically either, i.e. for example, they had more frequent problems with back pain after a longer period of employment at the EMRS.

In Sect. 2.3.3, it was shown that the type of injury is independent of the number of hours worked in the shift at the time of the injury. It is therefore likely that even, for example, a mandatory warm-up before the start of the shift would not affect the types of accidents at work at the beginning of the shift.

Section 2.3.4 showed that the probability of an accident at work (and therefore their number) does not depend on the age of the paramedic. This means that even after several years spent with the EMRS, more experienced paramedics are not able to avoid accidents at work. On the other hand, despite many years of stress, they are not more prone to injuries. However, since the last column of Table 4 shows that older employees have fewer injuries, it is possible that the case study with a larger amount of data will find a different conclusion, thus demonstrating interdependence.

In Sect. 2.3.5, it was found that the most risky period is halfway between 6 and 9 h of the shift. During this period, 31 work-related injuries occurred, which is 39% of all accidents in the monitored period of the twelve-hour shift. The second most risky period is the time at the beginning of the shift until 3 working hours where 32% of accidents occur. Explaining the reasons for increased injuries at these times is very difficult. The data were created by summing the data from day and night shifts. Thus, the highest number of accidents that occur between 6 and 9 working hours of the shift is the sum of the number of accidents that occurred between twelve and three o'clock in the afternoon with those that occurred in the same time period at night.

At the same time, day and night injuries can be of different nature and also have different origins. In both cases, the reasons for frequent injuries at the beginning of the shift may be, for example, that the crew is not asleep, has not had time to move, or is fully acquainted with the technology taken over. Reasons for injuries at the end of a shift may be fatigue, hunger, the need to go to the restroom. However, these reasons do not explain the decrease in accidents between 10 and 11 working hour, shown in Table 5. To better understand the increased accidents at these times, it would be necessary to obtain and analyze data for day and night shifts separately and also have information on injuries in individual hours of the shift.

4 Conclusion

Internal security management in an organization is a multidisciplinary field. It is a comprehensive approach with the application of especially technical, organizational and educational measures, the implementation of which prevents the emergence of identified risks, threats or damage to human health in the work process [11].

The main principles of workplace safety management are, in particular, the implementation of risk analysis and risk management, the review of the status and operation of safety measures with continuous and systematic improvement of the system and the elimination of identified deficiencies. It is always considered that employers and employees should prevent risks, not address the consequences.

The purpose of the submitted pre-case study in the field of Emergency Medical Services was to determine the origin and causal link of accidents at work. From the analysis of the principles of implementation of methods and internal security processes, the aim of the pre-case study was to determine whether it is possible to improve the current aspects of safety in the selected EMS of the Czech Republic, and to suggest possible measures. Based on the results described in Cpt. 3 and the proposed recommendations, the following research will be modified and extended to other subjects of the Emergency Medical Service of the Czech Republic so that the final result is applicable in practice. In order to increase the efficiency and validity of the results, it is absolutely necessary to obtain equally structured data from a larger number of EMS organizations in the Czech Republic.

The purpose of the submitted pre-case study in the field of EMS was to determine the origin and causal link of accidents at work. From the analysis of the principles of implementation of methods and internal security processes, the aim of the pre-case study was to determine whether it is possible to improve the current aspects of safety in the selected EMS of the Czech Republic and to suggest possible measures. Based on the results described in chapter 3 and the proposed recommendations, the following research will be modified and extended to other subjects of the Emergency Medical Service of the Czech Republic so that the final result is applicable in practice. In order to increase the efficiency and validity of the results, it is absolutely necessary to obtain equally structured data from a larger number of emergency medical service organizations in the Czech Republic.

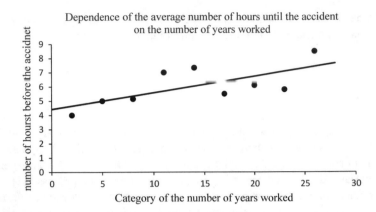

Graph 1 Average number of hours worked in a shift until the accident depending on the number of years worked. *Source* own

Acknowledgements Irena Tušer would like to thank AMBIS College, Prague for its support and Jiří Jánský thanks for support of the project DZRO K217, supported by the Ministry of Defence in the Czech Republic.

References

1. Act No. 262/2006 Sb. Labor Code, as amended
2. Adamoniene R (2018) Management presumptions and possibilities of human resources formation. In: Bekesiene S, Hoskova-Mayerova S (eds) Challenges to national defence in contemporary geopolitical situation (CNDCGS' 2018), pp 157–166
3. Bernstein S, Bernstein R (1999) Schaum's outline of theory and problems of elements of statistics: descriptive statistics and probability. McGraw-Hill, vii, 354. ISBN 0070050236
4. Government Regulation No. 201/2010 Coll. On the method of registration of accidents, reporting and sending records of accident, as amended
5. Hoskova-Mayerova S (2016) Education and training in crisis management, the European proceedings of social and behavioural sciences EpSBS, vol XVI, p 849–856. ISSN 2357–1330. https://doi.org/10.15405/epsbs.2016.11.87
6. Jánský J (2020) Teaching of mathematics in maple program. APLIMAT. In: 19th conference on applied mathematics, 4–6 Feb 2020, Bratislava
7. Kelemen M, Szabo S, Vajdova I (2018) Cybersecurity in the context of criminal law protection of the state security and sectors of critical infrastructure. In: Bekesiene S, Hoskova-Mayerova S (eds) Challenges to national defence in contemporary geopolitical situation (CNDCGS' 2018), pp 100–104
8. Lyon R, Sanders J (2012) The swiss bus accident on 13 March 2012: lessons for pre-hospital care. Crit Care 16(4):138. https://doi.org/10.1186/cc11370
9. Kudlák A, Urban R, Hošková-Mayerová Š (2020) Determination of the financial minimum in a municipal budget to deal with crisis situations. Soft Comput 24(12):8607–8618. ISSN 1432–7643. https://doi.org/10.1007/s00500-019-04527-w
10. Malachova H, Oulehlova A, Kincl P (2017) SIMEX simulation tool—"Accident" crisis scenario and crisis management entities' exercise. In: Barath J, Dedera L, Ockay M (eds) Communication and Information Technologies (KIT), Slovakia, pp 83–89

11. Navrátil J, Sadovská V, Švarcová I (2019) Health risk assessment of combustion products from simulated residential fire. Stud Syst Decis Control 104:15–23. https://doi.org/10.1007/978-3-319-54819-7_2
12. Oulehlova A, Malachova H, Rezac D (2015) Use of simulation in cooperation training of critical infrastructure entities. In: Distance learning, simulation and communication 'DLSC 2015', pp 103–112
13. Oulehlova A, Malachova H, Rezac D (2017) Risks evaluation in preparation of crisis management exercise. In: Distance learning, simulation and communication 'DLSC 2017', pp 143–153
14. Potůček R (2020) Life cycle of the crisis situation threat and its various models. Stud Syst Decis Control 208:443–461. https://doi.org/10.1007/978-3-030-18593-0_32
15. Svarcova I, Ptacek B, Navratil J (2015) Psychological intervention as support in disaster preparedness. Crisis Manage Sol Crisis Situat 2015:317–320
16. Svarcova I, Hoskova-Mayerova S, Navratil J (2016) Crisis management and education in health. Euro Proc Soc Behav Sci 16:255–261 (2016). https://doi.org/10.15405/epsbs.2016.11.26
17. Tušer I, Navrátil J (2020) Evaluation criteria of preparedness for emergency events within the emergency medical services. Stud Syst Decis Control 208:463–472. https://doi.org/10.1007/978-3-030-18593-0_33
18. Tušer I, Bekešienė S, Navrátil J (2020) Emergency management and internal audit of emergency preparedness of pre-hospital emergency care. Qual Quant. https://doi.org/10.1007/s11135-020-01039-w
19. Tušer I, Hošková-Mayerová Š (2020) Traffic safety sustainability and population protection in road tunnels. Qual Quant:1–22. https://doi.org/10.1007/s11135-020-01003-8
20. Weiss NA (2017) Introductory statistics, 10th edn. In: Weiss CA (ed) Global edition. Pearson, Boston, 763, 73. ISBN 9781292099729

Printed in the United States
by Baker & Taylor Publisher Services